Exosomes and Extracellular Vesicles

R. Kenneth Marcus · David Lubman
Editors

Exosomes and Extracellular Vesicles

Analytical Challenges and Biological Implications

Springer

Editors
R. Kenneth Marcus
Department of Chemistry
Clemson University
Clemson, SC, USA

David Lubman
Department of Surgery
University of Michigan
Ann Arbor, MI, USA

ISBN 978-3-031-89269-1 ISBN 978-3-031-89270-7 (eBook)
https://doi.org/10.1007/978-3-031-89270-7

This Springer imprint is published by the registered company Springer Nature Switzerland AG
The registered company address is: Gewerbestrasse 11, 6330 Cham, Switzerland

To the people who matter ... None of this would have been possible without the love and support of Lisa and the rest of the Brady Bunch. I am grateful to the dedication and efforts of my many graduate students, without whose efforts this laboratory would not have succeeded.

—R. Kenneth Marcus

I would like to thank the support and patience of my family during my 40+ years here at the University of Michigan in Ann Arbor and the many excellent students and fellows from around the world who have worked in my group during this time.

—David Lubman

Preface

Extracellular vesicles (EVs) such as exosomes are membrane-enclosed nanovesicles secreted by all cells. The vesicles are secreted into the bloodstream and can act as messengers to change cellular behavior throughout the body. Their effects may be important in the development of several different diseases including various cancers, Alzheimer's, and other neurological diseases. They are also seen as a mechanism for the propagation of immunogenicity. The vesicles contain a rich cargo of proteins, RNAs, lipids, and metabolites that reflect the contents of the particular cell type of origin and are surrounded by a membrane that protects the contents from degradation during the circulation of the blood. This cargo can be related to the presence of a disease state and so may be used as a marker for detection and prognosis in the development of disease.

There are now numerous publications that have appeared on the use of the contents of exosomes and EVs as markers for disease prognosis. Based on their cellular uptake and cargo delivery attributes, many research institutes and commercial entities are becoming highly vested in their application towards the therapeutic treatment of several types of cancers. In this regard, exosomes and EVs are being extensively evaluated and developed as vectors for drug delivery (e.g., gene therapy). Thus, the study and utilization of exosomes and EVs are of increasing interest and importance across the spectrum of fundamental biology, clinical diagnostics, and biotherapeutics.

Despite the rising interest in the study of EVs and exosomes in biology and medicine, there are still many issues to be addressed towards gaining a better understanding and ultimate utilization which are discussed in the various chapters of the current volume. Chapters 1 and 2 set the stage, covering issues in clinical applications and needs for standardization in the use of EVs and exosomes. A major issue in the study and use of these vesicles is the ability to isolate them from diverse matrices (biofluids, cell cultures, etc.) and to better characterize them, especially for liquid assays. Chapters 3 through 7 cover various novel methods for isolation of the vesicles from biofluids, also discussing some of the more standard methods such as ultracentrifugation, size exclusion chromatography and polymer precipitation among others. Some of the new methods described include techniques based on hydrophobic interaction chromatography and affinity column separations. Other methods discussed include field flow fractionation as a candidate for separating different types of EVs and exosomes, peptide affinity precipitation, and microfluidic methods for separations, especially as applied to the analysis of their RNA content.

In terms of characterizing EVs and exosomes, Chapters 8 and 9 describe state-of-the-art optical methods for studying single vesicles which are typically between 20–150 nm and thus below the diffraction limit of light microscopes. These methods use techniques that can circumvent the diffraction limit and can image vesicles down to 20 nm. The methods illustrate the ability to study the size, shapes, and protein composition in individual vesicles, which will ultimately be important for identifying protein markers associated with a disease that can be associated with specific parent cells and the organs that generate them.

At this point, the ultimate method for characterizing these vesicles in terms of chemical detail is mass spectrometry, where applications to EV and exosome characterization are discussed in Chapters 10 through 13. These chapters describe the use of mass spectrometry to characterize the protein content of EVs including the development of standards for quantitative analysis, the use of MALDI-MS as a rapid method to characterize protein content and the detailed characterization of post-translational modifications including glycosylation and phosphorylation.

The field this book describes: Exosomes and Extracellular Vesicles: Analytical Challenges and Biological Implications is a rapidly expanding field of research. The 13 chapters herein cover certain facets of this field, though there are certainly other areas that are missing such as in the use of EVs as biomarkers for diseases and in the analysis of lipid and metabolomic contents. Nevertheless, we hope that the contents herein are of use to the uninitiated in the field who would like to learn more about analytical handling, isolation, and analysis of these potentially important vesicles and their potential for study in problems of relevance in fundamental biology/medicine, clinical diagnostics, and biotherapeutics.

Ann Arbor, MI, USA David Lubman
Clemson, SC, USA R. Kenneth Marcus
January 2025

Contents

About the Editors

R. Kenneth Marcus is currently Robert Adger Bowen of Chemistry at Clemson University. He received B.S. degrees in chemistry and physics from Longwood College in 1982 and his Ph.D. in analytical chemistry from the University of Virgina in 1986 under the direction of Prof. Willard W. Harrison. He has been at Clemson University since 1986 and as Full Professor since 1996. His research program has consistently centered on the development of novel analytical methodologies and technologies, beginning with an emphasis on plasma source mass spectrometry and evolving over the last 20 years to include novel stationary phases for the liquid chromatography separations of proteins. Most recently, his research group has focused on novel approaches to two-dimensional liquid chromatography (2D-LC) and methods for the isolation and purification of extracellular vesicles. These developments have led to the awarding of over one dozen US patents.

David Lubman is currently Professor Emeritus of Surgery at the University of Michigan Medical Center. He received his B.A. from Cornell University in 1975 and his Ph.D. in physical chemistry from Stanford University in 1979 under the direction of Prof. Richard N. Zare. He has been at the University of Michigan since 1983, as a Full Professor since 1991 and as the Maud T. Lane Professor from 2005. His present research interests involve developing new proteomic methods for searching for protein markers of cancer. These include 2D liquid mapping and mass spectrometry of tumor cells and serum and the development of protein microarray technology. Most recent studies involve the study of cancer stem cells, early detection of cancer, circulating tumor cells and exosomes, and changes in glycosylation during stages of cancer progression and therapeutic treatment. Various collaborations with clinical researchers include studies in ovarian, pancreatic, colon, HCC, GBM, bladder and breast cancer.

The Development of New Technologies and Methods to Accelerate the Discovery and Validation of EV/EVPs as Clinical Biomarkers: An Introduction

Jennifer C. Jones

Abstract Technological advancements are rapidly transforming extracellular vesicle and particle (EVP) research from an emerging field into valuable clinical tools. This chapter introduces the complex world of EVPs, encompassing membrane-bound vesicles and non-vesicular structures that serve as information-rich biomarkers in blood and other body fluids. Fundamental methodological challenges that have historically limited EVP research are being overcome with new technologies and the development of standardized reporting methods. Through selected examples of prostate cancer liquid biopsy EVP assays, we demonstrate how innovative technologies are enabling the transition from laboratory discoveries to validated clinical assays. This introduction previews some of the technological advancements featured in this book and demonstrates how these advancements collectively are overcoming analytical barriers and accelerating EVP biomarker discovery.

More than a century ago, French biologist René Quinton posited that blood represents the ocean, the "marine plasma," within us [1]. The analogy of exploring the "oceans within us" has never seemed more apt than the current state of liquid biopsies and biofluid research. Since the early days of medicine, researchers have sought to detect and monitor diseases with blood or urine tests. In recent years, an explosion of liquid biopsy-type assays has transformed the scope and impact of clinical assay, and new frontiers of extracellular biology have been discovered. Some of these new approaches to liquid biopsies focus on extracellular vesicles (EV) and non-vesicular extracellular particles. Broadly speaking, the term "EVP" has been used to refer to the collective assortment of lipid bilayer-bound extracellular vesicles along with other composite extracellular, macromolecular particles, including lipoproteins and other macromolecular structures without membranes such as exomeres [2], and supermeres [3], large lipoprotein complexes, as roughly illustrated in Fig. 1. As more is being learned about the heterogeneity of extracellular vesicles, it has become

J. C. Jones (✉)
Translational Nanobiology Section, Laboratory of Pathology, Center for Cancer Research, National Cancer Institute, National Institutes of Health, Bethesda, MD, USA
e-mail: jennifer.jones2@nih.gov

R. K. Marcus and D. Lubman (eds.), *Exosomes and Extracellular Vesicles*,
https://doi.org/10.1007/978-3-031-89270-7_1

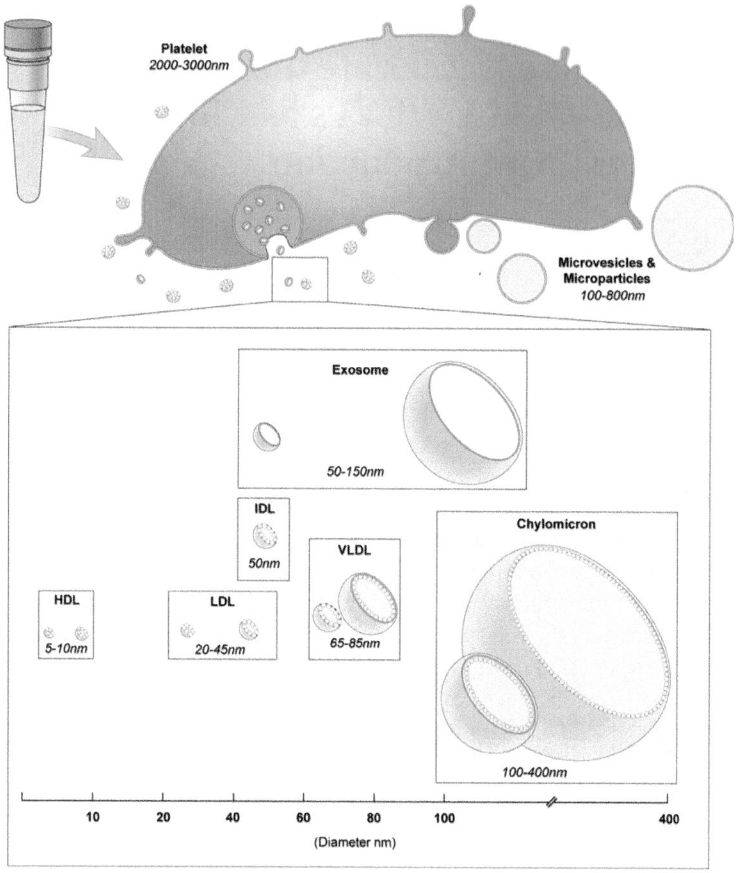

Fig. 1 A historical comparison of typical extracellular vesicle sizes to lipoproteins routinely assayed in clinical laboratories

well-recognized that there is an extensive diversity of types and sub-cellular origins of extracellular vesicles, extending far beyond the conceptualization of membrane-bound "exosomes" introduced in the past two decades [4, 5]. The field is only beginning to recognize the great diversity of different types of membrane-bound EVs, beyond canonical exosomes and shed microvesicles to include functional migrasomes [6], exophers [7], apoptotic bodies [8], and mitovesicles [9], which are each distinguished by their mode of biogenesis, cargo, and function [5].

From early reports identifying platelet-derived extracellular vesicles as "platelet dust" [10] to more recent practice-changing randomized controlled trials of EVP-based.

biomarkers of disease, great progress has been made in the study of EVPs and the use of EVPs to guide clinical decision-making [11–13]. One example of the rapid progression of basic EVP research to clinical applications is in the management of

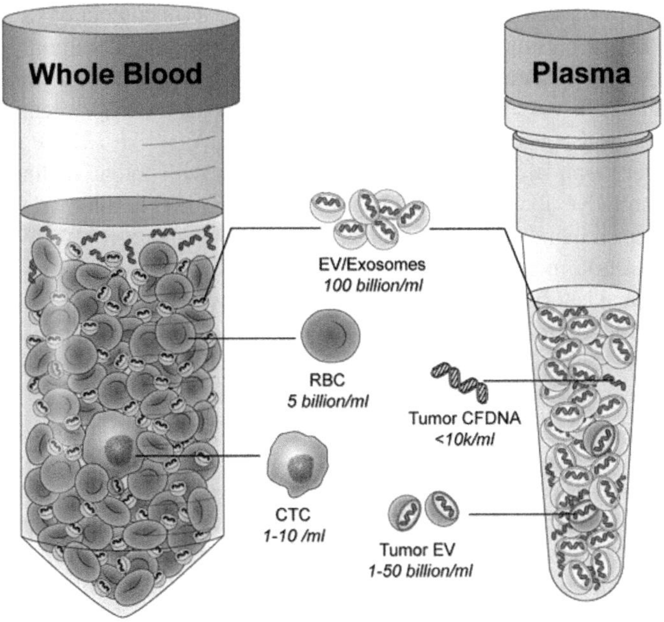

Fig. 2 Schematic of EVs admixed with other circulating "biomarkers" in blood, as one example of EVPs found in high concentrations in the blood, providing a robust source for liquid biopsy biomarkers. Selected other blood-borne biomarkers and components, with annotated concentrations: circulating tumor cells (CTC), red blood cells (RBC), cell-free DNA (CFDNA)

prostate cancer. EVPs were first identified in urine in 2004 [14], recognized as potential biomarkers for cancer in 2008 [15], and then found to carry small RNA signatures that distinguish high risk from low-risk prostate cancer (2013). With these studies, Skog and colleagues recognized the value of tumor-derived or tumor-associated EVPs as abundant (and informative packets of information relating to patient health and disease, as depicted in Fig. 2. Clinical laboratory tests developed based on this EVP miRNA signature have been validated in more than 3 prospective randomized clinical trials, and the use of such assays is transforming screening and treatment decision patterns for many patients [13, 16, 17]. Beyond this first example of prostate cancer-associated biomarkers, EVP-based biomarkers from bulk biofluid EVP isolates, are presently being discovered and validated for many other diseases.

Despite these rapid advancements in the field, much remains to be done to extend these advancements to realize the true potential for EVPs and their cargo in biomedical research. As a familiar example to patients who undergo blood tests for cholesterol and lipoprotein levels, LDL, HDL, and other lipoprotein parameters are routinely measured. However, standard clinical assays in use today for lipoprotein levels typically do not directly measure the particle number or size of those lipoproteins in a blood sample, but rather the tests measure content levels of specific lipoprotein-associated molecules that can readily be assayed in a clinical laboratory setting. For low density lipoprotein (LDL), the cholesterol content of LDL particles (LDL-C) [18] or plasma ApoB protein levels [19] are more commonly measured than the LDL particle count (LDL-P test) [20], which has been more technically challenging and expensive to perform. Technological barriers to implementing specific size, density, and concentration measurements for EVPs like LDL-P in biofluids have a barrier to implementing diagnostic approaches with potential clinical benefit, despite the relative abundance of EVs to other blood-bourne biomarkers (Fig. 2). Practically speaking, the fact that robust, reproducible, and cost-effective EVP size and concentration measurements remain elusive for characterizing most types of EVPs in clinical laboratories highlights the importance of the technological advances presented in this volume. In response to the growing need for robust, nanoscale single particle characterization methods, new methods are being developed and tested in the basic research and translational settings.

The importance of robust EVP size, concentration, density, and cargo measurements is not only of substantial importance in the clinical, translational setting, but also in the basic research setting. In order to investigate the molecular mechanisms for EVP assembly and release from cells, to delineate the full impact of EVPs and their cargo elements as mediators or mitigators of disease, and to systematically parse and classify distinctive EVPs from the spectrum of different types of macromolecules complexes produced by cells, a wide range of basic research endeavors investigating EVP biology requires the development and adaptation of technologies from disparate fields, intersecting together to bring single molecule and nanoscale sensitivity to complex extracellular molecular assemblies. Until the past decade, researchers had few, if any, biomedical research tools optimized for EVP characterization. An example of adaptation and development of improved tools for EV characterization in the clinical context of prostate cancer discussed above, stratification patient blood samples as either high or low for PSMA-positive EVPs by EV flow cytometry predicts tumor progression and modulates immune function following therapy [21]. In these studies, Lucien et al. leveraged many years of refinements to flow cytometry methods and instruments to resolve EVPs at single particle nanoscale resolution for classification [22, 23]. Further refinements of EV FCM instrumentation and single particle EVP methods continue to improve in terms of sensitivity and reproducibility [24], as is described in Chap. 7.

Fortunately, at this emerging EVP research frontier, EV flow cytometry is only one of many well-established biomedical research tools that have undergone transformative development and adaptation for the study of EVPs. The chapters that follow in this book detail a collection of other technologies, from mass spectrometry to affinity chromatography to field flow fractionation, that have undergone specialized transformations for the robust study of EVPs and their cargo. With these advanced EVP technologies, and with unified reporting frameworks to ensure robust interpretation of results across methods and institutions [25, 26], the biomedical research community is now able to dive more confidently and productively into the deep seas of translational liquid biopsies and the fundamental extracellular biology.

References

1. Quinton, R. (1912). *L'eau de mer milieu organique: Constance du milieu marin originel, comme milieu vital des cellulares, a travers la série animale.* Paris, France: LIBRAIRES DE L'ACADÉMIE DE MÉDECINE.
2. Zhang, H., Freitas, D., Kim, H. S., Fabijanic, K., Li, Z., Chen, H., Mark, M. T., Molina, H., Martin, A. B., Bojmar, L., Fang, J., Rampersaud, S., Hoshino, A., Matei, I., Kenific, C. M., Nakajima, M., Mutvei, A. P., Sansone, P., Buehring, W., … Lyden, D. (2018). Identification of distinct nanoparticles and subsets of extracellular vesicles by asymmetric flow field-flow fractionation. *Nature Cell Biology, 20*, 332–343.
3. Zhang, Q., Jeppesen, D. K., Higginbotham, J. N., Graves-Deal, R., Trinh, V. Q., Ramirez, M. A., Sohn, Y., Neininger, A. C., Taneja, N., McKinley, E. T., Niitsu, H., Cao, Z., Evans, R., Glass, S. E., Ray, K. C., Fissell, W. H., Hill, S., Rose, K. L., Huh, W. J., … Coffey, R. J. (2021). Supermeres are functional extracellular nanoparticles replete with disease biomarkers and therapeutic targets. *Nature Cell Biology, 23*, 1240–1254.
4. Cable, J., Witwer, K. W., Coffey, R. J., Milosavljevic, A., von Lersner, A. K., Jimenez, L., Pucci, F., Barr, M. M., Dekker, N., Barman, B., Humphrys, D., Williams, J., de Palma, M., Guo, W., Bastos, N., Hill, A. F., Levy, E., Hantak, M. P., Crewe, C., … Gololobova, O. (2023). Exosomes, microvesicles, and other extracellular vesicles-a Keystone Symposia report. *Annals of the New York Academy of Sciences, 1523*, 24–37.
5. Buzas, E. I. (2023). The roles of extracellular vesicles in the immune system. *Nature Reviews Immunology, 23*, 236–250.
6. Ma, L., Li, Y., Peng, J., Wu, D., Zhao, X., Cui, Y., Chen, L., Yan, X., Du, Y., & Yu, L. (2015). Discovery of the migrasome, an organelle mediating release of cytoplasmic contents during cell migration. *Cell Research, 25*, 24–38.
7. Cooper, J. F., Guasp, R. J., Arnold, M. L., Grant, B. D., & Driscoll, M. (2021). Stress increases in exopher-mediated neuronal extrusion require lipid biosynthesis, FGF, and EGF RAS/MAPK signaling. *Proceedings National Academic Science USA,* 118.
8. Kakarla, R., Hur, J., Kim, Y. J., Kim, J., & Chwae, Y. J. (2020). Apoptotic cell-derived exosomes: Messages from dying cells. *Experimental & Molecular Medicine, 52*, 1–6.
9. Melentijevic, I., Toth, M. L., Arnold, M. L., Guasp, R. J., Harinath, G., Nguyen, K. C., Taub, D., Parker, J. A., Neri, C., Gabel, C. V., Hall, D. H., & Driscoll, M. (2017). C. elegans neurons jettison protein aggregates and mitochondria under neurotoxic stress. *Nature, 542*, 367–371.
10. Wolf, P. (1967). The nature and significance of platelet products in human plasma. *British Journal of Haematology, 13*, 269–288.
11. Tutrone, R., Donovan, M. J., Torkler, P., Tadigotla, V., McLain, T., Noerholm, M., Skog, J., & McKiernan, J. (2020). Clinical utility of the exosome based ExoDx Prostate(IntelliScore) EPI test in men presenting for initial Biopsy with a PSA 2–10 ng/mL. *Prostate Cancer and Prostatic Diseases, 23*, 607–614.
12. McKiernan, J., Donovan, M. J., Margolis, E., Partin, A., Carter, B., Brown, G., Torkler, P., Noerholm, M., Skog, J., Shore, N., Andriole, G., Thompson, I., & Carroll, P. (2018). A prospective adaptive utility trial to validate performance of a novel urine exosome gene expression assay to predict high-grade prostate cancer in patients with prostate-specific antigen 2–10 ng/ml at initial biopsy. *European Urology, 74*, 731–738.
13. Margolis, E., Brown, G., Partin, A., Carter, B., McKiernan, J., Tutrone, R., Torkler, P., Fischer, C., Tadigotla, V., Noerholm, M., Donovan, M. J., & Skog, J. (2022). Predicting high-grade prostate cancer at initial biopsy: Clinical performance of the ExoDx (EPI) Prostate Intelliscore test in three independent prospective studies. *Prostate Cancer and Prostatic Diseases, 25*, 296–301.
14. Pisitkun, T., Shen, R. F., & Knepper, M. A. (2004). Identification and proteomic profiling of exosomes in human urine. *Proceedings National Academic Science USA, 101*, 13368–13373.
15. Skog, J., Wurdinger, T., van Rijn, S., Meijer, D. H., Gainche, L., Sena-Esteves, M., Curry, W. T., Jr., Carter, B. S., Krichevsky, A. M., & Breakefield, X. O. (2008). Glioblastoma microvesicles

transport RNA and proteins that promote tumour growth and provide diagnostic biomarkers. *Nature Cell Biology, 10*, 1470–1476.
16. McKiernan, J., Donovan, M. J., O'Neill, V., Bentink, S., Noerholm, M., Belzer, S., Skog, J., Kattan, M. W., Partin, A., Andriole, G., Brown, G., Wei, J. T., Thompson, I. M., Jr., & Carroll, P. (2016). A novel urine exosome gene expression assay to predict high-grade prostate cancer at initial biopsy. *JAMA Oncology, 2*, 882–889.
17. McKiernan, J., Noerholm, M., Tadigotla, V., Kumar, S., Torkler, P., Sant, G., Alter, J., Donovan, M. J., & Skog, J. (2020). A urine-based Exosomal gene expression test stratifies risk of high-grade prostate Cancer in men with prior negative prostate biopsy undergoing repeat biopsy. *BMC Urology, 20*, 138.
18. Friedewald, W. T., Levy, R. I., & Fredrickson, D. S. (1972). Estimation of the concentration of low-density lipoprotein cholesterol in plasma, without use of the preparative ultracentrifuge. *Clinical Chemistry, 18*, 499–502.
19. Sniderman, A., Vu, H., & Cianflone, K. (1991). Effect of moderate hypertriglyceridemia on the relation of plasma total and LDL apo B levels. *Atherosclerosis, 89*, 109–116.
20. Jeyarajah, E. J., Cromwell, W. C., & Otvos, J. D. (2006). Lipoprotein particle analysis by nuclear magnetic resonance spectroscopy. *Clinics in Laboratory Medicine, 26*, 847–870.
21. Lucien, F., Kim, Y., Qian, J., Orme, J. J., Zhang, H., Arafa, A., Abraha, F., Thapa, I., Tryggestad, E. J., Harmsen, W. S., Kosti, J., Ali, H., Lowe, V. J., Johnson, G. B., Kwon, E. D., Dong, H., & Park, S. S. (2022). Tumor-derived extracellular vesicles predict clinical outcomes in oligometastatic prostate cancer and suppress antitumor immunity. *International Journal of Radiation Oncology Biology Physics, 114*, 725–737.
22. Welsh, J. A., Arkesteijn, G. J. A., Bremer, M., Cimorelli, M., Dignat-George, F., Giebel, B., Gorgens, A., Hendrix, A., Kuiper, M., Lacroix, R., Lannigan, J., van Leeuwen, T. G., Lozano-Andres, E., Rao, S., Robert, S., de Rond, L., Tang, V. A., Tertel, T., Yan, X., … van der Pol, E. (2023). A compendium of single extracellular vesicle flow cytometry. *J Extracell Vesicles, 12*, Article e12299.
23. Welsh, J. A., Van Der Pol, E., Arkesteijn, G. J. A., Bremer, M., Brisson, A., Coumans, F., Dignat-George, F., Duggan, E., Ghiran, I., Giebel, B., Gorgens, A., Hendrix, A., Lacroix, R., Lannigan, J., Libregts, S., Lozano-Andres, E., Morales-Kastresana, A., Robert, S., De Rond, L., … Jones, J. C. (2020). MIFlowCyt-EV: A framework for standardized reporting of extracellular vesicle flow cytometry experiments. *J Extracell Vesicles, 9*, 1713526.
24. Kim, J., Xu, S., Jung, S. R., Nguyen, A., Cheng, Y., Zhao, M., Fujimoto, B. S., Nelson, W., Schiro, P., Franklin, J. L., Higginbotham, J. N., Coffey, R. J., Shi, M., Vojtech, L. N., Hladik, F., Tewari, M., Tigges, J., Ghiran, I., Jovanovic-Talisman, T., … Chiu, D. T. (2024). Comparison of EV characterization by commercial high-sensitivity flow cytometers and a custom single-molecule flow cytometer. *Journal of Extracellular Vesicles, 13*, Article e12498.
25. Lucien, F., Gustafson, D., Lenassi, M., Li, B., Teske, J. J., Boilard, E., von Hohenberg, K. C., Falcon-Perez, J. M., Gualerzi, A., Reale, A., Jones, J. C., Lasser, C., Lawson, C., Nazarenko, I., O'Driscoll, L., Pink, R., Siljander, P. R., Soekmadji, C., Hendrix, A., … Nieuwland, R. (2023). MIBlood-EV: Minimal information to enhance the quality and reproducibility of blood extracellular vesicle research. *Journal of Extracellular Vesicles, 12*, Article e12385.
26. Welsh, J. A., Goberdhan, D. C. I., O'Driscoll, L., Buzas, E. I., Blenkiron, C., Bussolati, B., Cai, H., Di Vizio, D., Driedonks, T. A. P., Erdbrugger, U., Falcon-Perez, J. M., Fu, Q. L., Hill, A. F., Lenassi, M., Lim, S. K., Mahoney, M. G., Mohanty, S., Moller, A., Nieuwland, R., … Witwer, K. W. (2024). Minimal information for studies of extracellular vesicles (MISEV2023): From basic to advanced approaches. *Journal of Extracellular Vesicles* 13, e12404.

The Need for Standardization of Extracellular Vesicle Characterization Methods

Bryant C. Nelson, Molly E. Wintenberg, Samuel W. Schaffter, and Wyatt N. Vreeland

Abstract Extracellular vesicles (EVs) are secreted from all prokaryotic and eukaryotic cells. The application of EVs in therapeutics, clinical diagnostics, and drug delivery is expanding rapidly and the EV field is set to capitalize on this expansion through the development of next-generation therapies for regenerative medicine and cell-free therapies. The relevant critical quality attributes (CQAs) for EV therapeutics have not been defined, yet there is a clear need to develop EV-specific characterization methods based on the experiences gained from the manufacturing of liposome- and lipid nanoparticle-based therapeutics. This chapter discusses potential EV CQAs and describes analytical methods and tools that could potentially be utilized to quantitatively characterize the selected attributes. We focus mainly on methods and tools for characterizing/measuring EV particle size and particle size distributions (PSDs), particle number concentrations (PNCs), surface charge/zeta potentials (ZPs), protein cargo and nucleic acid cargo.

1 Introduction

In the context of cell-free therapeutic modalities, extracellular vesicle (EV)-based therapeutics and diagnostics are set to become the next evolution in the development and application of lipid bilayer-based drug carriers. Starting from humble beginnings as "platelet dust [1]," EVs are lipid membrane-bound particles secreted from both

B. C. Nelson (✉) · M. E. Wintenberg · S. W. Schaffter · W. N. Vreeland (✉)
National Institute of Standards and Technology Material Measurement Laboratory, Gaithersburg, MD, USA
e-mail: bryant.nelson@nist.gov

W. N. Vreeland
e-mail: wyatt.vreeland@nist.gov

M. E. Wintenberg
e-mail: molly.wintenberg@nist.gov

S. W. Schaffter
e-mail: samuel.schaffter@nist.gov

R. K. Marcus and D. Lubman (eds.), *Exosomes and Extracellular Vesicles*,
https://doi.org/10.1007/978-3-031-89270-7_2

prokaryotic and eukaryotic cells (nonpathogenic and pathogenic). EVs are the established gatekeepers of cell-to-cell communication through the transfer of functional proteins (syntenin-1, ALIX, etc.) and coding and non-coding nucleic acids (mRNA, miRNA, DNA, etc.) from originator cells to recipient cells. EVs can be found in all biological fluids and have been repeatedly demonstrated to be biocompatible, minimally toxic and nonimmunogenic under a multiplicity of in vitro, in vivo and ex vivo settings. The repertoire of potential EV clinical applications is extensive and rapidly expanding. From regenerative medicine and advanced therapies (RMATs) utilizing stem cell-derived EVs to anticancer therapies based on the use of plant-derived EVs, EVs are becoming a significant presence in therapeutics and clinical diagnostics.

In the recent past, EVs were typically classified based on their postulated biogenesis pathway in combination with their relative particle size distributions (PSDs). EVs were classified into three major subtypes: exosomes (40 nm to 160 nm), ectosomes/microvesicles (50 nm to 1000 nm) or apoptotic bodies (500 nm to 2000 nm) [2] based on their formation via the endosomal sorting complex required for cargo transport (ESCRT) regulatory pathway [3, 4]. However, the International Society for Extracellular Vesicle (ISEV) research, currently recommends that unless the biogenesis of the EVs under study is confidently known, EVs should be subtyped as small EVs: sEVs (<200 nm in diameter) or large EVs: lEVs (>200 nm in diameter) [5]. Nevertheless, there is a significant overlap in size and size heterogeneity between and within the different EV subtypes. There also exists extensive EV heterogeneity (type and levels) within EV subtypes in terms of cargo content, protein surface receptors and lipid membrane content which makes isolation, purification and characterization of pure EV subpopulations analytically challenging [6–9]. Nevertheless, the use of EVs as potential therapeutic modalities for the treatment of cancers, inflammatory diseases, autoimmune diseases, viral diseases, cardiovascular diseases, renal diseases and neurodegenerative diseases in both preclinical and clinical trials has been widely reported and reviewed [10–14]. In addition to acting as direct moderators of cell-to-cell communication, EVs are intimately involved in the regulation of the immune system, angiogenesis, apoptosis, tissue regeneration and repair, antigen presentation and maintaining homeostasis [13]. Conversely, secreted EVs are also associated with viral pathogenicity, misfolding of proteins, neoplasia and cancer progression and metastasis [12]. These opposing biological activities of EVs are dependent upon the organism's disease or health status, and thus, the biological activities of EVs are not mutually exclusive.

The efficient and protected delivery of therapeutic levels of RNA transcripts (mRNA), miRNA and/or functionally active proteins to diseased recipient cells is correlated with many of the positive attributes of EV-based therapeutics and drug delivery vehicles. For example, mesenchymal stem cells (MSCs) are being tested to treat patients with inflammatory diseases due to the immunosuppressive activity of MSCs. However, numerous studies have demonstrated that it is the MSC-derived EVs that are the responsible factors for attenuating autoimmune and inflammatory responses. Accumulating evidence suggests that specific miRNAs, as well as other immunomodulatory factors within the MSC-EV secretome, are the main architects of the immunosuppressive activity [15–17]. Another example of the clinical potential of EV-based drug therapy is the targeted EV uptake into cells using functional engineered proteins. Chimeric antigen receptor (CAR) T cell therapy has been used to engineer T cells to target and kill cancer cells. CAR-T cells have been shown to secrete EVs containing CAR protein [18–20]. CAR containing EVs can then target cancer cells and release EV cargo that can kill target cells or inhibit cancer cell growth [19, 20]. CAR EVs may also overcome the limitations of CAR-T therapy by not inducing cytokine release syndrome and having improved solid tumor infiltration. Further research can improve understanding of the mechanisms of CAR EV activity and potentially lead to cancer therapies with CAR-T-derived EVs.

In addition to therapeutic and drug delivery applications, EVs are also heavily used as diagnostic particles for the detection of cardiovascular diseases, neurodegenerative diseases and, most prominently, for the early detection of cancers. Minimally invasive liquid biopsies on blood and/or urine specimens from cancer patients allow longitudinal sampling and treatment stratification as well as the ability to follow disease progression, response to treatment and disease recurrence [21, 22]. The key cargo in tumor-derived EVs that allow one to follow the progression of cancer consists of many different types of miRNAs, mRNAs with specific mutations, DNA, and tumor-derived proteins [23, 24]. Blood is the most common type of liquid biopsy sample. Of all the various types of RNA species found in blood, miRNAs are the most abundant, comprising approximately 76% of the total RNA components [25]. Importantly, the differential expression of tumorigenic vs. tumor suppressor miRNAs can be used as a potential strategy to detect cancer. One of the main challenges with using EV miRNAs as cancer biomarkers is their varying levels in blood which is dependent upon the individual patient status, disease state, originating cell type and the presence of non-tumor EV miRNAs. Because of this variability, definitive assessment of the presence or absence of cancer is often controversial [24]. One way to reduce diagnostic variability is to use multianalyte approaches for the early detection of cancer. These approaches involve the use of EV RNA in combination with circulating tumor DNA (ctDNA) and circulating tumor cells (CTCs) or EV RNA in combination with cell free DNA (cfDNA) and disease-specific protein measurements [21, 24].

Whether addressing their therapeutic and/or diagnostic potential, it is well recognized that EVs, especially sEVs are promising candidates for a wide range of drug delivery, therapeutic and diagnostic applications. However, the EV field is currently limited by the lack of robust, validated and fit-for-purpose analytical methods that can be used to characterize the potential critical quality attributes (CQAs) of EV therapeutics. These methods are needed to support timely drug product submissions and approvals. The shortage of EV-specific analytics is not surprising because EVs are the newest version of lipid-based therapeutic nanocarriers (after liposomes and lipid nanoparticles—LNPs) to receive global scientific attention and the relevant CQAs for EV-based therapeutics and medical products have not been determined. This chapter discusses selected EV CQAs and describes potential analytical methods and tools that could be utilized to quantitatively characterize the selected attributes. In this chapter, we focus on methods and tools for characterizing/measuring EV particle size and PSDs, particle number concentrations (PNCs), surface charge/zeta potentials (ZPs), protein cargo and nucleic acid cargo. The key standardization challenges inherent to the selected methods and tools are described in alignment with the most recent standardization guidelines reported by ISEV [5]. We also provide recommendations for overcoming method-specific technical challenges.

2 Methods and Tools for Characterizing EV PSDs

(*N.B.* unless noted otherwise, all sizes mentioned refer to a particle's equivalent spherical diameter).

2.1 *Introduction to Particle Size*

What is a particle's size? This question, *prima fascia,* seems very simple and straightforward, but upon a bit of cogitation, its simplicity becomes less evident. What aspect of a particle's size is important depends on what question is being asked. Before delving into the important aspects of EV particle size and how to measure it, a pedagogical exercise helps to frame the range of what size can mean.

Consider this relatively simple question: Which of the three following countries is the largest: the United States, China, or India [26]? The answer depends on what we can measure to determine size. A first answer might come from a cartographer, who can only measure land area, in which case the United States at 9.83M km^2 is roughly the same as China at 9.60M km^2; India is a little more than one-third the size of the United States at 3.30M km^2. A second answer is from a demographer, who can only count the number of people in an area. In this case, India's population of 1.37B people makes it slightly larger than China, with 1.25B inhabitants, and at 25% population of the other two countries, the United States is clearly the smallest. An answer from an economist, who can only know the amount of money a country produces in a year, would give yet a different answer to which country is "largest." The 2021 gross domestic product of the USA is $23.0T, which is significantly more than China's $17.7T; India's is by far the smallest at $3.2T. Thus, with this macroscale measurement of size we can get three different rankings of countries' size based on how we choose to measure size. This is the same when it comes to measuring the size of particles at the micro and nanoscales as will be explained in the coming sections.

Measurement of physical size for objects of a regular known geometry or shape on the macro (human) scale is straight forward; for a box we simply measure the height, width, and depth with a ruler. For a spherical ball, we can make similar measurements by measuring the distance between opposite sides of the ball to determine its diameter. For irregular shapes, the size is more difficult to define. Since EVs have a nearly uniform spherical shape, it is most convenient to measure size in terms of radius or diameter.[1] When the object size decreases to the micro and nano scale, we do not have physical rulers and we must rely on how the object interacts with other things in its environment to derive its size. Ancillary items carefully added into an EV's environment include light, electricity, body forces (aka G-forces) among others. How these ancillary additions change upon interacting with the particles, or the particles change upon interacting with them can yield information about size.

2.2 EV Particle Size Measurement Uses Two General Physical Phenomena

Current EV size measurement techniques rely on one of two general modalities: light scattering and interference *-or-* volume exclusion. The details of how these modalities are implemented are myriad, distinct, and have their own strengths, weaknesses, and biases. Most analytical strategies to measure EV particle size use some form of light scattering. Its deployment is so widespread that a basic understanding of its mechanism is warranted.

2.3 What is Light Scattering?

Light scattering is observed in almost every situation we observe with our eyes: It is the reason the sky is blue, sunsets are red, full-fat (whole) cow's milk and clouds in the sky are a bright white, whereas fat-free (skim) cow's milk can have a slightly blue hue. These commonly observed phenomena are due to how light of different colors (i.e., wavelengths) interact and hence scatter differently as they move through physical matter (the only situation where light doesn't scatter is when it moves through a complete vacuum). Light is a composite of electric and magnetic waves. Maxwell showed in 1861–1862 that electricity and magnetism are interdependent and that neither exists without the other [27]. When a light wave moves through condensed matter its electro-magnetic (EM) field will interact with the electrons in the propagating condensed matter. The greater the interaction between the EM light forces and the electrons, the more the electron will deviate from their natural non-excited state; if the deformation of the electrons is large enough, as they relax back to a less-excited state they will occasionally emit a photon of light of the same color as the EM light but in a random (and usually) different direction (this is known as elastic scattering where no energy is transferred to the scattering material, and the EM light wavelength remains unchanged). For most biological particles in

[1] For historical reasons, people with physics and/or engineering-based education will prefer "radius" whereas people with chemistry and biology-based education will prefer "diameter". To avoid miscommunication between various disciplines, it is important that the size of EVs always include which variant of the measurand is used.

aqueous solution, about one in one million photons are elastically scattered above the background scatter of the continuous medium, i.e. the scattering from the particle itself is around a 1 part-per-million event. This means light scattering is not a strong effect in contrast to other light-emitting phenomenon like fluorescence where 50% of EM photons or more undergo inelastic emission. Due to the low abundance of elastic scattering photons, the careful design and operation of optical equipment to measure elastic scattering events is required. All light scattering-based measurement techniques implement high intensity light sources in order to produce a detectable number of elastically scattered photons. For the last $\approx$ 40 years, lasers have become the dominant light source for scattering applications due to their high intensity and high collimation. Collimation is the degree to which all photons are moving parallel to each other; a collimated light beam will not expand as it travels away from its source (consider a collimated light source like a laser pointer, versus a non-collimated light source like an incandescent flashlight). Collimated light makes it simpler to confine the illumination light to a controlled volume and prevent it from obscuring the detection of any scattered light.

2.4 *How Do Light Waves Interact with Each Other?*[2]

Two light waves that coexist will couple together so that the observer will see only the composite combined properties. This combination results in a range of observed intensities bound by two limits depending on how the two waves interact. To use this phenomenon to derive particle size it is useful if the two light waves are *coherent*. Coherent means the two light waves have a constant phase relationship, i.e., the "distance" between the peaks of the light waves is constant. If the peaks of the light waves are exactly lined up, the light waves coupling is *constructive*, this represents the maximum magnitude observed. Conversely, if the peaks of one wave are exactly lined with troughs of the other wave the coupling is *destructive,* which produces the minimum magnitude observed. If two light waves are monochromatic and of equal intensity, *constructive* interference results in an observed light intensity 4X that of a single individual wave, whereas if the two identical waves interfere *destructively,* they "cancel" each other, and the observed intensity is zero. The phenomenon is shown in Fig. 1.

The above example illustrates limiting cases that only occur with coherent light, which is often produced with laser light sources. It is important to remember that coherent light, where the phase between the multiple light waves is constant can produce any magnitude between the limits of perfectly constructive and perfectly destructive, by changing the phase shift between the light waves. More common light sources (e.g., candles, light bulbs, etc.) do not produce *coherent* light, but *incoherent* light. Incoherent light is different from coherent light in that there is not

[2] * For pedagogical simplicity, many details of the precise physics have been omitted, but accuracy to the fundamental principles of interest are correct and true.

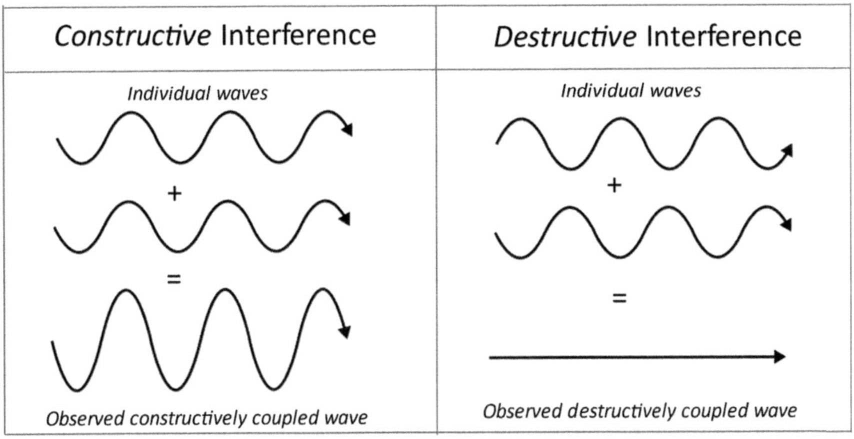

Fig. 1 Schematic of the observed outcome of two coherent light wave after constructive interference (left) or destructive interference (right). N.B. The constructively coupled wave is 2X the amplitude of the original interfering waves, the intensity of light is proportional to amplitude squared ($I \sim A^2$)

a particular correlation in the alignment of light wave phases; each light wave has a random phase relative to other light waves from that source. When incoherent light waves interact, the observed intensity is the simple arithmetic sum of the constituted light waves as is shown schematically in Fig. 2.

When a coherent light source interacts with a scattering particle (e.g., an EV), an emitted scattered light wave may *or* may not retain coherence with other scattered light waves. The level of coherence in scattered light waves is the primary observable

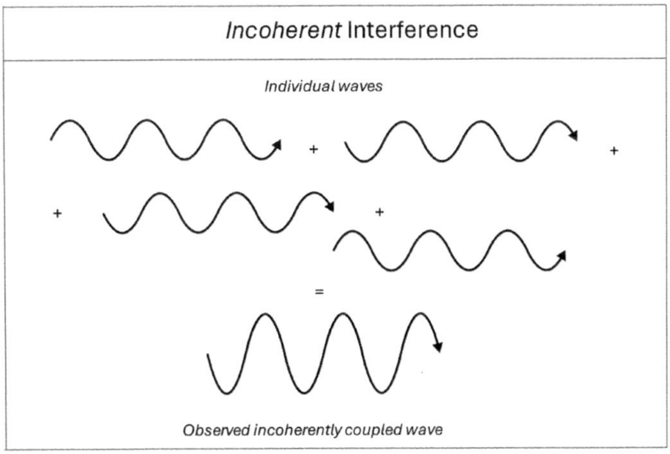

Fig. 2 Schematic of the observed outcome of four incoherent light waves after interference. N.B. it takes four incoherently coupled light waves to produce a composite wave with the same amplitude composite wave as two coherently constructively coupled light waves as shown in Fig. 1

(measurable) light scattering phenomenon used to determine the size of particles via the two dominant light scattering measurement modalities: Dynamic Light Scattering (DLS) and Static Light Scattering (SLS).

2.5 Measurement of EV Size: General Considerations

Light scattering-based measurements of EVs (and any other nano-sized colloidal particles) have one effect that is common that needs to be kept in mind when interpreting and evaluating light scattering-based methods. Many novice measurement efforts can be confused by and reach erroneous conclusions without a full appreciation of this fact. The vast majority of EVs scatter light in a so-called Rayleigh-Gans (RG) [28, 29] limit (all EVs with a radius <500nm). The RG limit is a solution for isotropic bodies derived from the more general Mie scattering equations that describe how spherical particles scatter light.. It is applicable to systems where the scattering particle radius is *both* smaller than the wavelength of scattering light *and* similar in refractive index to the continuous phase surrounding the particle. Most EV types satisfy these requirements in that their radius is generally between 10 and 300 nm and the wavelength of most light-scattering-based instruments is between 532 and 650 nm. Additionally, EVs are always in aqueous solution when isolated from their source fluid and most of the EV mass is also aqueous solution so the refractive indexes are similar.

RG scattering has one constant characteristic that can be extremely beneficial in some applications and present some limitations in others. This characteristic is that the magnitude of light scattered by a particle increases with particle radius in a *highly* non-linear manner, scattering intensity scales with particle radius to the sixth power (Eq. 1).

$$Intensity_{scatteredlight} \sim radius_{particle}{}^{6} \tag{1}$$

It is useful to explain how this can affect the measurement with a simplified thought experiment. Consider the measurement of two EV particles that are completely identical in all characteristics (intravesicular composition, refractive index, lipid composition, etc.) but one particle has 10 nm radius and the other 100 nm radius. The amount of scattered light produced by the 100 nm radius particle will be *one million times (!!!!)* more intense than that from the 10 nm radius particle. Effectively this means a very small number of large particles will produce an amount of light scatter that is vastly larger than light scattering from many small particles. This is useful when testing a sample for the presence of particle aggregation or agglomeration, where a small number of aggregates will produce a strong signal compared to the smaller particle from which the aggregates can form from. However, this characteristic can obscure and complicate the detection and measurement of small particles in a commonly encountered heterogeneous (in terms of EV size ranges) EV sample preparation. It is difficult to understate how this can impact measurement in a surprisingly non-intuitive manner.

2.6 PSD Measurements of EVs Using DLS

Dynamic Light Scattering (DLS), also known as Quasi-Elastic Light Scattering (QELS) or Photon Correlation Spectroscopy (PCS) is among the most widely used techniques to measure particle size. Instrumentation for the measurement is commonly available in laboratories, it is easy and simple for the user to perform, it can provide measurements quickly, and the sample can be recovered, without alteration, to be used in other applications. In general, DLS has a wide dynamic range for size measurement with most modern instruments able to measure particles from 2 nm to 20 μm (hydrodynamic diameter).

Like many measurement techniques, it doesn't directly measure the most commonly reported measurand, but rather another phenomenon that can be easily converted to the desired measurand. In the case of DLS, the technique directly measures the characteristic time of light scatter that exists in a *coherent* mode *vs.* an *incoherent* mode.

It is useful to explain the details of how the quick transition between coherent and incoherent scattering can yield a particle size. Consider a sample preparation of EVs in a solution state: these particles will undergo movement in a random walk manner and during that diffusion individual particles will move closer together and further apart in a random manner. However, smaller particles, which move more quickly in their random walk will move between being closer to and farther from neighboring particles relatively quickly. In contrast, large particles, which move more slowly under Brownian motion will move between being closer to and farther from neighboring particles more slowly. The time scale at which particles move close and far is what DLS probes and ultimately presents as hydrodynamic diameter or hydrodynamic radius. A short description of how this happens follows:

When two independently moving particles are close to each other, the coherence of the light waves they scatter will be very high. Alternatively, when the particles are far apart, the scattered light waves will have low coherence and can be considered incoherent as shown in Fig. 3. How quickly scattered light transitions from coherent scattering and incoherent scattering is directly related to the time it takes particles to move proximal and distal to each other, which is most often presented as the translational diffusion coefficient [30].

Perhaps the easiest technique to determine the time scale of some random fluctuation (in this case the transition between coherent and incoherent scattered light) is to measure its autocorrelation function; a technique used throughout physics. Qualitatively, as implemented in DLS, autocorrelation compares the change in intensity of scattered light as a function of time to the exact same intensity fluctuation shift by some time increment. As the time increment of the shift is increased, the original and time-shifted data sets will become less and less alike. One-half of the time shift

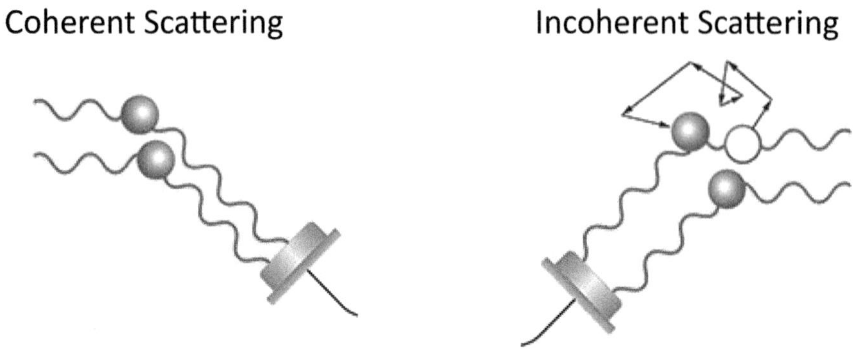

Fig. 3 Schematic of coherent constructive scattering (left) where particles are proximal and Incoherent scattering (right) where particles are distal. When the scattered light reaches the detector (gold disc) the intensities are different according to the coherence of the scattered light. (adapted from Wyatt Technologies)

required to go from completely correlated to completely uncorrelated is roughly the decay time. This is schematically depicted in Fig. 4.

This is essentially the average amount of time it takes the light scattered by a particle suspension to transition from coherent to incoherent scattering and can be

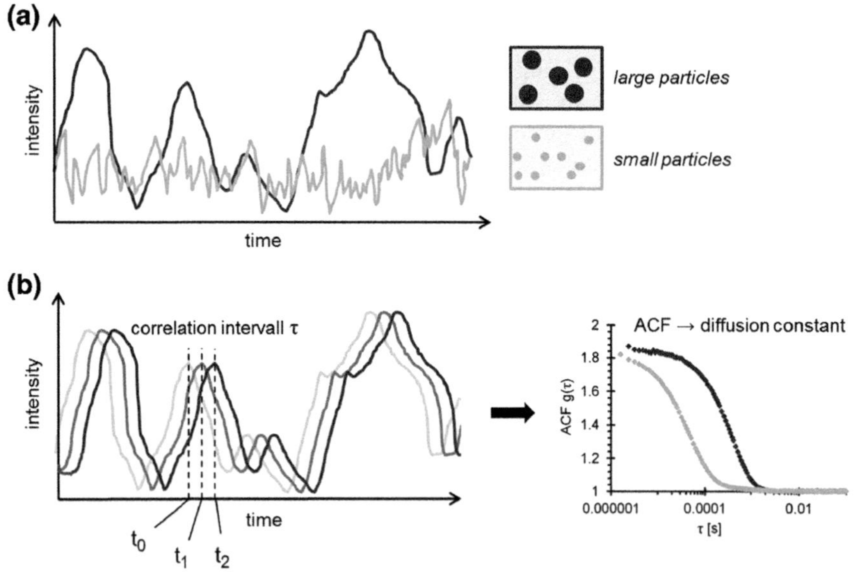

Fig. 4 Intensity fluctuations for particles (panel A) small particles (green) have a more rapid intensity fluctuation while large particles (blue) have a slower intensity fluctuation. Panel B shows the time shifted intensity fluctuations, where at larger times the fluctuations become less correlated and can be plotted on a semi-log axis to determine the characteristic correlation time [31]

directly transformed to the translational diffusion coefficient ($\mathcal{D}_{trans}$). $\mathcal{D}_{trans}$ has units of area/time, most often cm^2/sec, which is generally not intuitive to understand in common real-world characteristics, thus it is usually transformed into the more easily interpreted physical size equivalent hydrodynamic diameter via the Stokes–Einstein relationship as shown in Eq. 2.

$$d_h = \frac{k_B T}{3\pi \eta \mathcal{D}_{trans}} \tag{2}$$

where d_h is hydrodynamic diameter, k_B is Boltzmann constant, T is absolute temperature, 3π is a geometric constant for equating the size to a spherical shape, η is dynamic viscosity (not to be confused with the related kinematic viscosity). For accurate calculation of d_h from $\mathcal{D}_{trans}$ one must also know T and η. Generally, T can be easily measured and for aqueous solutions, η can be calculated from well-known viscosity-temperature relationships. Alternatively, it can be directly measured, but this is rarely done in practice, especially for common solvents. Further, during the measurement, diffusion must be the dominant source of particle movement without any significant electrophoretic movement or bulk movement.

It should also be appreciated that d_h is not strictly related to any physical size parameter of the scattering particle, but rather is the diameter of a solid sphere that has the same $\mathcal{D}_{trans}$ as the scattering particle. For EVs, which are generally spherical in shape this is a good approximation of physical reality. An important consideration is if EVs have significant coronal species on their periphery it can increase the measured d_h to be larger than the diameter delineated by its lipid envelope. Additionally, one must also always remember if the sample is heterogeneous in size, the DLS-measured size will be influenced more strongly by the larger particles. Finally, light scattering techniques have no particle-specific selectivity; dirt, dust, or non-EV particles will scatter light. EV-specific sample preparation procedures and laboratory cleanliness are of paramount importance.

2.7 PSD Measurements of EVs Using SLS

The other dominant light scattering modality used to size particles is SLS. SLS differs from DLS in that the average intensity of the scattered light over a ~1 s time is measured rather than the short time scales (10^{-6} s to 10^{-2} s) used in DLS. SLS is most often implemented in a multi-angle format called Multi-Angle Light Scattering (MALS). MALS is the same as DLS in that it measures how the particles' scattered light transitions between scattering modes. DLS measures the timescale of transition of coherent scattering (where the scattering light is phase-aligned) to incoherent scattering (where the scattering light has no phase alignment). MALS measures constructive to destructive transition of coherently scattered light as the observation angle changes. Legacy instruments are made of a single light intensity detector (photodiode, photoresistor, phototransistor, or photomultiplier tube) mounted on a goniometer that serially measures the scattered light at various angles. Thankfully, in the past few decades, the availability of high-quality, sensitive, wide-dynamic-range, and low-cost light detectors has made it more practical to position a plurality of detectors at various fixed angles around the scattering volume and measure the angular scattered light in parallel. This parallelization makes the measurement quicker, requires no manual detector positioning during the measurement, and allows implementation of MALS detectors inline with other detectors in various chromatography systems.

In 1908 Gustav Mie [32] provided a solution to the Maxwell equations applied to homogenous spheres. His mathematical *tour de force* created what has today become the seminal understanding that underpins myriad technology sectors. Mie theory provides the mathematics to calculate how electromagnetic radiation interacts (scatters) with objects. Mie theory is implemented in meteorology, communication, modern 3-D imaging, astronomy, medical diagnostics, particle sizing and in many other fields. The original paper is a turgid 69-page tome of dense mathematics that results in solutions sets of infinite series that can only be solved by modern computer systems. However, in the years after its publication some simplifications to the Mie theory resulted in more easily solved equations while maintaining sufficient accuracy to be quite useful. Perhaps the most widely used of these is the RG approximation which Robert Gans published in 1925 [33]. The two critical assumptions that allow a simplified solution are: (1) the refractive index of the particle and the surrounding medium are very similar *-and-* (2) the particle is equivalent to or smaller than the wavelength of light. Luckily for EVs and many other biologically important particles, these two assumptions are readily satisfied.

Angular dependence of scattered light: what causes it?

2.7.1 Isotropic Scattering

First, consider a particle that is much smaller than the wavelength of light. These particles can be considered "point scatterers", meaning that the scattered light can originate from any point on the particle and the emitted photon will have the same phase and intensity regardless of the scattering angle. This is shown in Fig. 5 where two coherent and collimated light waves move from left to right and interact with a small particle. If the light wave produces scattered photons at both the left-most and right-most edges of the particle, at a low angle of 15°, the scattered photons will be largely at the same phase alignment and produce a constant intensity at the detector as shown by the red-colored light waves. Similarly, if the other incoming light wave scatters a photon at both the leftmost and rightmost edges of the particle at 45° those photons will have roughly the same phase alignment as shown in the green-colored waves. Thus, the same light intensity is observed at both the 15° and 45° detectors. Effectively this means the amount of light scattered by small particles is independent of angle. In practice, particles with a radius less than 10% of the illuminating light's wavelength will have negligible variation with respect to angle. Hence the name isotropic which is from the Greek *iso* = same and *tropic* = to turn; meaning same for each turn or in this case angle.

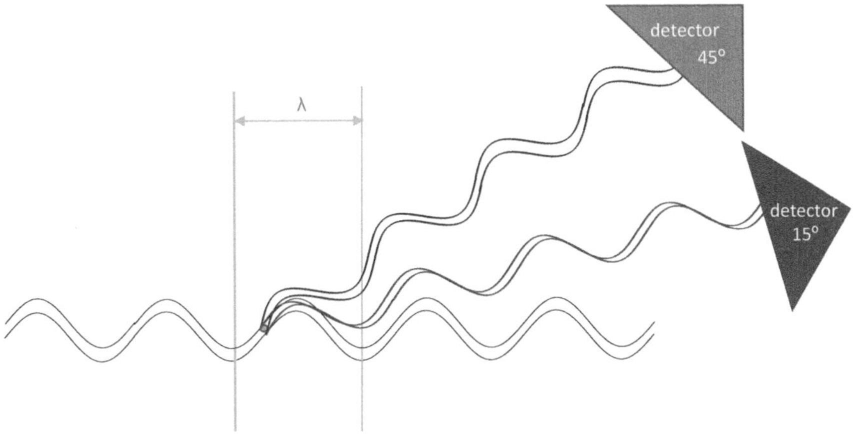

Fig. 5 Example of Isotropic scattering where the particle size is much smaller than the wavelength of light

2.7.2 Anisotropic Scattering

Now consider a particle with a radius that is equivalent to the wavelength of the illuminating light. For larger particles, the point where the scattered photon originates *as well as* the angle at which it scatters will impact its phase alignment and hence its measured intensity. This is shown in Fig. 6 showing the same scenario as Fig. 5, but with a larger scattering particle. Here if the incoming wave scatters photons from the leftmost side of the particle and the rightmost side of the particles at a low angle of 15°, the scattered photons will maintain phase alignment and arrive at the detector with mostly constructive interference and a relatively high intensity. In contrast, if the scattered photons are emitted at a higher angle (45° in this figure), their relative phases will be shifted by an amount proportional to the particle's radius and arrive at the detector with more destructive interference and hence a lower intensity. This is the basis of anisotropic scattering and how it can be used to measure particle diameter (or radius). The rate at which the light transitions from constructive to destructive interference with respect to scattering angle is proportional to particle size. Larger particles transition from constructive to destructive interference at lower scattering angles than smaller particles. Mathematical analysis of the attenuation with respect to angle allows for calculation of the particles' *radius of gyration.*

For the case of backscatter from anisotropic scatterers (scattering at angles greater than 90°), the scattering mechanism is more complicated due to multiple interferences from the illumination light and resonances and thus not suitable for a simple graphical explanation. For the purposes of SLS-based size measurements, it is sufficient to understand the forward scattering case and its angular attenuation. The interested reader is referred to the excellent and thorough text of van de Hulst [34] for a more detailed explanation of elastic light scattering.

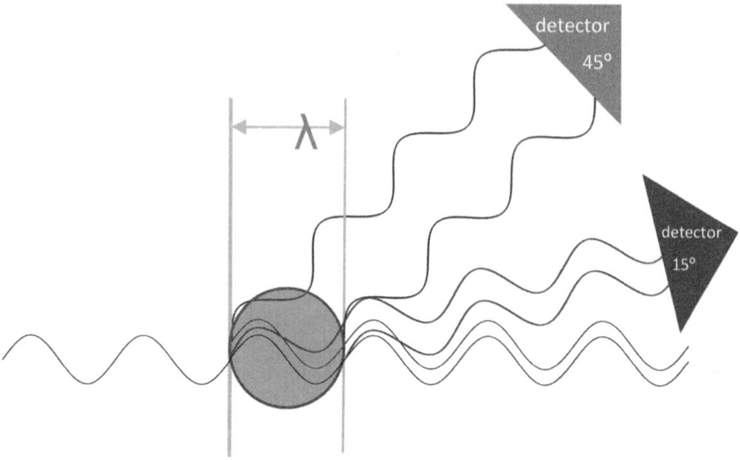

Fig. 6 Example of anisotropic scattering where the particle size is equivalent to the wavelength of light. In this case as the scattering angle changes, the scattered light transitions from constructive to destructive interference

2.8 Summary of Scattering Intensity-Based Size Measurements

(1) DLS measures the time-based change in scattering intensity at a fixed angle. The rate of this change is due to the speed at which particles move closer to each other or further from each other under diffusive motion. This is the particles' translational diffusion coefficient that is often transformed into a more intuitive hydrodynamic diameter (or radius). DLS does not have any selectivity for particle density or shape and only reports equivalent spherical diameter over a range extending from nm to several μm.

(2) SLS measures the time-average intensity of scattered light as a functional of angle. For particles that are sufficiently large in comparison to the wavelength of light (particle radius <10% light wavelength) the scattered light will transition from constructive interference to destructive interference as the scatter angle increases. The rate of this attenuation is proportional to the light wavelength. SLS can measure particle size in the range of $\approx$ 10 nm radius to $\approx$ 600 nm radius and is selective to particle density and general shape.

(3) Comparing DLS and SLS size measurements. The relative hydrodynamic size (DLS) to geometric size (SLS) can give some indication of particle architecture, i.e., is it a hollow sphere like a liposome or a solid sphere like a low-density lipoprotein (LDL) particle. The interested reader is referred to an excellent summary in *Macromolecules* [35]. This comparison can only be done over a size range where both DLS and SLS sizes are available (e.g., radii from $\approx$ 10 nm to $\approx$ 1000 nm).

(4) The scattering intensity for particles in the EV size range is *highly nonlinear* with slightly larger particles producing much higher intensities of scattered light.
(5) Light scattering methods "see" everything—this is both a strength and a limitation. Light scattering methods near universal sensitivity obviates the need for derivatization with fluorophores, antibodies, antigens, or other probes; the only requirement is that the particle has a different refractive index than the surrounding medium. However, light scattering measurements require clean, high purity samples because any particle contaminants in the samples will also scatter light..

3 Particle Size Distribution

Ensemble Measurements

The scattering-based measurements mentioned so far all measure the average plurality of particles and can provide an average size that is strongly influenced by any larger-sized particles that scatter light much more intensely. Batch-DLS (i.e. DLS measurement on a single solution with multiple species present) can provide coarse estimates of size distribution, but it requires several assumptions and is still constrained by the limited resolution of DLS in detecting distinct sizes in a sample composed of heterogeneously sized particles (e.g., EV preparation).

To obtain a more accurate estimate of PSD it is necessary to separate the constituent particles in a sample according to size, measure each separate fraction, and then reconstruct the PSD from these individual measurements. Ideally, this is done in a continuous manner with a chromatography system. In fact, for synthetic commodity polymers (i.e. plastics), this has been done for ~40 years with tandem size exclusion chromatography—light scattering (SEC-LS) systems where an SEC column provides size-based separation, and the LS provides a size measurement of the eluting stream. With an appropriately performing column, the eluting polymer at any instance in time is homogenous in size and the LS-measured size is not influenced by multiply-sized species. This allows for a more accurate construction of PSD compared to extrapolating distributions from bulk averages combined with various a priori assumptions.

The size-based separation provided by SEC columns for standard commodity polymers is not directly amenable to the separation of biological colloids like EVs because of their markedly increased surface activity and relative mechanical fragility. However, solid-phase chromatography approaches such as fast protein liquid chromatography (FPLC) based techniques can yield some benefit in determining PSDs. Consider a simple implementation where a crude or semi-enriched EV-containing fluid is loaded onto a SEC-type column (e.g., Izon) and fractions are collected in a drop-wise manner. These fractions can then be measured with other techniques to determine things like particle size, tetraspanin content, RNA sequences, and the like. However, one must be aware that the gel-column-based separation is not perfect and each collected drop can contain mixed populations of EVs (different sizes) that can confound subsequent downstream measurements. Further, EV membrane lipids are known to adsorb to the gel matrix and to carry over between samples on a given column. Lipid exchange between the stationary phase and the EVs can also be a concern.

3.1 Field Flow Fractionation

An ideal EV separation platform would have no interaction between the stationary phase and the EV sample and would not perturb the original EV sample's particle size or composition from column carry over between samples. An attractive platform exists that provides some of these advantages; it is called Field Flow Fractionation (FFF). This method separates particles using various physical forces without the involvement of a stationary phase. It was first proposed in a theoretical paper in 1976 by Calvin Giddings [36], though the hardware to realize the separation was not available at that time. All FFF techniques separate particles according to actions by two or more opposing physical forces as the particles move along a separation channel. The original publication presented different implementations, but the most widely used variant is Asymmetric Flow Field Flow Fractionation (AF^4).

The separation channel is generally of rectangular geometry that is several cm long, a few cm wide, and tens to hundreds of μm thick. In this rectangular channel a laminar flow is established that advects particles downstream. During this advection the particles are subjected to a force pushing them towards the bottom of the channel, simultaneously another force (of differing magnitude) pushes them away from the channel bottom. Usually, the later force is the ubiquitous isotropic diffusion of particles, and the former is a Stoke's drag force as is implemented in AF^4. In the other case, the downward force can be electrophoretic (Electrophoretic FFF (EFFF)), density/buoyancy (Sedimentation FFF) or thermal diffusivity (Thermal FFF). Currently, only AF^4 and EFFF are in significant use.

In all cases, the advective laminar flow adopts a parabolic distribution of flow velocities along the channel thickness between the upper and lower channel walls. For parabolic flows, the fluid elements in the channel center move at the highest velocity and the fluid elements near the top and bottom channel walls move more slowly. Thus, particles that reside further into the middle of the channel will advect down the channel at a faster rate than those that reside more proximal to the channel's top or bottom walls. This is the basis of FFF-based separations.

Some detail on the mechanism is helpful in understanding how FFF can separate particles (Fig. 7). Here the large downward arrow represents the method-specific body force (i.e. Stoke's drag or electrophoretic) the smaller upward blue arrows represent the upward diffusive movement. The diffusive movement is isotropic causing the particles to disperse across a large area, but to a first order approximation, it is only the upward component of the diffusion that impacts the separation. Here the green and blue spheres each occupy a "cloud" of space with a thickness that is determined by the ratio of separation force field to the upward component of diffusion. In all cases,

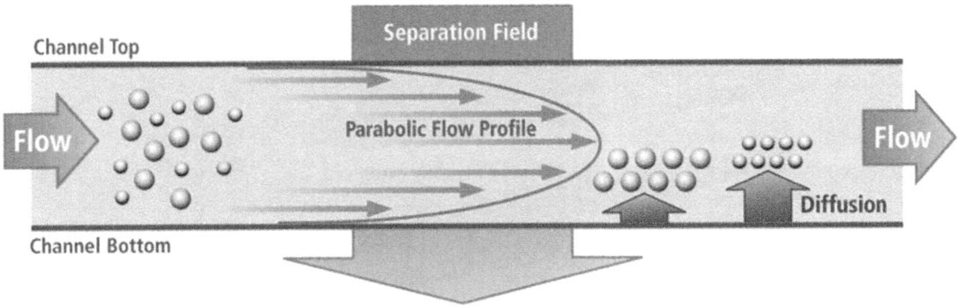

Fig. 7 Schematic of a general FF separation where the green particles occupy a thicker "cloud" and move through the channel more quickly compared to the blue particles which occupy a thinner "cloud" and move more slowly

the smaller particles will tend to diffuse into a larger "cloud" thickness while larger particles will occupy a thinner "cloud" over equivalent diffusion times. The magnitude of the opposing separation force field (Stoke's drag and/or electrophoretic) will then determine the relative thickness of the particles' cloud thickness and ultimately how quickly the particles advect along the channel.

3.1.1 Asymmetric Flow Field Flow Fractionation

Note: For pedagogical simplicity, it is easiest to explain the separation mechanism first and then describe how the particles are introduced into the channel (i.e., "injected").

In the most common implementation of AF^4, the separation field is a downward bulk fluid flow that imparts a downward Stoke's drag force on the particles that is counteracted by the upward component of the particles' diffusion. As particles increase in size, the magnitude of the downward Stoke's drag increases while the rate of upward diffusion decreases, so the larger particles occupy a relatively thin (and slowly advecting) cloud thickness. Whereas when particles decrease in size, the Stoke's drag decreases and the diffusive movement increases allowing the particles to occupy a greater thickness which results in faster advection of small particles through the channel.

An AF^4 channel is schematically shown in Fig. 8, where several layers are "sandwiched" together to form the channel. The bottom layer is a porous frit in a metal housing (grey and mottled black) that supports an ultrafiltration membrane (UF) (white) on the bottom of the channel. The next upward layer (blue) is a plastic spacer that defines the edges of the channel as well as the channel thickness. Finally, the top (transparent blue) is the impermeable upper layer that is the channel top. The top plate has three holes to allow fluidic access into and out of the separation channel, there is an additional hole in the bottom plate that allows is the outlet for fluid passing through the ultrafiltration membrane. The ultrafiltration membrane allows

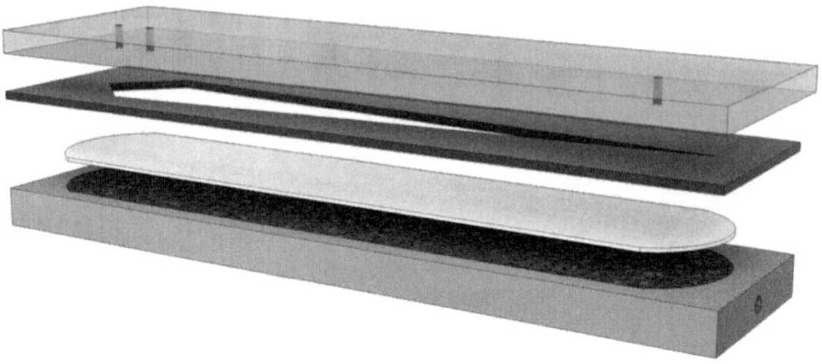

Fig. 8 A schematic of an AF^4 channel. The bottom most plate is a metal holder with an inlaid porous frit that supports an ultrafiltration membrane above it. The blue layer with a trapezoidal cut out is a spacer that defines the channel edges and thickness, and the topmost layer is an impermeable plastic block with three holes for fluidic access to the channel

the passage of bulk fluid flow in a downward direction but retains the particles in the separation.

Figure 9 shows a schematic of the separation of two particle species. At the start of the separation, all particles occupy a thin band near the ultrafiltration membrane channel bottom, while a high performance liquid chromatography pump delivers all the mobile phase to the upper left side of the channel. A flow meter measures the flow exiting the righthand side of the channel through both the exit at the channel top and at the channel bottom where fluid has passed through the ultrafiltration membrane. The flow meter is connected to a control valve that adjusts the relative flow rate between the top and bottom channel exits. This controls the magnitude of the downward cross-flow component and can be adjusted over the separation to increase the range of particle sizes that can be separated. The success of an AF^4 separation is largely determined by the appropriate selection of the cross-flow magnitude and rate of decrease during the separation. An optimum cross-flow and cross-flow ramp will sufficiently confine different particle sizes into distinct cloud thicknesses while still allowing access to sufficiently thick cloud heights so that particles elute before various dispersive mechanisms limit the separation efficiency. The flow conditions and particle cloud thickness distribution can be analytically calculated for perfectly well-behaved particles. Computer programs that simulate a separation can be a great aid in making initial estimates of "good" cross-flow rates and ramps if preliminary knowledge regarding the particles' physical characteristics is known. (However, current computer programs cannot account for the myriad subtleties and higher order effects that can impact a separation!!). An appropriate cross-flow will allow a continuously increasing size of particles to sequentially elute from the channel and onto a MALS detector where the particle size can be measured. The addition of a concentration detector (most often UV/Vis absorbance) can then be used to measure particle abundance at each size. The PSD can be reconstructed from these data.

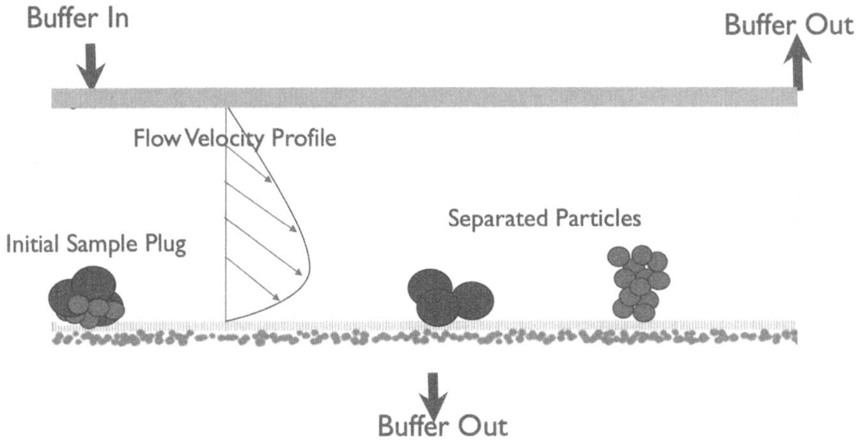

Fig. 9 A schematic of an AF^4 separation. The mobile phase's flow is a combination of both downward and rightward flows with the former forcing the particles in a size-selective manner to the bottom of the channel, while the later advect the particles through and out of the channel. This results in the smaller blue particles eluting before the larger red particles

3.1.2 Sample Injection

To introduce the sample as a narrow band proximal to the UF membrane, the right-hand upper fluid access port "changes direction" and serves as a fluid inlet port. Thus, during injection, fluid enters the channel from both the left upper and right upper ports and all the fluid exits through the bottom outlet underneath the UF membrane. While the flow is in this configuration, the sample is gently "sprinkled" in through the upper middle point. The relative right and left inlet flow rates are adjusted so there is a "stagnation" point under the sample inlet port on the UF membrane. Thus, the particles are advected with the fluid flow up to the membrane where they accumulate—this is the so-called "focusing" as shown in Fig. 10. Once the sample is introduced and focused, the flow configuration changes to "separation mode".

3.1.3 AF^4 Membrane, the Most Important Part of the Channel

Functionally the bottom of the separation channel is composed of an ultrafiltration membrane while the sides and top of the channel are impermeable. A perfect AF^4 membrane has the following characteristics:

(1) Completely impermeable to the particles being analysed
(2) Fully permeable to the separation's mobile phase and its components (e.g., salt ions, surfactants, micelles, etc.)
(3) No ionic, hydrophobic, or chromatographic interaction with the analytes

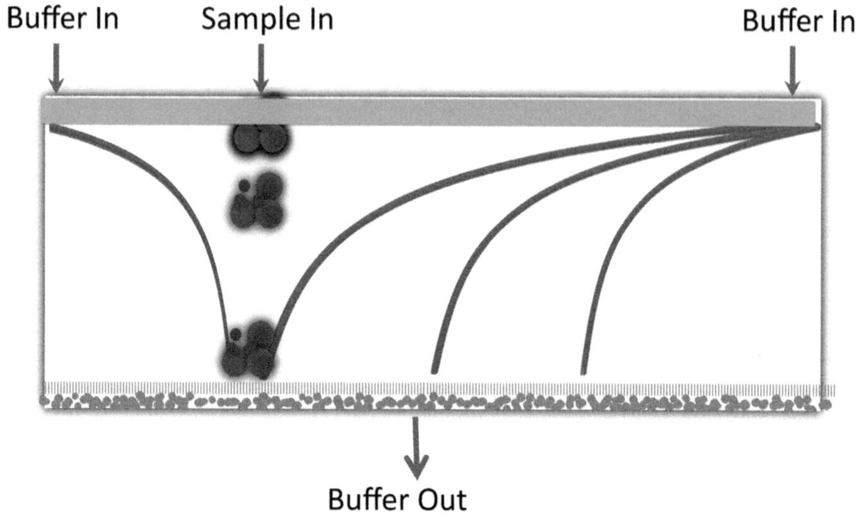

Fig. 10 A schematic of sample injection. Fluid flow is set up so that all fluid flows through the UF membrane. The relative flow rates on the left and right sides are adjected so the flow focuses under the middle sample inlet port as shown by the red flow lines. This results in the sample accumulating in a narrow region proximal to the UF channel bottom

Of course, no membrane can completely satisfy all these requirements. But rather membrane selection is an exercise in multiparameter optimization. The two most important characteristics of a membrane are: (1) its molecular weight cutoff (MWCO) and (2) its chemistry (i.e., material of construction).

The MWCO is not a strict demarcation where anything below the cut off freely passes through the membrane and everything above the cut-off is completely impermeable to the membrane. In fact, the membrane has absolutely no molecular weight (more correctly molar mass) selectivity (!!). The MWCO is rather the molar mass of a "normal" globular protein which is of a hydrodynamic size that is 90% retained from passing through the membrane. However, there is no standard definition of MWCO and some vendors may specify an 80% retention. Additionally, the flow rates and pressures for the MWCO determination are vendor and application-specific. Thus, one must empirically determine what membrane MWCO and chemistry sufficiently retain the analytes of interest in an AF^4 separation.

It is also important to recognize that membranes can show hysteresis as they are exposed to samples and various membrane "pores" get blocked and non-specific adsorption sites are occupied [37]. Further, the retention of the particles is dependent on both the injection "focusing flow" rate and "cross flow" rate. A modest flow rate may allow the particle to partition 100% in the separation channel whereas high flow rates may cause the particle to "squeeze" the membrane and be lost into the crossflow stream and not be measured by the MALS detector [38].

As well as concerns about sample retention, the area proximal to the ultrafiltration membrane has different properties than the average bulk solution of the channel.

During sample injection, the local concentration of the particles can increase by as much as 1000 X [39]. Any propensity particles may have towards aggregation may manifest at these locally high particle concentrations. Also, for aqueous buffers, the water molecules and hydrated buffer ions do not move through the membrane at the same rate. Ions in aqueous phase are surrounded by adherent water, making their effective size much larger than the ionic radius and thus they move more slowly through the membrane than free water molecules. Depending on the magnitude of the cross flow or focusing flow, the buffers' salt concentration and the membranes MWCO, the local increase in salt concentration can be significant. This so-called "membrane polarization" phenomenon is well known has been studied in water purification and filtration applications [40].

3.2 Individual Particle Measurement Systems

Thus far, the techniques explained all measure the average characteristics of a large ensemble of particles. Those particles may be directly taken from a biological source material where they are likely to be highly heterogeneous (e.g., EV sample preparation), or they can be purified and size-selected into smaller subfractions where those ensemble fractions are individually measured. Ideally, this fractionation will be done at a resolution where the subfractions are essentially homogeneous and unperturbed from their native condition. Currently, AF^4 is the separation method most amenable to these requirements.

In the past decade, only a few techniques have been applied to EVs and other colloidal particles that are capable of analysing individual particles in a serial or semi-serial manner, while the particles are in their pseudo-native aqueous environment. A few of these will be briefly described below. It should be noted this is not an exhaustive list and is selected from instrumentation accessible to the authors. The availability of technology and its implementation is constantly advancing and developing and these examples should only serve as a starting point for EV size and PSD analytics.

3.3 Nanoparticle Tracking Analysis (NTA) or Particle Tracking Analysis (PTA)

Nanoparticle Tracking Analysis (NTA), or more correctly Particle Tracking Analysis (PTA) became commercially available in 2003 from a company called NanoSight Ltd. and since then, other companies have developed instruments based on the same based methodology. The operational principle is perhaps the most direct measure of a particle's translational diffusion coefficient (and hence hydrodynamic size) since Robert Brown's famous jiggling pollen particle observation in 1827 [41].

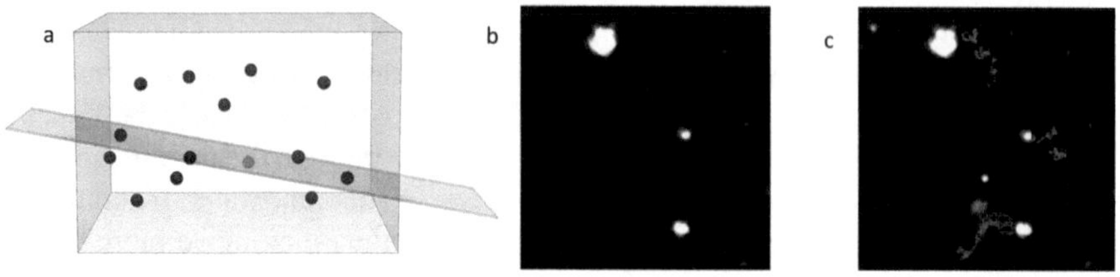

Fig. 11 **a** Schematic of a Particle Tracking Analysis detector where a laser sheet (light-red diagonal) illuminates some of the particles suspended in the flow cell. The light scattered from these particles is used to determine their location and movement over time to calculate a translational diffusion coefficient. **b** a camera image of the flow cell in (**a**) **c** the red lines show the location history of the particle over several sequential time frames

In PTA a quiescent fluid containing a solution of particles is illuminated with a thin sheet of laser light across two dimensions. A microscope camera is position in a sufficiently perpendicular direction to the laser sheet and records a series of images in time. As particles move into the laser sheet they will scatter light in all directions. The scattered light is recorded as a particle. This is shown in Fig. 11 where several blue balls are suspended in a cubic flow cell. The balls that are in the path of the light-red laser sheet will be illuminated and visible to the camera and produce an image like that in Fig. 11b. Sometime in the future, another image is recorded and compared to the first image. During the time between frames the particles will have diffused a certain distance and if the particles in the first image.

(Fig. 11b) can be correlated to the particles in the second image (Fig. 11c), the distance they have traveled is the random walk distance, which is proportional to the translational diffusion coefficient!!! If this process is repeated enough times the true average mean random walk distance can be measured and transformed into a hydrodynamic size, just as it is in a DLS measurement.

Operationally PTA requires the number of particles (particle concentration) in an image to be large enough to get a true representation of the particles, but not too large that the correlation of individual particles between frames become uncertain. The particle concentration is recommended by the instrument manufacturer, but in practice needs to be optimized according to the exact properties of the sample. Further, the intensity of the laser sheet illumination, the camera's exposure time (shutter speed) and gain (sensitivity) all must be adjusted to produce images of appropriate intensity for the camera and image analysis software. It is important to note that the intensity of light from a particle is NOT used to calculate its size, but rather serves to identify the particle's position in space, the particle size is determined solely from the frame-to-frame spatial displacement of the particle's image. It is also critical to remember that the non-linear scaling of scatter intensity with respect to particle size has a very strong effect on what particles can be effectively imaged. A modern 12-bit camera has 4096 different intensity values for a pixel in its digital range. With the radius to the 6th power scaling for scattering intensity, two particles varying by a factor of 4 in radius will, to a first order approximation, occupy the full dynamic range of the camera. This limit of dynamic range ultimately limits the size range of particles that can be measured under a given set of instrument setting. The range limit is higher than 4X since a saturated particle image can still be used for spatial localization but lowering the setting will cause the smaller particles to become invisible. Ultimately this means that PTA can only measure the larger particles in a sample. To successfully image smaller particles, the larger particle must be removed from the imaging, usually through filtration or sedimentation. The extreme sensitivity of the instrument setting to the detected particle sizes can affect the reported size of a measurement over a large range, thus extra care and operator experience is necessary to prevent erroneous measurements.

3.4 Resistive Pulse Sensing (RPS)

The other widely used single particle sizing platform for measuring PSDs is resistive pulse sensing (RPS). RPS is utilized for EV PSDs and is based on principles invented by Wallace Coulter in 1953, called the Coulter Principle [42]. In this method, two reservoirs of electrically conductive fluid, usually an aqueous salt solution, are connected by a short, small diameter tube. If the reservoirs contain particles and there is a bulk movement between the two reservoirs, occasionally one of these particles will enter the connecting tube and in doing so, displace a volume of the electrically conductive buffer solution. The change in the resistance (the reciprocal of the conductivity) will be proportional to the ratio of the volume of the particle to the volume of the channel. Thus, by monitoring the connecting channel's resistivity in time and serially shuttling a sufficient number of individual particles through the channel, a series of resistive electrical pulses, each proportional to the volume of the particle that produced it, is created and can be used to reconstruct the particle sizes and PSD.

RPS measurement requires that the particle volume be large enough relative to the connecting tube volume that it sufficiently perturbs the resistivity for a robust electrical measurement. However, the particle must also be small enough that it can freely move through the connecting tube without clogging or fouling the opening. Thus, a given tube volume can only measure particle size with RPS over a given volume range. This requires the use of tube volumes that are appropriately matched to the particles of interest or to dynamically adject the tube volume for a given particle volume; both of these approaches are available in current commercial instrumentation. Operationally, RPS is subject to fouling and/or clogging of the connecting tube and as such the electrical resistance cell is considered a consumable, is relatively low cost and routinely replaced after a set of measurements. It is also important to note that non-spherical particles with large aspect ratios can lead to erroneous measurements, especially if the long axis of the particle is aligned with the tube axis and is equivalent to or longer than the tube axis.

4 Methods and Tools for Characterizing EV PNCs

Perhaps one of the most useful plots to characterize an EV sample preparation would be a characteristic parameter on the x-axis (such as particle diameter, tetraspanin content, marker protein abundance, etc.) and the PNC on the y-axis. This requires the measurement of the number of particles that have a particular characteristic, and for the case of particle size, this is currently exceedingly difficult for EVs (i.e., due to the almost intractable problem of EV size heterogeneity) [9, 43–45]. This originates from the fact that each measurement platform has a unique set of biases and particle selectivity. Primary inter-technique studies using synthetic poly(styrene) latex bead standards of a single diameter generally only agree to a factor of 2 in particle count [46]. The addition of size heterogeneity in the original particle suspension as well as any compositional heterogeneity will only serve to increase the uncertainty of PNC with respect to size. Current state-of-the-art developments in particle metrology are focused on resolving these issues, but for the time being size-based PNCs should be considered informational and not quantitative.

5 Methods and Tools for Characterizing EV ZPs

The accurate measurement of the surface charge or zeta potential (ZP) of particles in a given solution environment is a quality attribute that can predict particle agglomeration/aggregation state, particle stability and particle-cell interactions that influence the biological uptake kinetics of the particles into recipient cells. So, what exactly is ZP and how is it measured? To be concise, ZP is the electrostatic potential that exists at the plane of the electrical double layer on the surface of particles with respect to the dispersion medium. The ZP value is strongly influenced by the solution environment (pH, ionic strength, valence and presence of multivalent ions, impurities, surfactants, etc.). Therefore, the solution environment must be exactly known and under tight experimental control for ZP measurements to have any meaning. Since ZP is a solution-state QA, it is typically calculated from particle electrophoretic mobility measurements in a specified medium. The measured electrophoretic mobility (U_E, instrument dependent) is converted into a ZP value using Henry's equation (Eq. 3):

$$U_{E=\frac{\varepsilon\zeta f(\kappa\alpha)}{6\pi\eta}} \tag{3}$$

where ε = the dielectric constant of the medium, ζ = calculated zeta potential (mV), $f(\kappa\alpha)$ = Henry's function [κ = the Debye parameter and α = the radius of the nanocarrier] and η = the viscosity of the medium [47]. If the measurement medium is polar in nature, then Henry's function (dimensionless) is assumed to be 1.5 based on the Smoluchowski approximation and if the medium is non-polar, then the function has a value of 1 based on the Hückel approximation [47–50]. Calculated ZP values for nanocarriers in polar solvents typically have a range of $\pm$ 30 mV, with values closer to |30| indicating greater overall nanocarrier solution state stability and ZP values closer to 0 indicating nanocarrier instability and greater tendency to aggregate [51, 52]. ZP values larger than ± 30 mV for particles are possible and mainly depend upon the composition of the medium and the employed measurement method. There exist two main techniques for determining the ZP of EVs: electrophoretic light scattering (ELS) and tunable resistive pulse sensing (TRPS). The method principles underlying both techniques are briefly described below. Note: RPS has been described previously under **Methods and Tools for Characterizing EV PSDs.**

5.1 *ZP Measurements of EVs Using ELS*

As the ELS acronym implies, electrophoresis is utilized to induce the movement of EVs in solution using a charged field with the movement (electrophoretic mobility) of the EVs detected, measured and deconvoluted based on laser light scattering. ELS is therefore based on an optical platform for measuring EV ZP. In brief, EVs in solution in a capillary cell containing two electrodes (both electrodes are in contact with the sample and a voltage is applied between the electrodes) are induced to move under the applied electric field. The mobility of the charged EVs in the cell is proportional to the ZP. Laser light from a discrete angle impinges onto the moving EVs and the resulting Doppler frequency shift in the scattered light is indirectly correlated with the EV speed. Most importantly, the Doppler frequency shift can be converted into EV electrophoretic mobility which allows one to use the Henry Equation (see Eq. 3) to calculate EV ZP values [53–55].

5.2 *ZP Measurements of EVs Using TRPS*

Tunable resistive pulse sensing (TRPS) is an electrical sensor-based technique that can be used to measure the PSDs, PNCs and ZPs of EVs in sample solutions. The technique is based on a non-optical measurement platform that is rapid and conceptually straightforward to implement. An EV sample is suspended in a conductive solution that passes through an orifice on a membrane that has electrodes on either side held at a defined electrostatic potential. The size of the orifice is tunable allowing EVs of different sizes to be analysed on a single particle basis. As the sample containing the EVs passes through the orifice, the ionic current between the two electrodes is reduced and this reduction in current is highly reproducible on a single particle basis. Each EV that passes through the orifice thus induces a change in resistance referred to as a "resistive pulse". However, the geometry of the orifice is not known thus requiring the analysis of calibration particles of known properties at the same time as the EVs. Each resistive pulse event corresponds to one EV passing through the orifice and the magnitude of the resistive pulse is proportional to the volume of the

EV which in turn allows one to calculate the diameter of the EV (assuming a spherical shape). The measurement of resistive pulses over a set time period corresponds to the PNC of EVs in a given sample volume. The EV electrophoretic mobility is calculated from the EV transit time through the pore. The EV ZP is then calculated from the electrophoretic mobility using Henry's Equation, with the appropriate approximation of Henry's function as described previously [52]. Alternatively, the EV ZP (electrokinetic surface charge) can be calculated by measuring the frequency of EV resistive pulses under various pressure conditions [56]. To obtain reproducible results, the pore size, pore stretch, pressure, temperature, and voltage of the system must be carefully optimized.

5.3 Key Challenges in EV ZP Measurements and Recommended Improvements

Several international documentary standards exist that carefully describe the requirements for making reliable ZP measurements on nanoparticles [47, 49, 50]. Though these documents are not specific for EV measurements, they are excellent starting points for the future development of documentary standards specific for EVs. In addition, there exists a National Institute of Standards and Technology (NIST) positively charged reference material (microcrystalline goethite) for electrophoretic mobility measurements [57]. Reliable secondary reference materials (e.g., carboxylated polystyrene beads or silica) can be purchased from several different vendors for validation of ZP measurements [47].

Regarding ELS-based ZP measurements on EVs, it is critical that the pH, buffer type and ionic strength of the conducting fluid be carefully optimized, controlled and reported to ensure reproducibility and accuracy of ZP results within and across platforms and laboratories [55, 58]. The pH of the conducting fluid has a large impact on the total electrical charge on EVS while the buffer type and ionic strength have large impacts on EV electrophoretic mobilities. It has recently been shown that the Ohshima approximation for Henry's function in the Henry Equation is more accurate than either the Smoluchowski or Hückel approximations for the determination of surface charge and consequently ZP values for sEVs [58]. It is recommended that users consider incorporating the Ohshima approximation when calculating ZP values from EV test samples known to predominantly contain sEVs.

An ISO documentary standard that reports the best practices and general operating procedures for determining the PSD, PNC and ZP of particles (including nanoparticles) in suspension using TRPS was recently published [59]. However, a documentary standard that outlines the best practices and general operating procedures for TRPS-based ZP measurements on biological nanoparticles (e.g., EVs) does not presently exist but would be useful. There exist several aspects of the TRPS technique that could be recommended for standardization and probably one of the most important aspects has to do with the use of calibration beads. TRPS utilizes stretchable pores that are not calibrated but must be calibrated to achieve accurate and reproducible EV electrophoretic mobility measurements. Standardized (as to type and diameter) calibration beads should be utilized for TRPS measurements to ensure that the specific instrument is operating within specifications before performing electrophoretic mobility measurements. As recommended by ISEV [5], another way to ensure TRPS measurement reproducibility among labs is to harmonize instrumental operating conditions reporting requirements. Some of the important reporting requirements for TRPS measurements are: instrument model, pore size, applied voltage, applied pore stretch, etc.

6 Methods and Tools for Characterizing EV Protein Cargo

The quantitative and reproducible measurement of EV protein content is a QA that is important for the discovery and validation of new biomarkers, developing new disease diagnostic applications, determining the biogenesis of EVs and EV subtypes, helping to identify the biological function of EVs and potentially identifying a universal biomarker of EVs. EVs from different cell types can contain both common and unique protein cargo based on cell type. To date, the total number of distinct proteins in all classes of EVs stands at 566,911 according to Vesiclepedia (http://www.microvesicles.org/). This is an enumeration of protein types from both published and non-published data. EV protein types range from surface receptors (MHCs) and membrane spanning proteins (integrins, annexins, tetraspanins) to luminal/cytosolic proteins (heat shock proteins, cytoskeletal proteins, ESCRT complexes and components, transport and fusion proteins, growth factors, cytokines, chemokines, etc.,). EV proteins can and do participate in a wide variety of important cellular pathways such as cAMP signaling, Ras signaling, axon guidance, RNA transport, proteasome activation, etc. The ability of EVs to transfer their protein cargo proximally and/or distally to recipient cells is critical for maintaining overall biological homeostasis in living organisms. The quantitative characterization of protein composition in EVs may lead to a greater understanding of the mechanisms of cell-to-cell interactions and protein functionality that drive the formation of disease. To achieve this lofty goal, it is necessary to have suitably fit-for-purpose and rigorous methods that can reproducibly measure the heterogeneous protein cargo found in EVs. The methods that are most often utilized in the EV field are mass spectrometry (MS)-based proteomics, Western blotting (WB) and enzyme linked immunoassay (ELISA). The general principles underlying the methods are briefly described below.

6.1 Protein Cargo Measurements on EVs Using MS-Based Proteomics

Mass spectrometry is a sensitive analytical technique capable of both qualitative and quantitative determination of a wide range of biomolecules (proteins, nucleic acids, small molecules, etc.) found in whole organisms and tissues as well as at the single cell level. MS-based proteomics is focused on investigation of the entire proteome of a biological system. MS-proteomic studies on EVs mostly follow the same protocols and workflows as MS-proteomic studies on tissue- and cell-based samples. MS-proteomics is the most comprehensive technique for the global analysis of proteins in complex biological samples and has demonstrated capacity to identify up to 10,000 proteins in a day [60–63]. Robust sample preparation in concert with ultrasensitive and stable MS instrumentation are key contributors to the reliability and high reproducibility of MS-proteomic methods.

What is actually measured when one performs MS-proteomics on EVs? MS is an analytical technique based on the measurement of charged analytes (in this case it is peptide ions formed from intact protein digestion) in the gas phase. Actually, it is the charge to mass ratio (*m/z*) of the peptide ions and peptide ion fragments that are detected and measured. How are the peptide ions, generated, detected and measured? A mass spectrometer generally consists of three components: an ion source (to convert molecules in solution into ions in the gaseous phase), a mass analyser (to separate the ions by their *m/z* values) and an ion detector (this could be Faraday cup, electron multiplier, photomultiplier tube, etc.).

A typical bottom-up or shotgun MS-proteomic workflow for EV studies involves: (1) obtaining a stable pellet or solution-based sample of highly purified EVs from the specified biological source; (2) lysis of the EV lipid membranes to extract the EV proteins; (3) denaturation, reduction and alkylation of the proteins; (4) digestion of the proteins into peptides using a sequence-specific protease such as trypsin; (5) analysis of the peptides using reversed phase liquid chromatography coupled to positive electrospray ionization mass spectrometry (quadrupole/time of flight (Q-ToF) or orbital trapping instruments are generally utilized); (6) peptides are identified and sequenced as precursor peptide ions (MS spectra produced) and undergo collision-induced dissociation to peptide fragment ions (MS/MS spectra produced) and finally, (7) EV protein identification via peptide spectral library generation/comparison and/or via EV protein sequence database searching. Untargeted (discovery-oriented) and targeted proteomic approaches can be utilized for quantifying EV protein levels. In both untargeted and targeted proteomics, EV proteins can be quantified based on the use of label-free or discrete labeling methods. There exist several different types of protein labeling methods. Tandem mass tags (TMTs) label peptides based upon chemical derivatization of free amines; isobaric peptides result that contain different numbers and combinations of ^{13}C and ^{15}N isotopes. Peptide quantitation is based upon mixing the labeled and unlabeled peptide samples together and analysing the resulting mixed sample using the appropriate proteomics workflow. Stable isotope labelling by using amino acids in cell culture (SILAC) is an in vitro method for labeling proteins during cell culture. In this labeling method, the cell medium is supplemented with ^{13}C and ^{15}N enriched arginine and lysine isotopes which are incorporated into the cellular proteins. Protein/peptide quantitation is based on mixing a non-isotope enriched cell culture sample with the heavy isotope-labeled culture sample and analysing the resulting sample using the appropriate proteomics workflow. It is noteworthy that both label-free and labeling methods enable relative (not absolute) quantitation of EV protein abundances.

6.2 *Protein Cargo Measurements on EVs Using Western Blotting*

Western blotting is widely used in biochemistry and molecular biology for the detection of specific proteins extracted from tissues or cells [64]. It can be used for both the qualitative identification and the absolute (using MS-based peptide measurements) and relative quantification of proteins. WB, as a proteomics technique, is generally slower, but more readily accessible in laboratories in comparison to MS-proteomics instruments and methods. Notably, WB is often used to validate potential EV disease biomarkers identified through MS-proteomics [65, 66]. WB is also commonly used to positively identify EVs and EV subtypes in EV sample preparations, but this is controversial because there does not exist a universal EV marker protein [5, 67–69]. EV investigations that are focused on the characterization of EV cargo proteins generally utilize WB as a first-line measurement, in combination with other orthogonal protein measurement tools [5].

A generalized workflow for measuring EV protein cargo using WB consists of (1) lysing the EVs in an EV sample preparation using an appropriate EV lysis buffer; (2) separation of the intact proteins in the sample using gel electrophoresis; (3) electrophoretic transfer of the separated proteins to a blot membrane, and finally 4() visualization of the separated proteins on the blot membrane via immunostaining [70]. For step (1) of the workflow, it is important to add a protease inhibitor to the lysis buffer to prevent loss of EV proteins. The lysis reagent should be compatible with the gel electrophoresis components and reagents. The protein separation mechanism in step (2) can be based on differences in protein isoelectric point, charge, or mass. Separation of the proteins can be performed under either native or denaturing buffer conditions. Denaturing requires boiling of the EV samples, and it is common to utilize sodium dodecyl sulfate–polyacrylamide gel electrophoresis (SDS-PAGE) to separate the denatured EV proteins. The SDS also helps to maintain the proteins in a denatured form by interacting with the hydrophobic and hydrophilic regions of the proteins and causing unfolding of the 3D structure. As an additional point, a reducing agent can be added to the sample buffer to break the protein disulfide bonds (S – S). The appropriate WB buffer conditions will have to be determined based on the protein under investigation and the activities of the primary and secondary antibodies used for immunostaining and protein detection. Following SDS-PAGE separation, the proteins are transferred (step 3) onto either a polyvinylidene difluoride (PVDF) or nitrocellulose blot membrane that should be blocked with bovine serum albumin (BSA) to prevent non-specific binding of non-EV proteins. Primary antibodies to the proteins of interest in combination with secondary antibodies tagged with fluorophores, colorimetric reagents, radioactive labels, or luminescent probes are used for detecting and quantifying the EV proteins of interest.

6.3 Protein Cargo Measurements on EVs Using ELISA

Enzyme-linked immunosorbent assay (ELISA) is a type of ligand-binding assay that can be used for either the qualitative or quantitative measurement of protein antigens and/or antibodies in a wide range of solution-based biological and clinical samples (urine, plasma, etc.,) [71]. ELISA's can also be automated and can be configured to perform high throughput screening assays. A typical ELISA is performed on either a 96- or 384-well polystyrene microtiter plate that contains the antigen test samples, antigen calibration curve, positive and negative controls, and other important assay controls (e.g., wells for sample blanks, interference samples, enzyme-only samples, substrate only, secondary antibody only, etc.) as required by the specific assay design. There exist four main ELISA formats for antigen/antibody measurements: direct, indirect, competitive, and sandwich. There exist comprehensive literature reviews describing the method principles for these various ELISA formats and thus that information will not be repeated here. However, since the sandwich ELISA is the most utilized format for quantifying proteins and is commonly utilized for EV protein measurements [72–75], the workflow for this specific assay is briefly described.

The sandwich ELISA consists of the following sequence of steps: (1) lyse the EVs in an EV sample preparation using an appropriate EV lysis buffer; (2) prepare the surface of the microtiter plate with the capture antibody; (3) block the remaining exposed surfaces of the wells of the microtiter plate to prevent non-specific binding; (4) apply the lysed EV test samples containing the antigen of interest, controls and blanks to the microtiter plate; (5) wash plate to remove any unbound antigen; (6) add the primary antibody to all of the wells in the plate to form a sandwich that contains the antigen between the capture antibody and the primary antibody; (7) add enzyme-linked secondary antibody; (8) wash plate to remove any unbound enzyme-linked secondary antibody; (9) add enzyme substrate to produce a measurable product; (10) quantify the signal produced from enzyme–substrate product using an appropriate plate reader. Like the WB workflow, there exist some specific considerations for measuring proteins in EV preparations. In step (1) of the workflow, it is important to add a protease inhibitor to the lysis buffer to prevent loss of EV proteins. After EV lysis, it is important to remove the lysis reagent from the EV preparation to prevent antigen–antibody binding interferences. For step (2), the added capture antibody is a primary antibody that binds to a specific epitope of the antigen. It is important to add a consistent and known amount of capture antibody to the plate so that the measured antigen response from the detection antibody is equivalent to the amount of antigen in the test samples and calibrants. The primary antibody in step 6 binds to a different epitope on the antigen than the capture antibody in step 2. In step 10, depending on the assay design and level of analytical sensitivity required for EV protein measurement, the assay readout may utilize colorimetric, fluorometric or luminescence detection.

6.4 Key Challenges for Protein Cargo EV Measurements and Recommended Improvements

EV protein cargo content (amount and identity) is a potential critical quality attribute (CQA) [76, 77] for emerging EV-based therapeutic applications and drug delivery systems. Three commonly utilized methods for measuring the protein cargo content of EVs as described previously are MS-proteomics, WB, and ELISA. Each method demonstrates excellent analytical measurement sensitivity with estimated limits of detection (LODs) for protein determinations of 50 pg, 50 pg, and 100 pg/mL, for MS-proteomics (utilizing orbital trapping MS) [78], WB (utilizing a conventional format, not single cell measurements) [79], and ELISA (utilizing an in silico approximation based on UV/Vis colorimetric detection) [80], respectively. Collectively, these three protein measurement methods are recognized globally as reliable, but technically complex methods to implement due to the number of variables that need to be controlled and reported at each step of their respective workflows. Standardization of technical steps across laboratories might not be feasible due to laboratory-specific protocols and preferences, but harmonization of the measurement protocols might be possible, especially at the pre-analytical EV sample preparation stages. Many of the measurement challenges for all three methods are concentrated at the EV sample preparation step.

For example, protein cargo measurements in EVs are complicated by the current lack of reproducibility and standardization of EV isolation and purification methods [2, 7, 8], the almost intractable heterogeneity of EV subtypes [6, 9] and the potential proteome false-identifications and misinterpretations emanating from EV protein corona signals overlapping with native EV protein cargo signals [81, 82]. Isolation of non-EV components like albumin, lipoproteins, and protein aggregates in conjunction with putative EVs due to non-ideal isolation methods challenges the accurate identification and subsequent quantitation of EV proteins. Two potential solutions to this measurement gap are to develop standardized and validated isolation methods and/or to combine and use multiple isolation methods in series that are based on orthogonal separation principles (e.g., SEC + affinity chromatography, etc.,) to produce non-contaminated EVs. EV size, cargo and biogenesis heterogeneity make it challenging to obtain pure EV populations for protein cargo measurements. One recently developed solution to the heterogeneity problem utilizes single vesicle flow cytometry in combination with multiparametric analysis to discriminate among EV populations based upon lipid membrane order and EV size [83]. The authors postulated that their "EV fingerprinting" method could be extended to clinical samples for EV biomarker analyses in future developments. There are currently no well-developed methods for completely removing the EV protein corona layer without disturbing and introducing artifacts into the native EV protein layer. Identification and quantitation of the proteins in the EV protein corona is an active area of research.

One special challenge for protein cargo measurements which is especially significant for MS-proteomic measurements concerns the presence of high abundance EV proteins or sample proteins that interfere with or mask the detection of low abundance EV proteins. The protein dynamic range in EVs, like their originator cells, span approximately seven orders of magnitude [84], far exceeding the measurement capabilities (linear dynamic range) of the highest resolution MS instruments. It is not technically possible to measure all the proteins in a single sample. The most direct approach for fixing this technical challenge is to employ protein depletion strategies to remove the highly abundant proteins from the sample preparation.

For reproducible and accurate WB-proteomic measurements, there exist a few unique challenges that should be discussed. First, the specificity of the antibodies for binding EV proteins in the sample preparations must be very high. If the antibody specificity is not high, then the EV protein measurements will suffer from interferences due to cross reactivity with non-EV proteins. It is recommended that antibodies be validated for the detection of specified EV proteins. Another challenge for WB is due to the inherently low concentration (low copy numbers) of EV-cargo proteins in EVs. Low protein concentrations make it difficult to detect the proteins using conventional WB. Low protein concentrations also contribute to increased variability in protein quantitation. One potential solution for low EV protein concentrations is to use advanced EV purification procedures (e.g., aptamer-based purification, etc.) to increase the purity of the EV sample preparation by isolating native EVs from contaminating proteins prior to EV lysis. Another potential solution is to utilize high throughput WBs (DigiWest) that are bead-based and only require 1 μg of protein for sample analysis [85]. To reduce the variability in EV protein quantitation, it is possible to normalize protein expression levels by including a structural (housekeeping) or total protein measurement when performing WBs [86].

Like WB, ELISA's for EV protein cargo measurements is challenged by the need to utilize highly specific antibodies for the detection of EV proteins and the need to concentrate EV sample preparations to increase EV protein concentrations. ELISA workflows can incorporate sample preparation strategies that are similar to the strategies proposed for WB previously.

7 Methods and Tools for Characterizing EV Nucleic Acid Cargo

In addition to the protein content, EVs also contain forms of nucleic acids including various types of DNA and RNA molecules [87, 88] with essential roles in intracellular communication and gene expression regulation [89]. EV populations are heterogeneous as the cargo composition and size is context-dependent and altered by various environmental and cellular signals [90, 91]. Additionally, nucleic acids from EVs often vary from their corresponding cellular counterparts in both type and concentration. In particular, RNA species within EVs can differ substantially depending on cell type and the environment. Types of EV RNA range in size from small RNAs [89, 92] like miRNA and piwi-interacting RNA (piRNA) to longer RNAs [93, 94] including mRNA, rRNA, and long non-coding RNA (lncRNA). The quantitative and reproducible measurement of EV RNA content is crucial for the discovery and validation of new biomarkers, understanding the biological functions of EVs, and possibly for the development of new disease diagnostics and therapeutic technologies [95, 96]. Analytical techniques like high-throughput next generation sequencing of RNA (RNA-seq) and reverse transcription quantitative polymerase chain reaction (RT-qPCR) are established methods for reproducible measurements of heterogenous EV RNA cargo. RNA-seq of EV RNA enables the broad discovery of novel transcripts, RNA species, and possible biomarkers, in addition to providing comprehensive characterization of EV RNA expression profiles and functional annotation of roles in biological processes. RT-qPCR, a lower throughput orthogonal method, is a sensitive, quantitative tool for targeted measurements of select EV RNA with established MIQE guidelines throughout the entire method [97–99]. The general principles underlying both methods are briefly described below.

7.1 RNA Cargo Measurements on EVs Using RNA-Seq and Transcriptomics

The selection of an optimal analytical technique for EV RNA cargo measurements depends on whether the scope of the study is exploratory or targeted. As previously discussed, RNA-seq and transcriptomic analysis will provide broad insight into the population of RNA species and their relative abundance in EVs. Similar to MS-based proteomics, RNA-seq is a method for investigating the identity and abundance of every transcript within the transcriptome of a biological system at a specific moment in time. Transcriptomic studies on EVs mostly follow the same experimental protocols and workflows as those used on tissue- and cell-based samples with minor modifications [100]. Due to the transient nature of RNA, cDNA reverse transcribed from EV RNA is actually what is measured via next-generation sequencing (NGS), with RNA sequence and abundance subsequently inferred from DNA reads.

A typical RNA-seq workflow for EV studies involves (Fig. 12): (1) obtaining a stable pellet or solution-based sample of highly purified EVs from the specified biological source (Fig. 12a); (2) extracting RNA from the purified EV sample using a procedure optimized for the specific EV RNA type (Fig. 12b); (3) characterizing extracted RNA using quality control measurements to determine quality, purity, and concentration of each sample. Typical quality control measurements include capillary electrophoresis for quality (Bioanalyzer 2100 or equivalent), UV absorbance measurements with a spectrophotometer for purity, and fluorometric measurements for concentration (Qubit fluorometer or equivalent); (4) preparing cDNA libraries using pure, intact RNA as the input (Fig. 12c); (5) characterizing the quality of individual cDNA libraries using a similar workflow as the RNA workflow described in (3), to determine library size and presence of indices prior to sequencing (Fig. 12c); (6) pooling individual cDNA libraries in equimolar concentration according to their indices prior to sequencing to ensure equal read distribution across all samples; (7) NGS of cDNA libraries using settings (single- or paired-end, read length, and instrument capacity) to achieve desired read depth and aim of the study (Fig. 12d); (8) lastly, bioinformatic analysis should be performed to process data from raw reads to transcript-level expression with quality filtering steps throughout the pipeline. Additional downstream bioinformatic analyses of hierarchical clustering and principal component analysis, differential expression analysis, and functional annotation can provide insight into data correlation of expression sets, expression differences between sample groups, and context of biological function of EV RNA.

Similar to EV protein workflows, specific considerations must be made to accurately measure EV RNAs with RNA-seq and transcriptomic analysis. In step (2) of the workflow, the RNA extraction procedure selected must be optimized to acquire the desired RNA type (small RNAs, total RNA, etc.). Several methods are well characterized, but improper selection could introduce bias and unwanted variation into RNA expression patterns [101]. Quality control measurements, described in step 3, are essential to ensure samples remain intact and pure and should be performed after

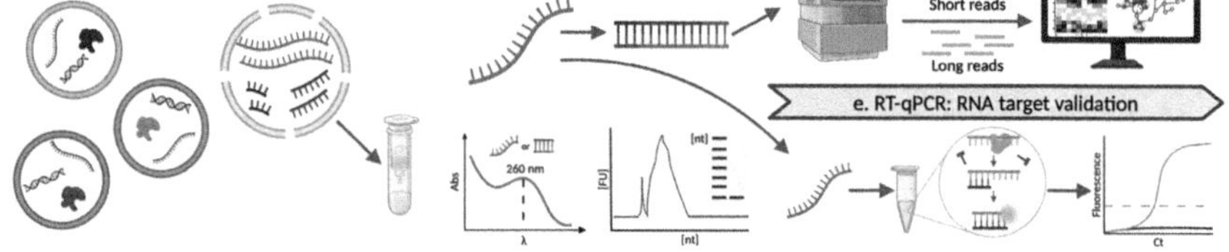

Fig. 12 Schematic of the RNA-seq workflow for the measurement of EV nucleic acids. A typical workflow includes: **a** isolation and characterization of EV sample, **b** isolation and size selection of EV RNA, **c** quality control measurements and cDNA library preparation for **d** NGS and bioinformatic analysis or **e** validation of EV RNA targets with RT-qPCR. Selection of RNA-seq vs. RT-qPCR depends on the aims of the study. For analysis of EV DNA, the workflow is similar, but RT steps are unnecessary for either method. *Created in BioRender. Romantseva, J. (2024)* https://BioRender.com/q21q393

each step in the workflow. Prior to cDNA library prep in step 4, a size selection step is often performed on small RNA extracts and rRNA and globin mRNA depletion is performed on total RNA to ensure compatibility with library preparation input guidelines, deeper sequencing, and minimal loss of low abundance transcripts [96, 102]. Similar to RNA extraction methodology, RNA type, sample number, output read depth, and sequencer compatibility should be considered when selecting a cDNA library preparation method in step 4. All cDNA libraries for a given study should be prepared simultaneously to minimize batch effects [103]. In step 7, consideration should be taken to achieve approximate read depths of 10–25 million reads per sample for total RNA from EVs and 5–10 million for small RNAs like miRNAs [96, 104].

7.2 RNA Cargo Measurements on EVs Using RT-QPCR

RT-qPCR is a sensitive analytical method with established MIQE guidelines often used in validation studies of novel biomarkers and specific transcriptional signatures [97]. Briefly, RT-qPCR consists of the conversion of EV RNA into cDNA via RT followed by the amplification and quantification of specific target and reference sequences. Advantages of this method include accurate detection of low copy number or highly diluted transcripts of interest, quantitative measurements of nucleic acid abundance, and easy adaptability for in house and clinical settings [96, 105, 106]. Similar to RNA-seq, several considerations of input type and experimental design must be made prior to initiating a workflow. These include accurate primer design [97], the selection of non-specific dye-based or sequence-specific hydrolysis probes for quantification [107], and reaction experimental design with either one-step or two-step approaches with respect to RT and subsequent qPCR reaction [108–110]. Additionally, special techniques must be employed for measurements of miRNA with RT-qPCR due to similar lengths of input RNAs and primers [96, 111].

A typical RT-qPCR workflow for EV studies involves (Fig. 12): (1) identifying a target RNA sequence or biomarker for detection and quantification; (2) selecting appropriate experimental design including dye-based or hydrolysis probes and one-step or two-step RT-qPCR reactions to enable accurate measurements of input material; (3) following the design parameters described in the MIQE guidelines, sequence specific primers should be designed to bind to the amplicon region of the input RNA sequence; (4) obtaining a stable pellet or solution-based sample of highly purified EVs from the specified biological source; (5) extracting RNA from the purified EV sample and performing quality control measurements described earlier (Fig. 12c); (6) preparing the reaction mixture for either one-step or two-step RT-qPCR in an RNase free environment with appropriate assay controls for DNA contamination in the input RNA sample and extraneous nucleic acid contamination and formation of primer dimers (Fig. 12e); (7) running the assay on a real-time thermal cycler with the appropriate cycling protocol based on the RT-qPCR kit and instrument specifications

[109, 110]; (8) analysing the cycle threshold (Ct) of the samples compared to appropriate negative controls. For quantification of abundance, sample Cts are compared to Cts of a standard curve of known RNA concentrations. For differential abundance analysis across samples, Cts are compared to the Cts of a reference RNA present at a constant amount across samples. Considerations described in the RNA-seq section including RNA extraction methods and quality control measurements also apply to RT-qPCR. Additionally, pure EV RNA with no contamination is required for accurate amplification and quantification. For example, DNase treatment can be added following EV RNA extraction to remove co-isolated DNA from sample [110].

7.3 Key Challenges for Nucleic Acid Cargo EV Measurements and Recommended Improvements

EV RNA has the potential to aid the development of emerging EV-based therapeutic applications like the discovery of novel biomarkers. RNA-seq/transcriptomics and RT-qPCR methods provide quantitative measurements of RNA abundance and identity in EV samples. Similar to protein cargo, RNA cargo measurement methods are reliable, but challenging to implement due to the technical complexity of controlling numerous variables at each step in the workflow. Standardization of technical workflows across laboratories may not be fully achievable due to varying sample types and laboratory-specific protocols, however the field would benefit from harmonization of measurement protocols. Methodological variability throughout experimental and computational workflows and the collection of sample metadata annotations are challenges in comparing EV RNA measurements between different studies and for developing standards. For example, workflows must be optimized for specific RNA types (small or long RNA), and sample type (cell line and biofluid type) to achieve accurate measurements [96, 112]. Also, the large heterogeneity of EV populations intertwined with RNA diversity inflicts a steep challenge that limits the route to standardization. The low abundance of RNA in EVs further complicates establishing a consensus of acceptable variation for diverse sample and RNA types [87]. A recommendation to minimize variability and achieve comparability between studies is to develop standardized and validated protocols for RNA-seq and RT-qPCR specific to target RNA type. Due to heterogeneity of clinical samples and diverse research aims, protocols will likely need to be standardized for specific RNA types.

For continued progress, the field will need to address additional key challenges. For example, methods to improve sensitivity will need to be developed to detect and quantify low abundance EV RNA. RNA cargo measurements can also suffer from RT and PCR biases prior to quantification with RT-qPCR and RNA sequencing. For the RNA-seq method, unique molecular identifiers could be used to combat PCR bias by filtering PCR artifacts and duplicate reads out during data analysis [113]. Challenges specific to RT-qPCR include limited reference transcripts for use in differential abundance analysis and streamlined design of functional primers.

The further identification of reference transcripts with exploratory RNA-seq and continual updating of databases like miREV [105] could improve accurate detection of RNA cargo with RT-qPCR. To overcome these challenges, the community could update the MIQE guidelines for EV-specific applications like biomarker discovery and implement transparent reporting of results including both successes and failures.

8 Conclusions and Future Perspectives

The current analytical methods for characterizing the potential CQAs of EV therapeutics are not EV-specific. The first problem is that the CQAs for EV therapeutics have not been established because there are no guidelines or good manufacturing practice protocols for the reproducible production of large-scale batches of EVs. The secondary problem is that current methods for producing therapeutic EVs yields EVs that are inherently heterogeneous in terms of size, cargo content and biogenesis. And finally, the traditional methods that are used to isolate EVs are not able to resolve the EVs into pure populations Hence, it is extremely challenging to identify the EV CQAs that are relevant for clinical applications and that should be characterized. Nevertheless, the current methods that are used to characterize the CQAs of other types of lipid-based particles can, with certain modifications, be utilized to characterize EVs. However, these methods, need to be modified and validated specifically for the detection of EVs. We have described and discussed the most popular methods for characterizing the PSDs, PNCs, ZPs, protein cargo and nucleic acid cargo content of EVs. The most critical challenge for reproducible EV detection and/or quantitation is the pre-analytical EV sample preparation step. Low EV concentrations and/or the presence of contaminating non-EV particles, viruses, lipoproteins, and protein aggregates complicate the accurate measurement of potential EV CQAs. We are now witnessing rapid advancements in the development of high throughput, automated methods for the highly specific measurement of EV CQAs in low concentration samples. For example, there are significant technical innovations in methods for EV protein cargo measurements. The DigiWest high throughput Western blot platform was previously described [85], but recently, a high throughput, automated ELISA platform for single molecule (protein) detection was developed. Enzyme-linked single molecule array (eSimoa) uses paramagnetic beads for single EV isolation and detection on a single platform [114–116]. The platform enables the discrimination and characterization of EV subtypes and the detection and quantitation of low abundance, surface and luminal EV proteins. The technology for EV protein identification and quantitation uses machine learning algorithms to reduce the measurement variability for single EV data.

Acknowledgements The authors would like to acknowledge John Elliott, Pawel Jaruga and Vytas Reipa for thoroughly reviewing the chapter during its preparation.

NIST Disclaimer Certain commercial equipment, instruments, and software are identified in this paper to specify an experimental procedure as completely as possible. In no case does the identification of equipment, instruments, or software imply a recommendation or endorsement by the National Institute of Standards and Technology nor does it imply that the equipment, instruments, or software are necessarily the best available for the purpose.

References

1. Wolf, P. (1967). The nature and significance of platelet products in human plasma. *British Journal of Haematology, 13*, 269–288.
2. Nelson, B. C., Maragh, S., Ghiran, I. C., Jones, J. C., DeRose, P. C., Elsheikh, E., et al. (2020). Measurement and standardization challenges for extracellular vesicle therapeutic delivery vectors. *Nanomedicine-UK, 15*, 2149–2170.
3. Henne, W. M., Buchkovich, N. J., & Emr, S. D. (2011). The ESCRT pathway. *Developmental Cell, 21*, 77–91.
4. Juan, T., & Fürthauer, M. (2018). Biogenesis and function of ESCRT-dependent extracellular vesicles. *Seminars in Cell & Developmental Biology, 74*, 66–77.
5. Welsh, J. A., Goberdhan, D. C., O'Driscoll, L., Buzas, E., Blenkiron, C., & Bussolati, B., et al. (2024). Minimal information for studies of extracellular vesicles (MISEV2023): From basic to advanced approaches (vol. 13, e12404, 2024). *Journal of Extracellular Vesicles*, 13.
6. Bordanaba-Florit, G., Royo, F., Kruglik, S. G., & Falcon-Perez, J. M. (2021). Using single-vesicle technologies to unravel the heterogeneity of extracellular vesicles. *Nature Protocols, 16*, 3163–3185.
7. Buschmann, D., Mussack, V., & Byrd, J. B. (2021). Separation, characterization, and standardization of extracellular vesicles for drug delivery applications. *Advanced Drug Delivery Reviews, 174*, 348–368.
8. Hartjes, T. A., Mytnyk, S., Jenster, G. W., van Steijn, V., & van Royen, M. E. (2019). Extracellular vesicle quantification and characterization: Common methods and emerging approaches. *Bioengineering (Basel)*, 6.
9. Willms, E., Cabanas, C., Mager, I., Wood, M. J. A., & Vader, P. (2018). Extracellular vesicle heterogeneity: Subpopulations, isolation techniques, and diverse functions in cancer progression. *Frontiers in Immunology, 9*, 738.
10. Gurunathan, S., Kang, M. H., Jeyaraj, M., Qasim, M., & Kim, J. H. (2019). Review of the isolation, characterization, biological function, and multifarious therapeutic approaches of exosomes. *Cells-Basel*, 8.
11. Herrmann, I. K., Wood, M. J. A., & Fuhrmann, G. (2021). Extracellular vesicles as a next-generation drug delivery platform. *Nature Nanotechnology, 16*, 748–759.
12. Kalluri, R., & LeBleu, V. S. (2020). The biology, function, and biomedical applications of exosomes. *Science, 367*, 640-+.
13. Kumar, M. A., Baba, S. K., Sadida, H. Q., Marzooqi, S. A., Jerobin, J., & Altemani, F. H., et al. (2024). Extracellular vesicles as tools and targets in therapy for diseases. *Signal Transduction Target*, 9.
14. Murphy, D. E., de Jong, O. G., Brouwer, M., Wood, M. J., Lavieu, G., Schiffelers, R. M. et al. (2019). Extracellular vesicle-based therapeutics: natural versus engineered targeting and trafficking. *Experimental Molecular Medicine*, 51.
15. Harrell, C. R., Fellabaum, C., Jovicic, N., Djonov, V., Arsenijevic, N., & Volarevic, V. (2019). Molecular mechanisms responsible for therapeutic potential of mesenchymal stem cell-derived secretome. *Cells-Basel*, 8.
16. Harrell, C. R., Jovicic, N., Djonov, V., Arsenijevic, N., & Volarevic, V. (2019). Mesenchymal stem cell-derived exosomes and other extracellular vesicles as new remedies in the therapy of inflammatory diseases. *Cells-Basel*, 8.

17. Pers, Y. M., Bony, C., Duroux-Richard, I., Bernard, L., Maumus, M., & Assou, S., et al. (2021). miR-155 contributes to the immunoregulatory function of human mesenchymal stem cells. *Frontiers in Immunology*, 12.
18. Calvo, V., & Izquierdo, M. (2022). T Lymphocyte and CAR-T cell-derived extracellular vesicles and their applications in cancer therapy. *Cells-Basel*, 11.
19. Fu, W. Y., Lei, C. H., Liu, S. W., Cui, Y. S., Wang, C. Q., & Qian, K. W., et al. (2019). CAR exosomes derived from effector CAR-T cells have potent antitumour effects and low toxicity. *Nature Communication*, 10.
20. Pagotto, S., Simeone, P., Brocco, D., Catitti, G., De Bellis, D., & Vespa, S., et al. (2023) CAR-T-derived extracellular vesicles: A promising development of CAR-T anti-tumor therapy. *Cancers*, 15.
21. Yu, W., Hurley, J., Roberts, D., Chakrabortty, S. K., Enderle, D., Noerholm, M., et al. (2021). Exosome-based liquid biopsies in cancer: Opportunities and challenges. *Annals of Oncology, 32*, 466–477.
22. Hatano, K., & Fujita, K. (2021). Extracellular vesicles in prostate cancer: A narrative review. *Translation Andrology Urology*, 10
23. Shimada, Y., Yoshioka, Y., Kudo, Y., Mimae, T., Miyata, Y., & Adachi, H. et al. (2023). Extracellular vesicle-associated microRNA signatures related to lymphovascular invasion in early-stage lung adenocarcinoma. *Science Report-UK*, 13.
24. Irmer, B., Chandrabalan, S., Maas, L., Bleckmann, A., & Menck, K. (2023). Extracellular vesicles in liquid biopsies as biomarkers for solid tumors. *Cancers*, 15.
25. Huang, X. Y., Yuan, T. Z., Tschannen, M., Sun, Z. F., Jacob, H., & Du, M. J. et al. (2013). Characterization of human plasma-derived exosomal RNAs by deep sequencing. *BMC Genomics*, 14.
26. The World FactBook.
27. Maxwell, J. C. V. I. I. I. (1865). A dynamical theory of the electromagnetic field. *Philosophical Transactions of the Royal Society of London, 155*, 459–512.
28. Rayleigh, L. X. (1881). On the electromagnetic theory of light. *The London, Edinburgh, and Dublin Philosophical Magazine and Journal of Science, 12*, 81–101.
29. Bohren, C. F., & Huffman, D. R. (1983). *Absorption and scattering of light by small particles*. Wiley.
30. Daune, M. (1999). *Molecular biophysics: Structures in motion*. Oxford University Press, Oxford.
31. Falke, S., & Betzel, C. (2019). Dynamic light scattering (DLS). In A. S. Pereira, P. Tavares & P. Limão-Vieira (eds.), *Radiation in Bioanalysis: Spectroscopic Techniques and Theoretical Methods*, Springer International Publishing, Cham, pp. 173–93.
32. Mie, G. (1908). Beiträge zur Optik trüber Medien, speziell kolloidaler Metallösungen. *Annalen der Physik, 330*, 377–445.
33. Gans, R. (1925). Die molekulare Lichtzerstreuung in festen isotropen Körpern und Flüssigkeiten. *Annalen der Physik, 382*, 317–324.
34. Light scattering by small particles. By H. C. van de Hulst. New York (John Wiley and Sons), London (Chapman and Hall), 1957. Pp. xiii, 470; 103 Figs.; 46 Tables. 96s. *Quarterly Journal of the Royal Meteorological Society* 1958;84:198–99.
35. Tande, B. M., Wagner, N. J., Mackay, M. E., Hawker, C. J., & Jeong, M. (2001). Viscosimetric, hydrodynamic, and conformational properties of dendrimers and Dendrons. *Macromolecules, 34*, 8580–8585.
36. Giddings, J. C., Yang, F. J. F., & Myers, M. N. (1976). Flow-field-flow fractionation: A versatile new separation method. *Science, 193*, 1244–1245.
37. Cuperus, F. P., & Smolders, C. A. (1991). Characterization of Uf membranes—membrane-characteristics and characterization techniques. *Adv Colloid Interfac, 34*, 135–173.
38. Marioli, M., & Kok, W. T. (2019). Recovery, overloading, and protein interactions in asymmetrical flow field-flow fractionation. *Analytical and Bioanalytical Chemistry, 411*, 2327–2338.

39. Bian, J., Gobalasingham, N., Purchel, A., & Lin, J. (2023). The power of field-flow fractionation in characterization of nanoparticles in drug delivery. *Molecules, 28*, 4169.
40. Gray, G. T., McCutcheon, J. R., & Elimelech, M. (2006). Internal concentration polarization in forward osmosis: Role of membrane orientation. *Desalination, 197*, 1–8.
41. Brown, R. XXVII. (1828). A brief account of microscopical observations made in the months of June, July and August 1827, on the particles contained in the pollen of plants; and on the general existence of active molecules in organic and inorganic bodies. *The Philosophical Magazine, 4*, 161–73.
42. Coulter, W. H. (1953). Means for counting particles suspended in a fluid, INdividual, United States.
43. Nieuwland, R., Siljander, P. R. M., Falcon-Perez, J. M., & Witwer, K. W. (2022). Reproducibility of extracellular vesicle research. *European Journal Cell Biology*, 101.
44. Vestad, B., Llorente, A., Neurauter, A., Phuyal, S., Kierulf, B., Kierulf, P., et al. (2017). Size and concentration analyses of extracellular vesicles by nanoparticle tracking analysis: A variation study. *J Extracell Vesicles, 6*, 1–11.
45. Wakker, S. I. V., Meijers, F. M., Sluijter, J. P. G., & Vader, P. (2023). Extracellular vesicle heterogeneity and its impact for regenerative medicine applications. *Pharmacological Reviews, 75*, 1043–1061.
46. DeRose, P. C., Benkstein, K. D., Elsheikh, E. B., Gaigalas, A. K., Lehman, S. E., & Ripple, D. C. et al. (2022). Number concentration measurements of polystyrene submicrometer particles. *Nanomaterials-Basel*, 12.
47. ASTM, 2022. E2865 – 12 (Reapproved 2022) Standard Guide for Measurement of Electrophoretic Mobility and Zeta Potential of Nanosized Biological Materials, ASTM.
48. Hunter, R. J. (1981). *Zeta potential in colloid science: Principles and applications*. Academic Press.
49. ISO, 2012. ISO 13099-1 Colloidal Systems — Methods for Zeta-Potential Determination — Part 1: Electroacoustic and Electrokinetic Phenomena.
50. ISO, 2012. ISO 13099-2 Colloidal Systems — Methods for Zeta-Potential Determination — Part 2: Optical Methods.
51. Clogston, J. D., & Patri, A. K. (2011). Characterization of nanoparticles intended for drug delivery. In S. E. McNeil (ed.) *Methods in Molecular Biology*, pp. 63–70.
52. Vogel, R., Pal, A. K., Jambhrunkar, S., Patel, P., Thakur, S. S., & Reátegui, E. et al. (2017). High-resolution single particle zeta potential characterisation of biological nanoparticles using tunable resistive pulse sensing. *Science Report-UK*, 7.
53. Mendivil-Alvarado, H., Limon-Miro, A. T., Carvajal-Millan, E., Lizardi-Mendoza, J., Mercado-Lara, A., & Coronado-Alvarado, C. D. et al. (2023). Extracellular vesicles and their zeta potential as future markers associated with nutrition and molecular biomarkers in breast cancer. *International Journal Molecular Science*, 24.
54. Wang, Y. J., Zhang, L., Li, Y. J., Chen, L. J., Wang, X. L., Guo, W., et al. (2015). Exosomes/microvesicles from induced pluripotent stem cells deliver cardioprotective miRNAs and prevent cardiomyocyte apoptosis in the ischemic myocardium. *International Journal of Cardiology, 192*, 61–69.
55. Midekessa, G., Godakumara, K., Ord, J., Viil, J., Lättekivi, F., Dissanayake, K., et al. (2020). Zeta Potential of extracellular vesicles: Toward understanding the attributes that determine colloidal stability. *ACS Omega, 5*, 16701–16710.
56. Vogel, R., Anderson, W., Eldridge, J., Glossop, B., & Willmott, G. (2012). A variable pressure method for characterizing nanoparticle surface charge using pore sensors. *Analytical Chemistry, 84*, 3125–3131.
57. NIST, 2011. SRM 1980—Positive Electrophoretic Mobility standard from NIST.
58. Tamrin, S. H., Phelps, J., Nezhad, A. S., & Sen, A. (2023) Critical considerations in determining the surface charge of small extracellular vesicles. *Journal Extracellular Vesicles*, 12.
59. ISO, 2023. ISO 13319-2 Determination of particle size distribution — Electrical sensing zone method. Part 2: Tunable resistive pulse sensing method.

60. Gholami, A. M., Hahne, H., Wu, Z. X., Auer, F. J., Meng, C., Wilhelm, M., et al. (2013). Global proteome analysis of the NCI-60 cell line panel. *Cell Reports, 4*, 609–620.
61. Kelstrup, C. D., Bekker-Jensen, D. B., Arrey, T. N., Hogrebe, A., Harder, A., & Olsen, J. V. (2018). Performance evaluation of the Q exactive HF-X for shotgun proteomics. *Journal of Proteome Research, 17*, 727–738.
62. Kim, M. S., Pinto, S. M., Getnet, D., Nirujogi, R. S., Manda, S. S., & Chaerkady, R. et al. (2014). A draft map of the human proteome. *Nature, 509*, 575-+.
63. Sinitcyn, P., Richards, A. L., Weatheritt, R. J., Brademan, D. R., Marx, H., & Shishkova, E. et al. (2023). Global detection of human variants and isoforms by deep proteome sequencing. *Nature Biotechnology, 41*, 1776-+.
64. Towbin, H., Staehelin, T., & Gordon, J. (1979). Electrophoretic transfer of proteins from polyacrylamide gels to nitrocellulose sheets—procedure and some applications. *P Natl Acad Sci USA, 76*, 4350–4354.
65. Li, M. Y., Zhao, C., Chen, L., Yao, F. Y., Zhong, F. M., & Chen, Y. et al. (2021). Quantitative proteomic analysis of plasma exosomes to identify the candidate biomarker of Imatinib resistance in chronic myeloid leukemia patients. *Front Oncology,* 11.
66. Liu, T. C., Liu, N., Wang, Y. Z., Li, T., & Zhang, M. (2023). Differential expression of coagulation pathway-related proteins in diabetic urine exosomes. *Cardiovasc Diabetol*, 22.
67. Mizenko, R. R., Brostoff, T., Rojalin, T., Koster, H. J., Swindell, H. S., & Leiserowitz, G. S. et al. (2021). Tetraspanins are unevenly distributed across single extracellular vesicles and bias sensitivity to multiplexed cancer biomarkers. *Journal Nanobiotechnology*, 19.
68. Théry, C., Witwer, K. W., Aikawa, E., Alcaraz, M. J., Anderson, J. D., & Andriantsitohaina, R. et al. (2018). Minimal information for studies of extracellular vesicles 2018 (MISEV2018): A position statement of the international society for extracellular vesicles and update of the MISEV2014 guidelines. *Journal Extracellular Vesicles*, 7.
69. Kowal, J., Arras, G., Colombo, M., Jouve, M., Morath, J. P., Primdal-Bengtson, B., et al. (2016). Proteomic comparison defines novel markers to characterize heterogeneous populations of extracellular vesicle subtypes. *P National Academic Science USA, 113*, E968–E977.
70. Kowal, E. J. K., Ter-Ovanesyan, D., Regev, A., & Church, G. M. (2017). Extracellular vesicle isolation and analysis by western blotting. *Methods in Molecular Biology, 1660*, 143–152.
71. Engvall, E., & Perlmann, P. (1971). Enzyme-linked immunosorbent assay (Elisa) quantitative assay of immunoglobulin-G. *Immunochemistry, 8*, 871–0.
72. Driedonks, T. A. P., Jiang, L., Gololobova, O., Liao, Z., & Witwer, K. W. (2024). ELISA-based detection of immunoglobulins against extracellular vesicles in blood plasma. *Journal Extracellular Biology, 3*, Article e129.
73. Iha, K., Tsurusawa, N., Tsai, H. Y., Lin, M. W., Sonoda, H., & Watabe, S. et al. (2022). Ultrasensitive ELISA detection of proteins in separated lumen and membrane fractions of cancer cell exosomes. *Anal Biochemistry*, 654.
74. Jorgensen, M., Baek, R., Pedersen, S., Sondergaard, E. K., Kristensen, S. R., & Varming, K. (2013). Extracellular Vesicle (EV) Array: Microarray capturing of exosomes and other extracellular vesicles for multiplexed phenotyping. *Journal Extracellular Vesicles*, 2.
75. Yuyama, K., Sun, H., Fujii, R., Hemmi, I., Ueda, K., & Igeta, Y. (2024). Extracellular vesicle proteome unveils cathepsin B connection to Alzheimer's disease pathogenesis. *Brain, 147*, 627–636.
76. Adlerz, K., Patel, D., Rowley, J., Ng, K., & Ahsan, T. (2020). Strategies for scalable manufacturing and translation of MSC-derived extracellular vesicles. *Stem Cell Research*, 48.
77. Gandham, S., Su, X. Y., Wood, J., Nocera, A. L., Alli, S. C., Milane, L., et al. (2020). Technologies and standardization in research on extracellular vesicles. *Trends in Biotechnology, 38*, 1066–1098.
78. Levy, M. J., Washburn, M. P., & Florens, L. (2018). Probing the sensitivity of the Orbitrap Lumos mass spectrometer using a standard reference protein in a complex background. *Journal of Proteome Research, 17*, 3586–3592.

79. Ye, F. M., Smith, P. B., Wu, C. F., & Chiu, D. T. (2013). Ultrasensitive detection of proteins on western blots with semiconducting polymer dots. *Macromol Rapid Comm, 34*, 785–790.
80. Zhang, S. Y., Garcia-D'Angeli, A., Brennan, J. P., & Huo, Q. (2014). Predicting detection limits of enzyme-linked immunosorbent assay (ELISA) and bioanalytical techniques in general. *The Analyst, 139*, 439–445.
81. Heidarzadeh, M., Zarebkohan, A., Rahbarghazi, R., & Sokullu, E. (2023). Protein corona and exosomes: New challenges and prospects. *Cell Communication Signal*, 21.
82. Tóth, E. A., Turiák, L., Visnovitz, T., Cserép, C., Mázló, A., & Sódar, B. W. et al. (2021). Formation of a protein corona on the surface of extracellular vesicles in blood plasma. *Journal Extracellular Vesicles*, 10.
83. von Lersner, A. K., Fernandes, F., Ozawa, P. M. M., Jackson, M., Masureel, M., Ho, H., et al. (2024). Multiparametric single-vesicle flow cytometry resolves extracellular vesicle heterogeneity and reveals selective regulation of biogenesis and cargo distribution. *ACS Nano, 18*, 10464–10484.
84. Zubarev, R. A. (2013). The challenge of the proteome dynamic range and its implications for in-depth proteomics. *Proteomics, 13*, 723–726.
85. van Maanen, J. C., Bach, F. C., Braun, T. S., Giovanazzi, A., van Balkom, B. W. M., Templin, M., et al. (2023). A combined western and bead-based multiplex platform to characterize extracellular vesicles. *Tissue Engineering. Part C, Methods, 29*, 493–504.
86. Diller, T., Thompson, J., & Steer, B. (2021). Biological validation of a novel process and product for quantitating western blots. *Journal of Biotechnology, 326*, 52–60.
87. O'Brien, K., Breyne, K., Ughetto, S., Laurent, L. C., & Breakefield, X. O. (2020). RNA delivery by extracellular vesicles in mammalian cells and its applications. *Nature Reviews Molecular Cell Biology, 21*, 585–606.
88. van Niel, G., D'Angelo, G., & Raposo, G. (2018). Shedding light on the cell biology of extracellular vesicles. *Nature Reviews Molecular Cell Biology, 19*, 213–228.
89. Xiong, Q., & Zhang, Y. (2023). Small RNA modifications: Regulatory molecules and potential applications. *Journal of Hematology & Oncology, 16*, 64.
90. Lee, Y. J., Shin, K. J., & Chae, Y. C. (2024). Regulation of cargo selection in exosome biogenesis and its biomedical applications in cancer. *Experimental & Molecular Medicine, 56*, 877–889.
91. Liang, Y., Huang, S., Qiao, L., Peng, X., Li, C., Lin, K., et al. (2020). Characterization of protein, long noncoding RNA and microRNA signatures in extracellular vesicles derived from resting and degranulated mast cells. *J Extracell Vesicles, 9*, 1697583.
92. Kim, V. N. (2005). Small RNAs: Classification, biogenesis, and function. *Molecules and Cells, 19*, 1–15.
93. Fang, Y., & Fullwood, M. J. (2016). Roles, Functions, and Mechanisms of Long Non-coding RNAs in Cancer. *Genomics, Proteomics & Bioinformatics, 14*, 42–54.
94. Zhou, R., Chen, K. K., Zhang, J., Xiao, B., Huang, Z., Ju, C., et al. (2018). The decade of exosomal long RNA species: An emerging cancer antagonist. *Molecular Cancer, 17*, 75.
95. Spornraft, M., Kirchner, B., Haase, B., Benes, V., Pfaffl, M. W., & Riedmaier, I. (2014). Optimization of extraction of circulating RNAs from plasma–enabling small RNA sequencing. *PLoS ONE, 9*, Article e107259.
96. Gratz, C., Schuster, M., Brandes, F., Meidert, A. S., Kirchner, B., Reithmair, M., et al. (2024). A pipeline for the development and analysis of extracellular vesicle-based transcriptomic biomarkers in molecular diagnostics. *Molecular Aspects of Medicine, 97*, Article 101269.
97. Bustin, S. A., Benes, V., Garson, J. A., Hellemans, J., Huggett, J., Kubista, M., et al. (2009). The MIQE guidelines: Minimum information for publication of quantitative real-time PCR experiments. *Clinical Chemistry, 55*, 611–622.
98. Huggett, J. F., Foy, C. A., Benes, V., Emslie, K., Garson, J. A., Haynes, R., et al. (2013). The digital MIQE guidelines: Minimum information for publication of quantitative digital PCR experiments. *Clinical Chemistry, 59*, 892–902.
99. Huggett, J. F., Whale, A. S., De Spiegelaere, W., Nour, A. A., Bae, Y. K., Benes, V., et al. (2020). The digital MIQE guidelines update: minimum information for publication of quantitative digital PCR experiments for 2020. *Clinical Chemistry, 66*, 1012–1029.

100. Tesovnik, T., Jenko Bizjan, B., Sket, R., Debeljak, M., Battelino, T., & Kovac, J. (2021). Technological approaches in the analysis of extracellular vesicle nucleotide sequences. *Front Bioeng Biotechnol, 9*, Article 787551.
101. Eldh, M., Lotvall, J., Malmhall, C., & Ekstrom, K. (2012). Importance of RNA isolation methods for analysis of exosomal RNA: Evaluation of different methods. *Molecular Immunology, 50*, 278–286.
102. Shin, H., Shannon, C. P., Fishbane, N., Ruan, J., Zhou, M., Balshaw, R., et al. (2014). Variation in RNA-Seq transcriptome profiles of peripheral whole blood from healthy individuals with and without globin depletion. *PLoS ONE, 9*, Article e91041.
103. Hartl, C., & Gao, Y. (2020). Clarifying the effect of library batch on extracellular RNA sequencing. *Proceedings of National Academic Science USA, 117*, 1849–1850.
104. Hildebrandt, A., Kirchner, B., Meidert, A. S., Brandes, F., Lindemann, A., Doose, G., et al. (2021). Detection of atherosclerosis by small RNA-sequencing analysis of extracellular vesicle enriched serum samples. *Frontiers Cell Development Biology, 9*, Article 729061.
105. Hildebrandt, A., Kirchner, B., Nolte-'t Hoen, E. N. M., & Pfaffl, M. W. (2021). MiREV: An online database and tool to uncover potential reference RNAS and biomarkers in small-RNA sequencing data sets from extracellular vesicles enriched samples. *Journal of Molecular Biology, 433*, Article 167070.
106. Nolan, T., Huggett, J., Sanchez, E., Sanders, R., Redshaw, N., & Wilkes, T. (2013). *Good practice guide for the application of quantitative PCR (qPCR).*
107. Holland, P. M., Abramson, R. D., Watson, R., & Gelfand, D. H. (1991). Detection of specific polymerase chain reaction product by utilizing the 5′—3′ exonuclease activity of Thermus aquaticus DNA polymerase. *Proceedings of National Academic Science USA, 88*, 7276–7280.
108. Wacker, M. J., & Godard, M. P. (2005). Analysis of one-step and two-step real-time RT-PCR using SuperScript III. *Journal of Biomolecular Techniques, 16*, 266–271.
109. Lee, H., He, X., Le, T., Carnino, J. M., & Jin, Y. (2020). Single-step RT-qPCR for detection of extracellular vesicle microRNAs in vivo: A time- and cost-effective method. *American Journal of Physiology. Lung Cellular and Molecular Physiology, 318*, L742–L749.
110. Bracht, J. W. P., Gimenez-Capitan, A., Huang, C. Y., Potie, N., Pedraz-Valdunciel, C., Warren, S., et al. (2021). Analysis of extracellular vesicle mRNA derived from plasma using the nCounter platform. *Science and Reports, 11*, 3712.
111. Androvic, P., Valihrach, L., Elling, J., Sjoback, R., & Kubista, M. (2017). Two-tailed RT-qPCR: A novel method for highly accurate miRNA quantification. *Nucleic Acids Research, 45*, Article e144.
112. Zong, H., Yu, W., Lai, H., Chen, B., Zhang, H., Zhao, J., et al. (2023). Extracellular vesicles long RNA profiling identifies abundant mRNA, circRNA and lncRNA in human bile as potential biomarkers for cancer diagnosis. *Carcinogenesis, 44*, 671–681.
113. Kivioja, T., Vaharautio, A., Karlsson, K., Bonke, M., Enge, M., Linnarsson, S., et al. (2011). Counting absolute numbers of molecules using unique molecular identifiers. *Nature Methods, 9*, 72–74.
114. Cheng, C. A. (2024). Before translating extracellular vesicles into personalized diagnostics and therapeutics: What we could do. *Molecular Pharmaceutics, 21*, 2625–2636.
115. Cheng, C. A., Hou, K. C., Hsu, C. W., & Chiang, L. C. (2024). Ultrasensitive and high-resolution protein spatially decoding framework for tumor extracellular vesicles. *Advance Science*, 11.
116. Wang, X., Ogata, A. F., & Walt, D. R. (2020). Ultrasensitive detection of enzymatic activity using single molecule arrays. *Journal of the American Chemical Society, 142*, 15098–15106.

EV Isolation from Diverse Matrices via Capillary-Channeled Polymer (C-CP) Fiber Platforms

R. Kenneth Marcus, Carolina Mata, and William F. Pons

Abstract Without argument, the most essential aspect of discerning the fundamental biochemistry, development of clinical applications, or the biotherapeutic implementation of extracellular vesicles (EVs) lies in the ability to efficiently and cost-effectively isolate those vesicles from their host media with high yields and purity. This general challenge is exacerbated by the fact that EVs, and more specifically exosomes, must be isolated from an extremely wide diversity of matrices, including mammalian body fluids (serum, urine, saliva, breast milk), plant materials, and a wide range of cell culture media. A final challenge lies in the fact that the sample/matrix volumes can range from microliters to milliliters, and even liters based on the application at hand; covering fundamental studies, clinical diagnostics, and gene therapy vector production. Described in this chapter are developments from the Clemson University laboratory in the use of capillary-channeled polymer (C-CP) fibers as stationary phases for the isolation of exosomes via a hydrophobic interaction chromatography (HIC) elution method. The novel-shaped fibers, when packed into a column format, provide 1–4 μm open channels for efficient solution flow and exosome interaction with the modestly hydrophobic polyester (PET) and nylon-6 fiber surfaces. The general strategy involves an injection of the sample into a concentrated salt solution (2 M $(NH_4)_2SO_4$), wherein matrix salts, small ionic compounds, and genetic material from the sample matrix pass directly through the column or spin-down tip structures. Proteins initially adsorbed to the fiber surfaces are eluted through a reduction in the salt content and the addition of small percentages of organic solvent. Adsorbed exosomes are ultimately released with complete removal of salt content and a further increase in the organic solvent strength. C-CP fiber columns and spin-down tips provide high sample throughput on standard high performance liquid chromatography platforms or through small-format (tabletop) centrifuge processing. Specific applications and results are presented across a wide diversity of sample matrices, suggesting distinct advantages relative to more commonly applied EV isolation methods. Very high throughput and separation quality combine with comparatively

R. K. Marcus (✉) · C. Mata · W. F. Pons
Department of Chemistry, Biosystems Research Complex, Clemson University, Clemson, SC, USA
e-mail: marcusr@clemson.edu

R. K. Marcus and D. Lubman (eds.), *Exosomes and Extracellular Vesicles*,
https://doi.org/10.1007/978-3-031-89270-7_3

inconsequential materials costs to offer practical solutions to challenging exosome separations.

1 Introduction

Whether the point of the effort is a better understanding of fundamental biochemical processes, the development of clinical diagnostic tests, or the delivery of therapeutics, the heart of any utilization of extracellular vesicles (EVs) or more specifically exosomes, is the effective isolation of the target species from their host matrix [1–8]. In the case of fundamental studies [9], effective isolation is required to allow unambiguous cause-and-effect assessment and attribution. When considering diagnostic assays [10, 11], matrix alleviation is needed to present the active entity to the biospecific agent in a concentrated form and in the absence of potential interferents. Finally, with reference to biotherapeutic (vector) applications [12–14], the presence of remnant media species not only lessens the efficacy of the dosing but also increases the potential for catastrophic side effects. Throughout this volume, the many aspects of exosome science are covered by experts in their respective fields, with aspects of exosome isolation holding key positions in each of the end applications. Likewise, chapters deal with specific modalities of isolation and purification which can be applied across multiple disciplines. In this chapter, novel polymer materials, capillary-channeled polymer (C-CP) fibers, are described as immobilization phases towards a diverse set of initial host matrices and end uses of the derived exosomes. The practical aspects of their application and utility are placed into context with more established methods, highlighting areas of opportunity across the areas of fundamental biochemistry, clinical diagnostics, and biotherapeutics.

As in the case of separations in realms of protein chemistry, proteomics, and protein-based biotherapeutics (e.g. monoclonal antibodies (mAbs)) [15, 16], there are some fairly universal performance targets that are relevant to exosome isolation. Indeed, the goals of the two operations, mAb and EV purification, have many parallels. Table 1 presents what might be a typical performance target list generated by researchers in the three aforementioned areas of exosome application. The precise order of importance will vary greatly on a per-case basis, but as presented, allow for some level of grouping. The first four entries (a–d) represent what might be considered the physical aspects of the isolation process; simply put, how efficiently can one extract and purify the exosomes from their native (host) matrix? Clearly, these quantities are highly dependent on the host matrix's identity, with relevant targets dependent on the end application. The next four items (e–h) refer specifically to the retention of the biological function of the vesicles; literally, do they retain the fundamental biochemical and biophysical qualities of relevance to the intended use? The final four entries (i–l) represent the practical aspects of the isolation process, which must always be balanced against the first eight characteristics. In very simple terms, do the ends justify the means? Again, the parallels to the chemico-physico separation involved with protein study and biotherapeutic production are profound. The

Table 1 Target characteristics/metrics describing EV separations

a. Particle count (total)
b. Particle concentration (in extractant)
c. Particle yield (versus native medium)
d. Particle purity (#EV per microgram latent protein)
e. Vesicular structure retention
f. Surface protein retention
g. Genetic cargo retention
h. Bio-functionality/cellular uptake
i. Processing time and complexity
j. Specimen size (minimum and maximum)
k. Capital costs
l. Supplies costs

following paragraphs describe the most common modalities of exosome isolation, providing a basic discussion of processes and practical implementation.

Ultracentrifugation is generally considered to be the 'gold standard' in the isolation of extracellular vesicles. Ultracentrifugation is a density-based isolation method by which larger particulates like cells and cell debris are sedimented first, and smaller particulates like microvesicles and EVs are sedimented with increasing centrifugal forces according to their density [17]. Ultracentrifugation has been used to isolate EV populations that are heterogenous in size (50–200 nm) [18], from various biofluids such as urine, serum, and plasma [17]. However, in addition to the shear stress applied to the vesicle membranes, matrix-related low- and very low-density lipoproteins (LDL and VLDLs), cellular debris, and other species of similar density are found to coelute with the target EVs, impacting purity and subsequent application to downstream processes [19].

Polymer precipitation techniques are often used in commercial exosome isolation kits, including ExoQuick-TC (System BioSciences) and Total Exosome Isolation Kit (Invitrogen); these methods isolate exosomes using the aggregation of polyethylene glycol (PEG) to exosomes and lead to their subsequent precipitation. These PEG-exosome aggregates can then be isolated by low-speed centrifugation [20]. Challenges with polymer precipitation include PEG impurities in the final elution, and poor solubility of precipitated exosomes, leading to sample loss and issues with compatibility for downstream applications [21]. Due to its extended incubation times and limited throughput, this method becomes costly when isolating exosomes from a variety of samples [22]. Finally, specificity is a concern, as this is a separation method based on surface charge, allowing similarly charged species to co-precipitate [23].

Size-exclusion chromatography (SEC) is an effective size-based method for isolating EVs from biofluid matrix components and is available in both spin-down and HPLC column formats. This method successfully removes free proteins from samples, although, as a size-based approach, the coelution of similarly sized particles is the main concern. A major contributor, specifically is the coelution of lipoproteins

(LDLs/HDLs) and sEVs/exosomes, which can interfere with downstream applications [24]. Furthermore, final EV isolate protein concentrations are typically greater than that of UC [25]. Another concern of this approach is throughput; this approach results in extensive dilution of eluates due to on-column diffusion, resulting in diminished exosome isolate concentrations [25]. An alternative approach to this issue is the pairing of SEC and ultracentrifugation, which results in highly pure exosome eluates, although this pairing sacrifices particle concentration and greatly increases separation times and reduces throughput [26].

Filtration methods such as tangential flow filtration (TFF) allow for bulk separation of EVs using size-based methods. By flowing solvent tangentially across the surface of a membrane with desired pore sizes, EVs are isolated from large sample volumes with greater batch-to-batch reproducibility than other separation methods [27]. The constant flow of solvent minimizes membrane fouling effects but contributes to large dilution factors making smaller-scale EV isolation a challenge [28]. TFF yields EVs with lower amounts of matrix-related proteins, but similar to ultracentrifugation, shear stress is a concern and impacts the quality of recovered vesicles for downstream applications [28].

The remainder of this chapter describes the implementation of C-CP fiber stationary phases in the isolation of exosomes from very diverse matrices, employing columns designed for operation in standard high performance liquid chromatography (HPLC) systems or spin-down tips which can be processed using tabletop microcentrifuges with very short spin/processing times. While the potential advantages of the materials lie first in the realm of practicalities (items i-l in Table 1), the performance figures of merit relative to overall exosome recoveries, purities, and retention of basic biochemical functions are equal to or better than the more established methods. While applications to date have centered on sample quantities of relevance to fundamental biochemistry and clinical diagnostics, scale-up to volumes of relevance to vector production are envisioned and indeed underway.

2 C-CP Fiber Exosome Separation Platforms

2.1 *Introduction and Background to C-CP Fiber Separations*

The concept of using natural and polymeric fibers as chromatographic media for liquid-phase separations extends back almost 75 years, with the concepts and applications summarized by Marcus [29, 30]. Indeed, there are both kinetic and thermodynamic reasons for their application over conventional porous bead, monolithic, or membrane stationary phase formats. Those other formats, most specifically porous beads, far and away dominate the HPLC separation space [15, 16, 31, 32]. The primary advantages here lie in the use of single-micron diameter silica beads, where high surface area-to-volume ratios affect high phase ratios to drive the enthalpically driven equilibria processes. The use of small-diameter particles also provides for very efficient column packing and uniform mobile phase flow patterns. However, highly porous phases suffer when applied to large/macromolecules, where intra-pore particle hold-up results in poor mass transfer characteristics as revealed in the classic van Deemter equation C-term [33, 34]. For this reason, phases of nominal surface porosity (so-called superficially porous) find greater application in the area of protein separations [31, 35, 36]. A complicating factor as well is the de facto size selectivity imparted by porous phases, wherein molecules of increasing size are excluded from pore access. This concept lies at the heart of the technique of size exclusion chromatography (SEC). In one vision of ideal chemical separation, there would be no sized-based (entropic) aspects to the process, being only driven by solute-surface (enthalpic) interactions, which would be the case for non-porous phases. Such separations would admittedly be limited in terms of dynamic binding capacities by reduced phase surface areas. In principle, natural and polymeric fibers provide the potential to affect chemical separations based solely on enthalpic interactions between solutes and the fiber surfaces, without potential complications of size-based entropic effects.

Marcus and co-workers at Clemson University have developed the use of a family of channeled (rather than round cross-section), synthetic polymer fibers, termed capillary-channeled polymer (C-CP) fibers, as stationary phases for the HPLC separation of proteins [37–42] and bionanoparticles [43–46]. The shaped fibers were developed by the Tennessee-Eastman company to take advantage of their inherent ability to wick various liquids in single-fiber and non-woven composite formats [47, 48]. The fibers are melt-extruded from commodity polymers, including poly(ethylene) terephthalate (aka polyester, PET), polypropylene (PP), and nylon-6 such that each fiber includes micron-sized channels running along its periphery. The Clemson group found that pulling the fibers through standard and microbore HPLC column structures resulted in a self-alignment of the fibers, effectively creating a monolith of thousands of parallel, 1–4 μm open channels which affect efficient solvent/solute transport [38, 49–51]. As a result, columns provided very efficient fluid transport, with the non-porous fiber structure resulting in no size-based effects and thus, excellent mass transfer characteristics for macromolecules such as proteins [52]. Protein separations could be affected by reversed phase (RP) [37, 38], ion exchange (IEC) [53–55], hydrophobic interaction (HIC) [39, 40], and a variety of affinity-based interactions [42, 56] based on the chemical nature of the native fibers and through readily-implemented surface modifications.

Beginning with discussions with collaborator T. Bruce (Clemson University Light Imaging Facility (CLIF), the scope of application of the C-CP fiber phases was dramatically expanded to the world of extracellular vesicles and exosomes. Based on the basic characteristics of the vesicles, sizes of 50–200 nm and relatively hydrophobic surface chemistry, it was proposed that HIC affected on PET fibers, having a moderately strong hydrophobic yet slightly anionic character, could be applied to isolate exosomes in an HPLC column format [43]. To this point, LC of EVs had not been demonstrated, with SEC-type separations only just becoming known. In HIC, solutes are injected in a high concentration of an antichaotropic salt (e.g. 2 M $(NH_4)_2SO_4$) such that ionic and highly-polar species pass unretained, with an inverse gradient used to successively elute species of increasing hydrophobic character. In the case of exosomes, matrix proteins would elute prior to the vesicles. The method worked on initial application and has been refined over the course of the last 5 years to include isolation of EVs from more than 30 matrices, using the original HPLC column format [43, 44, 57–63] as well as a micropipette tip-based, spin-down processing approach [41, 45, 64–67]. The remainder of this chapter provides a foundation for the overall C-CP fiber separation approach, practical and novel aspects of the approaches relative to more common isolation methods, and prospects for future application.

2.2 C-CP Fiber Column and Spin-Down Tip Assembly

The method of C-CP fiber microbore HPLC column assembly has remained constant over the last decade. In general, the target aspects are to provide the greatest fiber surface area (i.e., number of fibers) for solute interaction while not perturbing the hydrodynamic flow characteristics, which manifest in poor peak shapes and high backpressures. The basic steps in column assembly are shown in Fig. 1a, with the most common format being 0.76 mm i.d. × 30 cm long polyether ether ketone (PEEK) tubing [38]. Based on prior experience, the number of fibers (350–450 depending on fiber geometry) needed to achieve optimum performance (an interstitial fraction of $\varepsilon_i = 0.6$–0.65) are accumulated on a circular cylinder (e.g., slide carousel) and removed as a loop which is pre-shrunk by submersion in hot water [38, 51, 68]. As the loop is pulled through the column, the shaped fibers interdigitate and self-align along the column axis, forming thousands of open parallel channels. As the length of the pulled-fiber bundle is longer than the column tubing, the ends are trimmed, and the column is completed with standard HPLC fittings. This assembly process is accomplished in <5 min. At this point, the column is mounted on the HPLC system and flushed with DI-H_2O and acetonitrile (ACN) to remove any anti-static treatment from the fiber spinning process; the column is now ready for analytical-scale separations. These ~0.1 mL bed volume microbore C-CP fiber columns contain ~85 mg of fiber, having a commercial fiber cost of <$0.25!

The process for assembling the C-CP fiber spin-down tips depicted in Fig. 1b is quite analogous to the HPLC column, except that the construct is a more pliable 0.8 mm i.d. fluorinated ethylene propylene (FEP) tubing [66, 69]. In this case, due to the difference in inner diameter, greater numbers of fibers are employed to achieve similar interstitial fractions. Here, successive washings in hot water and ACN are performed prior to insertion into the tubing. Once the fibers are placed in the tubing, the bundle is pulled back through the column to leave a ~5 mm gap at the end, and the tubing is sliced with a surgical knife to a length of 1.5 cm; this process is repeated to make ~20 tips. The open end of the fiber tip is affixed to the end of standard 200 μL low-retention micropipette tips with a drop of household adhesive to provide stability. Rather than manual aspiration, the spin-down tips are adapted for microcentrifuge processing by first fitting the low-retention tip into a 1 mL tip

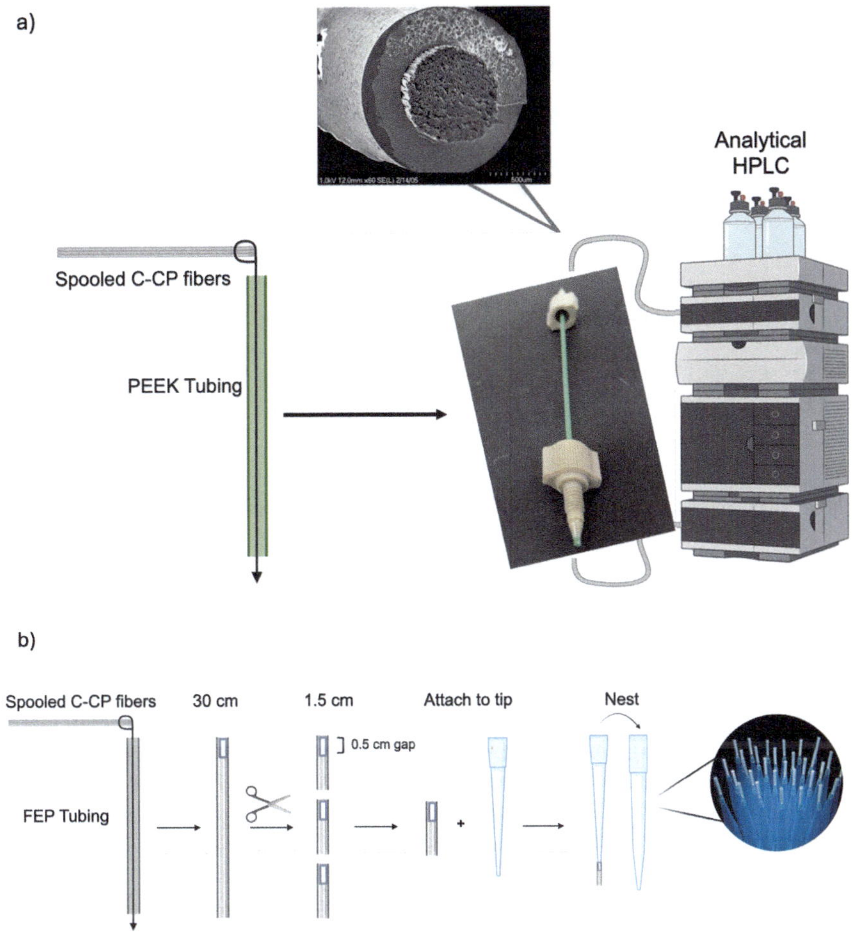

Fig. 1 Implementation of C-CP fibers in **a** HPLC column and **b** spin-down tip formats

for added support, drilling a hole into the screw top of a 15 mL conical centrifuge tube, and inserting the nested tip into place.

2.3 C-CP Fiber Column and Spin-Down Tip Methodologies

One of the most attractive aspects of the C-CP fiber processing of exosomes is the fact that the manipulations are completely in line with other common processes in analytical or biochemical laboratories. As depicted in Fig. 2, the initial processing of all samples except in the case of bovine milk is simple passage through a 0.2 μm syringe filter. (Processing of bovine milk first requires skimming and acetic acid

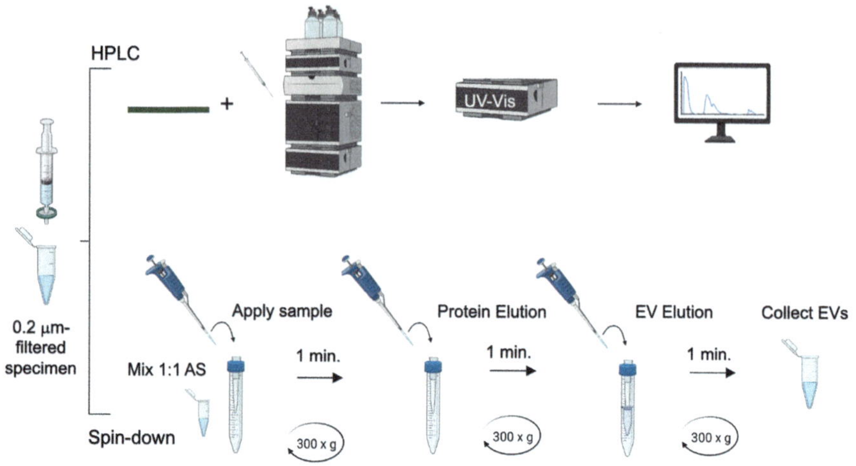

Fig. 2 General procedures for effecting HPLC and spin-down tip separation/isolation of exosomes

treatment to remove the high density of fats and protein agglomerates.) The HPLC process simply involves the placement of the fiber column in the normal solvent flow stream and injection of sample volumes of 4–100 μL in the 2 M ammonium sulfate (AS) mobile phase. At this point, ionic and polar compounds pass unretained, and the reverse salt gradient is initiated in the presence of an organic modifier: glycerol, ACN, or Tween-20 [43, 59, 70]. The gradient can be of a linear form or, most efficiently, a pair of gradient steps that successively remove proteins from the stationary phase by reducing the salt content by 50%, and then the EV isolates by reducing the salt content to zero. This general scenario has been found to be universal across all sample matrices, with the organic modifier chosen based on the end application, as discussed in detail in subsequent sections. As depicted in the diagram, in-line absorbance detection is readily implemented and is a remarkably simple yet powerful means of exosome quantification. In-line multiangle light scattering (MALS) is also practically employed alone or in conjunction with absorbance detection providing greater analytical information such as EV sizing and number densities. The C-CP fiber column approach offers rapid processing (<10 min) of sample volumes composed of up to 10^{12} exosomes; a number sufficient for many biochemical studies [63, 71]. In addition to collecting the target EV fraction, comprehensive profiling of the sample medium can be done by collecting the injection and protein elution fractions. Finally, it should be mentioned that injection of samples at the intermediate mobile phase composition results in the sole retention of the target exosomes, with all small molecules, salts, and proteins passing un-retained [63]; increasing the process throughput and immobilization of greater numbers of vesicles.

Exosome extraction via the spin-down tip method provides a lower-cost, straightforward means of handling volume-limited samples of up to 100 μL. Here, the filtered sample (and indeed multiples for most microcentrifuges) is mixed 1:1 in 4 M AS (to achieve 2 M salt content), aliquoted into a tip and spun at 300 × g for 1 min [66]. An equal volume of the protein elution solvent (lower salt plus organic) is added and spun for 1 min, and the receiving vial is then changed, and an equal volume of the exosome elution solvent (no salt plus organic) is spun through. Here again, initial application in the mixed-solvent mode would result solely in the retention of the EVs. At this point, the sample is available for evaluation via absorbance detection, MALS, nanoflow cytometry, Western blots, etc. or use in biological/clinical assays or other downstream processing.

2.4 Scope of C-CP Fiber EV Separations

The methods described above have been applied to a wide variety of isolation challenges, perhaps touching the vast majority of matrices of relevance in fundamental biochemistry, clinical diagnostics, and vector production. Importantly, in every case, save bovine milk as referenced above, the same general methodologies involving an HIC capture/elution protocol have been followed. In some cases, either HPLC columns or spin-down tips have been applied, and in some, both platforms have been utilized. In general, the HPLC method allows initial confirmation of proper solvent composition choices, with column separations being utilized in cases where in-line detection or high exosome binding capacities are desired. The spin-down tips are most applicable for cases where sample volumes are limited and the use of HPLC will result in excess dilution. It also provides a lower-cost, more parallel means of processing.

Table 2 presents a listing of the various host matrices from which exosomes have been isolated utilizing C-CP fibers. The first listing represents those matrices in which researchers have fundamental biochemical profiling and clinical diagnostics interest [44, 45, 57, 59, 62, 66, 72]. The range of matrices each presents a different set of potential 'chemical' challenges. For example, urine is a high-salt, low-protein matrix, while serum/plasma pose high protein loads, with the presence of lipoproteins being a major challenge. The C-CP fiber column method has demonstrated reductions in lipoprotein carryover of >97% [57, 62]. Human breast milk has moderately high fat and protein content and is of fundamental interest in studies of immunity transfer between mother and child [57]. Bovine milk is a primary focus as a means of bulk, natural vector isolation for biotherapeutics delivery. The high fat/protein content of bovine milk presents an incredible challenge, though, for methods that rely on the size/density isolation of EVs as native fats and casein aggregates present themselves in the same fraction as the target exosomes. Recent efforts in this laboratory point the way to high efficiency isolation following simple skimming and acetic acid treatment of the raw milk [58].

Plant-derived EVs (PDEVs) represent an interesting opportunity towards bulk exosome isolation for therapeutic applications [64]. The challenges in these applications are the incredible diversity in structural (physical) makeup and the need to process the material in such a way that cellular and extracellular species can be distinguished. While much work remains regarding the latter aspect, C-CP fiber spin-down tips provide ready homogenate processing with high yields. Finally, mammalian and other eukaryotic cell cultures present the possibility for controlled production of exosomes in large quantities towards applications as biotherapeutic delivery vectors and use in basic biochemistry studies [41, 60, 63]. While human embryonic kidney (HEK) cells are the most prominent means of targeted EV production, exosomes are a natural by-product in the production of monoclonal antibodies (mAbs) from Chinese hamster ovary (CHO) cells. Studies here have shown that exosomes are produced in cultures in a time-dependent manner, just as mAbs. In fact, their production could be used as a monitor of CHO cell culture health. Ultimately, as EVs exist in the mAb processing waste stream, their successful harvest adds a value-added component to the overall process.

3 Examples of Basic C-CP Fiber and Spin-Down Tip Separations and General Figures of Merit

3.1 C-CP Fiber Column Separations

The initial demonstration of the C-CP fiber column isolation of exosomes was from a commercially-sourced EV sample and the supernatant from a *D. discoideum* culture [43]. Based on previous experiences in the HIC separation of proteins, a simple linear gradient protocol was adopted, with the respective samples injected into a 2 M AS mobile phase, followed by a 20 min. gradient to 30% ACN. In these chromatograms, a distinct injection peak of salts and polar molecules was followed by a broad band of eluting proteins, and then a discrete band of what were shown to be exosomes. The entire process was shortened by injecting the sample at an intermediate solvent composition wherein the proteins passed unretained, with the EVs eluted upon subsequent continuation of the gradient. Key findings in this initial effort included exosome validation via scanning and transmission electron microscopy (SEM/TEM), the ability to inject samples under conditions that only resulted in exosome retention, and the demonstration that in-line absorbance detection (actually

Table 2 Matrices sampled for EV isolation via C-CP fiber column and spin-down tip methods

a. Human urine, serum/plasma, saliva, cervical mucus, and seminal fluid
b. Human breast and bovine milk
c. Plant materials (20 varieties of fruits and vegetables)
d. Mammalian cell cultures (CHO, HEK293, SKOV-3, and CaOv-3)
e. Eukaryote amoeba (*D. discoideum*)

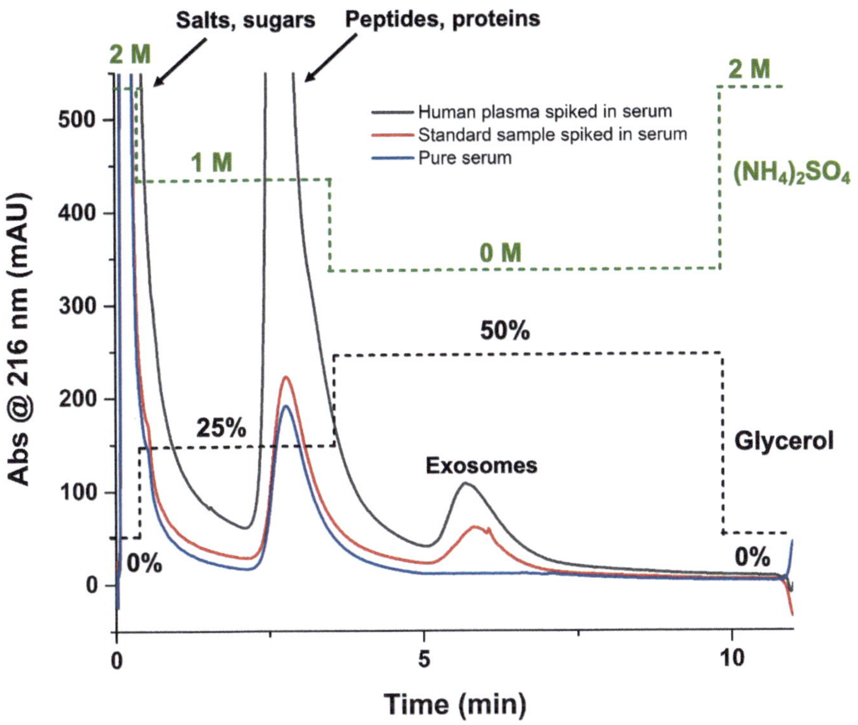

Fig. 3 PET C-CP fiber HIC chromatograms of pure bovine serum (blue line), a 50:50 mixture of pure exosome-depleted bovine serum spiked with a standard exosome sample (red line), and a 50:50 mixture of pure exosome-depleted bovine serum spiked with a human plasma sample (black line). Mobile phase flow rate of 0.5 mL min^{-1}, 20 μL aliquot injections. Dashed lines depict the program aspects of the 100% buffer A (2.0 M $(NH_4)_2SO_4$ solution dissolved in PBS (green line); pH = 7.4) and 100% buffer B (50% glycerol v/v dissolved in PBS) (black line) [59]

scattering by the vesicles) could be readily applied for post-column quantification. The latter point is perhaps one of the most impactful aspects of this research effort.

Based on the initial efforts, all subsequent work has concentrated on the use of step gradients to process exosomes; injection under high-salt conditions, protein elution at intermediate solvent strength, and EV elution in the final step to a zero-salt condition [59]. This general protocol is depicted in the chromatograms of Fig. 3 for the case of exosome isolation from human plasma, wherein glycerol is used as the organic modifier. While the initial efforts employed ACN, there was concern (ultimately proven not relevant) that vesicles could not survive those solvent conditions. Glycerol was chosen here as it is also known to be a solvent compatible with the cryogenic storage of EVs. In the case of human blood, there should be no EVs present in serum but would remain in the plasma fraction. The chromatogram in blue, serum, shows only the injection peak and a subsequent peak corresponding to proteinaceous species following the first gradient step. The chromatogram in red represents serum that has been spiked with a commercial EV isolate. In this case, there is a clear peak occurring following the onset of the final gradient step, representing the added exosomes. The final chromatogram in black represents the spiking of serum with plasma. As can be seen, the protein elution peak response is greatly increased due to the contribution of the plasma proteins, with the exosomes present in the plasma producing an appreciable response as expected following the last gradient step. The composite of these chromatograms set the groundwork for all subsequent work on HIC EV isolations; use of step gradients rather than linear gradients, much reduced (<10 min.) processing times, and the idea that alternative organic solvents might be applied.

The ability to use different organic solvent modifiers has many ramifications in terms of developing application-at-hand exosome isolation strategies. In addition to ACN and glycerol, recent efforts have demonstrated that modest concentrations of the non-ionic detergent Tween-20, can be used very effectively. While in-depth comparisons of the different solvent systems towards different detection scenarios and end uses are currently underway, some very simple guiding principles are evolving. First, while ACN would naturally be expected to have deleterious effects on the vesicular structure of EVs, concentrations of up to 40% ACN in PBS buffer can be used without damage. In fact, for many applications, this system is ideal as once the fraction is collected, the organic content can be simply evaporated, leaving the isolates in a nearly pristine 1 × PBS matrix. Glycerol, in general, provides slightly higher overall particle yields but is more difficult to deal with when applying subsequent processes including cell-lysing prior to proteomic mass spectrometric analysis. It is also problematic in light scattering detection scenarios. In these cases, low-g spinning to separate the viscous phase is readily applied, though it is not 100% efficient. That said, if the intention is to cryogenically store the isolates for later study, glycerol is a near-ideal matrix. Finally, recent efforts with Tween-20 have been very promising [70]. In this case, recoveries as high as those for glycerol can be found. Natural questions exist in the fact that detergents may lyse the vesicles, but the elution strengths used in these separations, <0.5%, have not shown to be detrimental to the exosome integrity. That said, detergent solutions may pose challenges in terms of light scattering detection as well as the need to remove them prior to intentional lysing for proteomics analysis. Very importantly, it remains to be determined what role the elution matrix (and potential remnants) has on the cellular uptake of the exosomes. In general, the respective organic solvent systems must be paired with the means of detection and ultimate use of the exosomes.

3.2 *Isolation of Exosomes from Lipoproteins in Blood Plasma*

Perhaps one of the greatest hinderances to the advancement of the fundamental biochemistry of exosomes in humans is the inability to separate them from lipoproteins (e.g., low-density lipoproteins, LDLs) present in the plasma. The challenge lies in their very similar size and densities. As such, separations based on those qualities result in EV isolates polluted with the proteins. In the most general proteomics evaluations, the presence of the lipoproteins in great excess to the exosome cargo and surface proteins creates great spectral complexity and dynamic range limitations. To assess the ability of the C-CP fiber HIC method to segregate LDLs from exosomes, studies were performed using tandem, post-column absorbance and fluorescence detection to differentiate fluorescently-labeled lipoproteins from native exosomes using the standard glycerol step-gradient method [57].

Depicted in Fig. 4 are the HIC chromatograms for a commercial exosome isolate, a commercial fluorescently-labeled LDL, and a mixture of the two based on absorbance and fluorescence detection [57]. To be clear, both the exosomes and the LDLs should produce responses in the absorbance channel, while only the LDLs should register in the fluorescence channel. What is seen in Fig. 4a is the expected response for the exosome sample, with peaks representing the protein and EV elution fractions. Interestingly, but not surprisingly, the LDL injection also exhibits two peaks; protein and EV. This is not unexpected as the LDLs are themselves derived from blood, rich in exosomes; i.e., they are not pure. Finally, the absorbance chromatogram of the mixture looks much like a linear combination of the two individuals. The fluorescence chromatograms of Fig. 4b provide clear evidence of the successful separation of the two species. As expected, the fluorescence chromatogram of the EV standard is effectively a blank, while those for the LDL and mixture samples show an appreciable fluorescence response in the protein fraction and a minor response in exosome elution step, again reflective of the fact that LDL sample contains EVs that themselves retain a fluorescent tag or have adsorbed LDL molecules. Subsequent MS proteomics analysis of the exosome elution fraction of the mixture sample clearly demonstrated that the composition was virtually, entirely that of exosome-related proteins. In fact, of the total proteomic profile, only 0.3% of the harvested proteins could be attributed to the Apo-B protein content, which would be the most expected LDL signature species. Indeed, these results directly mirrored proteomics analysis of human blood plasma samples performed by the Lubman research group and the University of Michigan, wherein 100 μL samples yielded 10^{11} particles mL^{-1} extraction efficiencies in 15 min. chromatographic runs [62]. More importantly, the proteomics results indicated that, as compared with several currently used separation methods, the PET C-CP HIC method yields higher purity EVs with minimal contamination from serum proteins and apolipoproteins.

3.3 Comparison of C-CP Fiber Spin-Down Separations with Standard EV Isolation Methods

As described previously, the use of the C-CP fiber spin down tip approach to processing has its most direct relevance in the realm of fundamental biochemistry, where exosomes are to be extracted from samples of limited volume; i.e., biofluids.

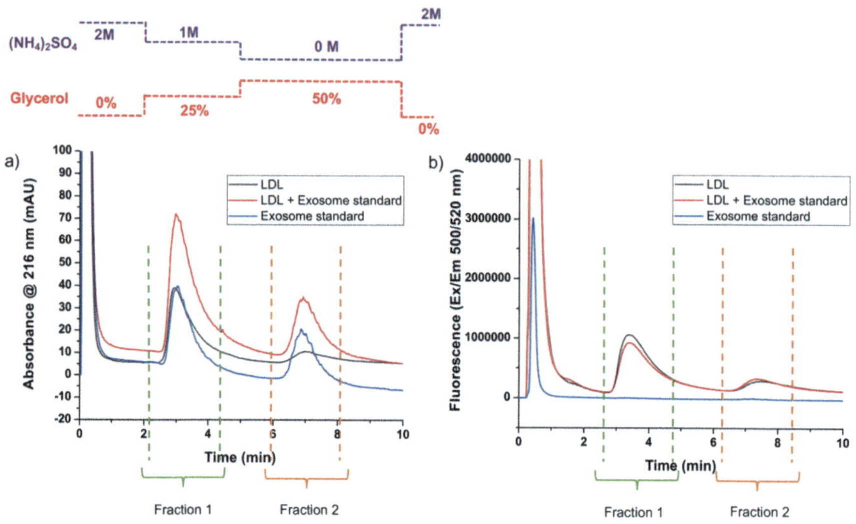

Fig. 4 Chromatograms of LDL, plasma-based exosome standard, and LDL + exosome standard isolation. **a** UV absorbance at 216 nm; **b** Fluorescence excitation at 500 nm and emission at 520 nm. Step gradient of 25% glycerol with 1 M $(NH_4)_2SO_4$ (3 min) (fraction 1) and 50% glycerol (5 min) (fraction 2) were performed. Separations were performed with a mobile phase flow rate of 0.5 mL min^{-1}. Injection volume = 40 µL. LDL concentration = 0.25 mg mL^{-1} and exosome standard density = 2.9×10^{12} particles mL^{-1} [57]

A benchmarking study was performed between the fiber tips (using both the ACN- and glycerol-modified elution solvents, two common polymeric precipitation kits (exoEasy™ (Qiagen) and ExoQuick™ (System Biosciences)), and ultracentrifugation (UC) as the gold standard [65]. Each of the other approaches was implemented exactly as directed by the respective manufacturers for the processing of a singular, human urine, starting material. The respective isolates were subjected to physical and chemical evaluations commonly employed to characterize exosome extraction quality. The respective isolation and test methods are depicted graphically in Fig. 5: TEM to assess the vesicular structures, optical absorbance to determine the EV extract concentrations, microBCA analysis to determine latent protein content, ELISA to assess the tetraspanin protein content and nanoparticle tracking analysis (NTA) to perform particle sizing.

Table 3 presents a summary of the findings of the comparative analysis of a human urine sample as well as some practical figures of merit in terms of processing. As can be seen, before any critical comparison of the separation characteristics, the C-CP fiber spin-down tip method holds a very distinct set of practical advantages versus the other methods in terms of the required sample size and overall processing times. Indeed, while the tips require only 0.2 mL of sample, the other methods require 40–200 × that volume, just to affect the separation. While parallel sample processing is an advantage for the fiber tips over HPLC processing, each of these methods also affords that possibility. The times here are for singular samples.

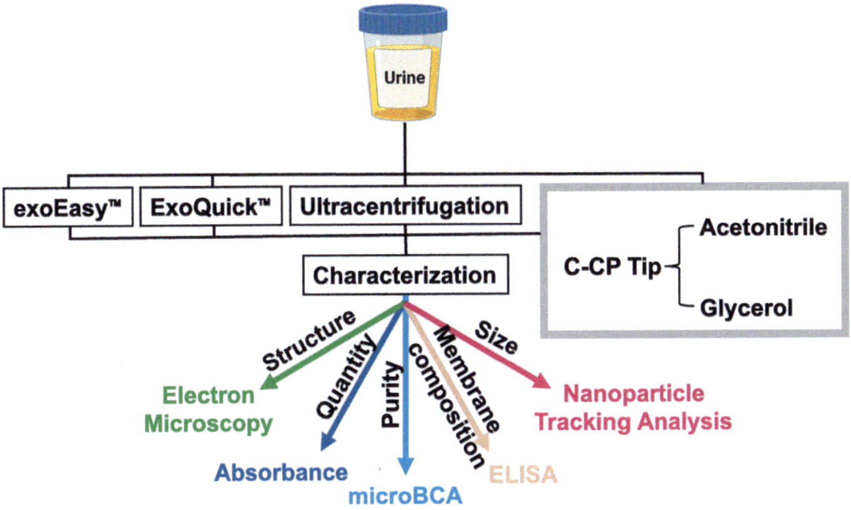

Fig. 5 Diagrammatic depiction of the parallel processing of a human urine sample via standard methods and C-CP fiber spin-down tips using ACN and glycerol as organic elution solvent modifiers, along with the means of the EV isolate characterization

The more quantitative aspects of the comparison point in favor of the spin-down tips as well, with the fibers yielding far higher capture efficiencies than the other methods. It is important to keep in mind that the concentrations here are in the final extractant volumes and that the primary volumes for the commercial packages and the UC methods are multiple times larger, so the actual recoveries in those are far less efficient. In many respects, the comparison among particle sizes is difficult to discern, though the fiber methods and the exoEasy™ do appear to recover more discrete populations. The ExoQuick™ and UC methods generally provide just a broad continuum of sizes which may reflect a good deal of exosome/exosome and exosome/protein agglomeration. The assessment of the characteristic testraspanin proteins present in the vesicular membranes basically reflects the respective total particle counts but may suggest some difficulties in the immunolabelling process for the tip-derived exosomes, which may be affected by the organic solvents present in the isolate fractions. The final entries in the table speak to the obtainable purity of the exosome isolates in terms of the latent protein content. In this case, a microBCA test was employed. It must be emphasized that this test is actually a reflection of amino-acid content, relative to the use of a bovine serum albumin (BSA) standard, and so both free proteins and those embedded in the vesicle surfaces will register a response. As seen, only the fiber spin-down tips yield EV purities that lie above the accepted standard of 3×10^{10} particles per μg of protein [73, 74]. Ultimately, the C-CP fiber tip method holds the promise of providing > 40 × higher recovery efficiencies and 10 × purity levels than the other methods, using far lower sample volumes and shorter processing times, with supply costs of <$1 per tip.

Table 3 Comparison of EV isolation figures of merit using C-CP fiber spin-down tip method with those of standard methods for a common human urine sample matrix

		C-CP fiber tips				
Characteristic	Units	Acetonitrile	Glycerol	ExoEasy™	Exo quick™	Ultra-centrifugation
Primary sample volume	mL	0.2	0.2	8	10	45
Nominal processing time	Minutes	5	5	35	60	130
Concentration (via absorbance)	Particles mL^{-1}	3.7×10^{12}	4.0×10^{12}	$1.0\ 10^{12}$	9.2×10^{11}	1.6×10^{12}
Particle sizing (via NTA)	Maximum population (nm)/general distribution	100 and 180 nm/ distinct populations	60 and 120 nm/ broad continuum	70 nm/broad continuum 100–200 nm	Multiples across broad continuum 80–200 nm	Multiples across broad continuum 60–300 nm
Protein surface activity (via ELISA)	Relative luminescent units (RLU) CD81; CD63	2.4×10^4 2.1×10^4	3.3×10^4 3.9×10^4	2.3×10^4 1.8×10^4	3.1×10^4 1.4×10^4	3.0×10^4 2.7×10^4
Latent protein concentration (via microBCA)	μg mL^{-1}	17	21	54	57	443

3.4 General Figures of Merit for C-CP Fiber Column and Tip Separations

As the C-CP fiber column and spin-down tip methods have shown a very broad range of applicability across diverse matrices, and provide a variety of elution environments that can be tailored to specific end applications, it is good to summarize their general operational and performance characteristics. Presented in Table 4 are those metrics as have been published for two representative matrices, human urine and HEK293 cell culture supernatants. In the case of the HPLC column format, the figures of merit concentrate on the use of the platform to process larger volume samples, looking to harvest large numbers of exosomes [63, 71]. For the spin-down tips, the focus is on the ability to affect rapid, high efficiency isolation [41, 65, 66]. That said, characteristics such as EV purity and binding capacity per unit structure size (e.g. 30 vs 1 cm) are directly transferable. Likewise, the cost of the stationary phase materials is inconsequential in both formats, being <$1 per unit in both cases.

The sample sizes in the two cases are set by two parameters, the reality for isolate dilution in the HPLC application limiting absorbance detection limits at the low end, and the total binding capacity for 1 cm fiber tips limiting the upper range of sample volumes. The processing times in both cases are considered excellent in comparison to competing methods. The ability to achieve similar quantification values for the

Table 4 General figures of merit for EV isolations via liquid chromatography and spin-down tip methods

		C-CP fiber HPLC columns		C-CP fiber Spin-down tips	
Characteristic	Units	Human urine	HEK293	Human urine	HEK293
Sample sizes	μL	20–1000	20–1000	10–100	10–100
Process time	Minutes per sample	10	10	5; w/multiple samples in parallel	5; w/multiple samples in parallel
Typical EV concentrations (via absorbance)	Particles mL^{-1}	$0.2–1.2 \times 10^{11}$	$0.3–1.5 \times 10^{11}$	$\sim 5 \times 10^{10}$	$\sim 1 \times 10^{11}$
Latent protein concentration (via microBCA/ Bradford)	$\mu g\ mL^{-1}$	–	–	<500	<100
EV purity	EVs per μg protein	–	–	$0.9–2 \times 10^{11}$	$0.3–2 \times 10^{12}$
Binding capacity	Particles	$\sim 4 \times 10^{12}$	$3–5 \times 10^{12}$	–	–
Throughput	Particles min^{-1}	$\sim 1 \times 10^{11}$	$\sim 9 \times 10^{10}$	–	–

respective matrices speaks well for the scalability of the approaches and the general applicability of the absorbance quantification approach, whether employed in-line in the HPLC case or performed on standalone microvolume absorbance spectrometers. As described previously in detail, the spin-down tip method has shown very high levels of quantified exosome isolate purities. While not determined explicitly, the clearly high level of protein and EV elution bands in the HIC HPLC method (e.g., Fig. 3) suggest an equally high level of selectivity. In the case of bulk EV recoveries, the fiber columns of less than 0.5 mL bed volumes have exceedingly high binding capacities, on the order of 10^{12} particles. This number of vesicles is above what would be desired for basic biochemistry studies of a given population. Likewise, that number can be isolated on time scales of less than 10 min. Needless to say, such high binding capacities and throughput, coupled with the exceedingly high levels of product purity, point to exciting possibilities in terms of column scaleup for preparative applications.

4 Novel Aspects of C-CP Fiber Exosome Separations

The examples of C-CP fiber platform separations presented in the previous sections demonstrate characteristics that can be directly compared to more common protocols. The fiber platform also affords modes of utilization that offer different levels of implementation and greater information content that is not readily achieved with those methods. Concepts regarding the ability to immobilize exosomes for further interrogation, facile processing for diverse detection modalities, and hybridized methods wherein exosome isolation is used in tandem with other chromatographic columns to gain further insights into complex systems are described here.

4.1 Fluorescence Exosome Identification and On-Fiber Imaging

The concept of using fluorescently labeled antibodies to identify the presence of biomarker proteins on the surface of exosomes is very powerful, but not new. Pools of previously isolated EVs can be incubated with the desired immunofluorescent tags and counted using flow cytometry-type methods. Those sorts of experiments can involve the use of Rayleigh scattering to provide particle count/sizing information, with one or more immunolabels employed to identify the presence of multiple chemical moieties/characteristics. This general concept is illustrated in Fig. 6 for a bovine milk specimen processed on a C-CP fiber column using ACN as the organic solvent modifier, wherein 30 μL of the extract (post-ACN evaporation) is sampled. The NanoFCM (Nottingham, Nottinghamshire, UK) Nanoflow analyzer provides

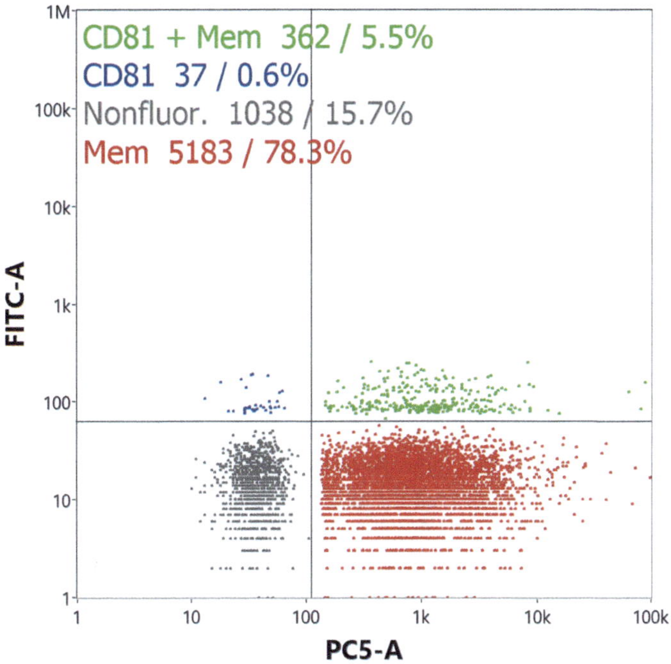

Fig. 6 Nanoflow cytometry analysis of milk-derived EVs fluorescently labeled with membrane label Memglow (PC5 axis) and fluorescent anti-CD81 antibodies (FITC axis)

for particle counting/sizing via light scattering, and provides two channels (for fluorescence monitoring). Presented is a 4-quadrant plot of EVs that have been labeled with both the anti-CD81 and MemGlow labels for the presence of CD81 tetraspanin proteins and a lipid bilayer membrane structure, respectively. The FITC (y-axis) refers to the anti-CD81 Abs, and the PC5 (x-axis), MemGlow. The lower left is the negative population and consists of vesicles that were unlabeled, 'under-labeled' (below threshold), or unbound labels in solution. It is clear the vesicles favor the uptake of MemGlow, with a total of ~81% of detected particles showing positive fluorescence. In the CD81-positive quadrants, less total fluorescence is seen. CD81, a membrane protein, is supposed to be present at a density of only 1–10 repeats per EV, resulting in a lower total fluorescent response in comparison to MemGlow. However, ~6% of vesicles were positive for this protein, and 4.5% of the total vesicles demonstrated a positive response for both CD81 and MemGlow probes. It is hypothesized that the CD81 negative portions still contain a significant portion of CD81 positive vesicles that did not generate sufficient response to exceed the S/N threshold and be detected. A control experiment of only fluorescent labels in 1X PBS was also conducted for both labels, showing the unbound fluorescent label background to be very low, suggesting the fluorescent signal was due to fluorescent amplification.

The processing of EV-containing matrices using the C-CP fiber spin-down tips provides a ready means of immobilizing exosomes on a surface for further interrogation. Analogous to the cytometry experiment, immobilized EVs can be labeled and imaged on an individual basis using confocal fluorescence microscopy [67]. In this way, the physical structure of the vesicles is maintained, with the capability to practically employ more labels (colors) than the typical 3–4 employed in flow systems. The possibility also allows for subsequent probing on-fiber or following elution. Figure 7 illustrates the basic concept for the case of labeling cells for the presence of the CD81 tetraspanin protein and common biomarker upregulated in the case of ovarian cancer (OC), CA125.

The isolation method was the same as presented in Fig. 2, wherein 100 μL aliquots of the test samples were processed in a more or less standard overnight staining cycle (which was reduced to 5 h in subsequent images), and the fibers extracted from the tips. The images here are from 20 × 20 μm regions of single fibers. As expected, each of the sample types shows positive images reflective of exosomes containing the CD81 protein, with those EVs isolated from the CaOv-3 cell culture supernatant, and the urine of an OC patient reading positive for CA125. In each case, the overlays are self-consistent for both proteins. Also, as expected, the isolates from the *d. Discoideum* culture only read positive for CD81. What was unexpected was the co-existence of both proteins in the isolate of exosomes commercially sold as being from 'healthy' human urine. As these products are usually taken from pooled specimens, this was not the case here. It is fully expected that this sort of assay could see appreciably higher throughput based on different immunolabeling approaches, as again, the actual separations take <5 min. It is believed that this general approach could have wide applicability in both the fundamental biochemistry and clinical diagnostics laboratory.

4.2 Coupling with Diverse Detection Modalities

Beyond the concept of on-fiber interrogation, the use of spin-down processing necessitates the use of a separate, discrete means of isolate characterization, as in all other common EV isolation methods. The use of a liquid chromatography platform and C-CP fiber column processing affords a great deal of flexibility in terms of eluate quantification and characterization. The initial demonstration of the methodology, employing a standard HPLC absorbance detector, has been incredibly successful as a multiple of quantification approaches can be employed, including standard Beer's Law calibration curves or the use of standard additions [43–45, 59].

As in the case of traditional areas (e.g. pharmaceuticals) where HPLC is extensively employed, many different approaches to detection and characterization are also available, including optical fluorescence, refractive index, and multi-angle light scattering (MALS) methods. Perhaps most information-rich, is the coupling of in-line MALS detection, including following passage through an absorbance detector. In doing so, the quantification aspects of absorbance are complemented with a second

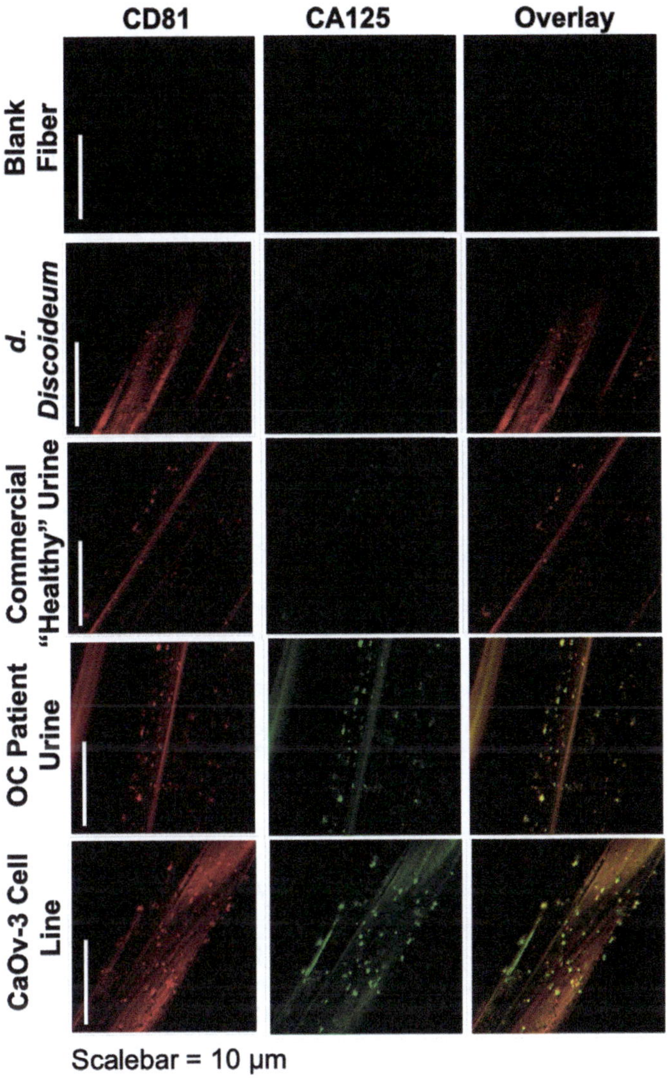

Fig. 7 Confocal fluorescence images of C-CP fibers from process blank and test specimens following standard immunolabeling procedure. Red color depicts the presence of the CD81 tetraspanin protein on exosome membranes, with the green color representing the presence of CA125 [67]

means of quantification along with particle sizing information [61]. Figure 8 presents the absorbance and MALS responses for a 100 μL injection of CHO cell supernatant, subjected to a standard glycerol-based HIC separation. Following calibration with standards, the particles can be quantified and sized based on the chromatographic responses with very high levels of precision and accuracy. For example, based on

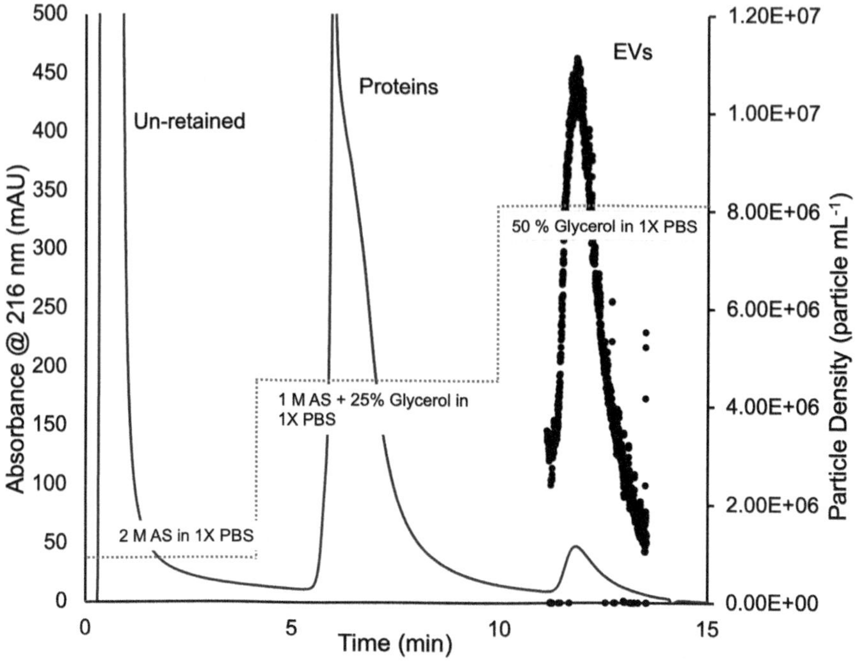

Fig. 8 Overlay of averaged HIC separations across 5 PET C-CP columns. Absorbance is measured at 216 nm, where 2 M ammonium sulfate (AS) is introduced from 0–4 min and unretained species flow through. From 4 to 10 min, a step is taken to 1 M AS and 25% glycerol in 1X PBS, where proteins are eluted. Finally, 50% glycerol in 1 X PBS is introduced from 10 to 15 min, and EVs are eluted [61]

absorbance values, the concentrations varied by <5%RSD across triplicate injections over 5 different columns, with the average values between the two detection modes of 3.6×10^8 particles mL^{-1} and 3.7×10^8 particles mL^{-1} for absorbance and MALS being in excellent agreement. Not surprisingly, the MALS values showed higher overall variability. That said, the correlation between the integrated MALS particle counts and the integrated absorbance values agreed with an $R^2 = 0.999$. As a complement, for the MALS particle sizing, the median and 50th percentile sizes were in excellent agreement across the 5 columns. The ability to affect different in-line detection modalities to provide more comprehensive information about exosome isolates is considered to be one of the strongest attributes of the methodology.

4.3 Tandem Isolation of Exosomes and Monoclonal Antibodies

As stated multiple times in this chapter, the isolation of exosomes from any biological or cell culture matrix is challenged by the presence of matrix-related proteins. To this point, those proteins have been treated as incidental waste products. Of course, this can be no further from the truth in the operating biological system. A comprehensive picture of the state of any system would include a proteomic assessment. The C-CP fiber methods described here provide a ready means of harvesting native proteins from the host matrix. A unique situation is presented in the case of mammalian cell cultures which are typically used in the production of biotherapeutic proteins including monoclonal antibodies (mAbs). Most specifically, CHO cells are far and away the most common modality for the production of mAbs. While the growth of human embryonic kidney (HEK) cells is the most widely applied culture process for biotherapeutic exosome vector production, EVs are a natural by-product of CHO cell culturing. Indeed, as in all other systems, the rate of exosome production is a reflection of the overall stress, positive or negative, in a cellular system. To this end, harvesting of exosomes as a function of the culture time of HEK293S supernatants using the C-CP fiber spin-down tips showed an excellent correlation along the cell multiplication cycle, a time of system stress [41]. As the viable cell count increased, so too did the exosome recoveries. As the cell density reached a steady state (i.e., low stress), the exosome numbers actually decreased. In all cases, very high EV purities (»3 $\times$ 10^{10} particles μg^{-1} protein) were realized, with the daily sampling volumes being only 100 μL.

The concept that exosomes are both by-products and potential process monitoring elements in CHO cell culturing led to the development of a two-dimensional (2D) liquid chromatography method, permitting the isolation and quantification of exosomes and mAbs [60]. The processes were affected on a standard 2D-HPLC system, wherein a PET C-CP fiber column was employed in the first dimension to isolate EVs, and a protein A (ProA) modified polypropylene C-CP fiber column was employed in the second dimension to selectively capture immunoglobulin G (IgG). Both columns were employed for selective capture, elution, and absorbance detection for quantification. The chromatogram of Fig. 9a depicts the injection of 20 μL of CHO cell supernatant under protein non-retention conditions (1 M AS/25% ACN), where all proteins including the mAbs pass to the second (ProA) column, and the subsequent step elution of exosomes as in normal HIC separations. Of that protein effluent, IgG molecules are selectively retained and then released in a step gradient with 150 mM citric acid (pH = 2.6) as shown in the chromatogram of Fig. 9b. As depicted, the overall process of mAb capture and recovery occurs with very high efficiency and precision. Based on this relatively simple experimental setup, it is proposed that exosome production in cell cultures can be employed in process monitoring scenarios, both to access the growth process and to diagnose negative stresses within the culture due to pH, O_2, and feedstock variations. Likewise, this approach provides a direct line to developing fermentation conditions specifically

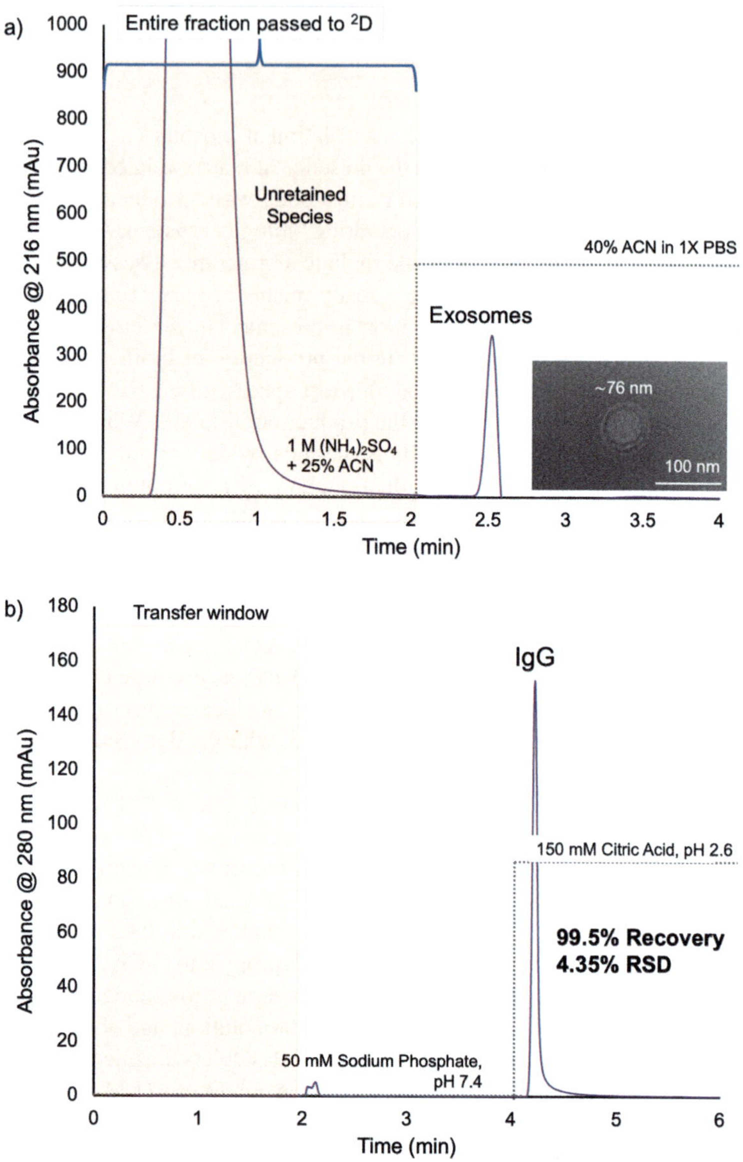

Fig. 9 Product chromatograms for the separate stages of the isolation of (a) exosomes and (b) IgG from a 20 μL injection of CHO cell culture supernatant. **a** ^{1}D HIC isolation of exosomes with absorbance detection. TEM image of a single exosome. **b** ^{2}D ProA isolation of IgG with column placed in the switch valve loop and absorbance detection. The orange region of each chromatogram represents the timeframe where the two columns are in-line with each other [60]

for the production of EVs. Finally, as has recently been demonstrated, reversing the column order sets the groundwork for performing standard mAb purification/recovery, with subsequent exosome recovery as a value-added product from what would normally be considered a process waste stream.

5 Conclusions and Paths Forward

As has been well documented for the last two decades, the isolation, purification and recovery of exosomes is an incredible challenge. The chemical and physical makeup of the nanovesicles and their likeness to many different matrix components only serve to complicate the issue. Add to this, the disparity in those matrix types. The C-CP fiber stationary phase, in either HPLC column or spin-down tip formats, provides a basis for straightforward processing with exceedingly low capital and consumable costs. High throughput in either mode, along with the ability to handle sample volumes from microliter to single milliliter, allows for tailoring in terms of scale. At this point, across the vast diversity of matrices, including human plasma, urine, saliva, cell cultures, human breast and bovine milk, and 20 plant types, EV yields and purities have surpassed all standards in the field. Finally, direct comparisons and literature reports suggest excellent figures of merit with respect to the generic harvesting of exosomes.

Moving forward, there are basic questions that remain to be answered. For example, is PET the best polymer fiber matrix? Current efforts are looking to the use of more hydrophilic nylon-6 as the stationary phase. More qualitative and quantitative evaluations of the respective organic elution modifiers (ACN, glycerol, and Tween-20) are needed, tailored more specifically to eventual means of exosome characterization and application. Questions about the ultimate processing capacity of C-CP fiber columns must be answered relative to the production of large quantities of vector materials. Current efforts in the Clemson University laboratory look to capabilities on the preparative scale, with the target of 10^{15} EVs on a single column. Past performance in just five years absolutely justifies these efforts.

Acknowledgements The efforts described here relative to the use of C-CP fibers for the isolation and purification of extracellular vesicles and exosomes have been funded by the US National Science Foundation under grant numbers CHE-1608663 and 2107882, and by the National Institutes of Health under grant number 1RO1GM1411347.

References

1. Adnan, S.-S., Benjamin, G. T., & David, R. W. (2023). Advances in extracellular vesicle isolation methods: A path towards cell-type specific EV isolation. *Extracellular Vesicles and Circulating Nucleic Acids, 4*(3), 447–460.
2. Cho, S., Yang, H. C., & Rhee, W. J. (2020). Development and comparative analysis of human urine exosome isolation strategies. *Process Biochemistry, 88*, 197–203.
3. Liangsupree, T., Multia, E., & Riekkola, M.-L. (2021). Modern isolation and separation techniques for extracellular vesicles. *Journal of Chromatography A, 1636*, Article 461773.
4. Patel, G. K., Khan, M. A., Zubair, H., Srivastava, S. K., Khushman, M., Singh, S., Singh, A.P. (2019). Comparative analysis of exosome isolation methods using culture supernatant for optimum yield, purity and downstream applications. *Scientific Reports, 9.*
5. Wang, W., Sun, H., Duan, H., Sheng, G., Tian, N., Liu, D., Sun, Z. (2024). Isolation and usage of exosomes in central nervous system diseases, CNS Neurosci. *Ther 30*, e14677.
6. Benayas, B., Morales, J., Egea, C., & Armisen, P. (2023). Yanez-Mo, Optimization of extracellular vesicle isolation and their separation from lipoproteins by size exclusion chromatography. *Journal of Extracellular Biology, 2*, Article e100.
7. Shami-shah, A., Travis, B. G., & Walt, D. R. (2023). Advances in extracellular vesicle isolation methods: A path towards cell-type specific EV isolation. *Extracellular Vesicles and Circulating Nucleic Acids, 4*, 447–460.
8. Dong, L., Zieren, R. C., Horie, K., Kim, C.-J., Mallick, E., Jing, Y., Feng, M., Kuczler, M. D., Green, J., Amend, S. R., Witwer, K. W., de Reijke, T. M., Cho, Y.-K., Pienta, K. J., & Xue, W. (2020). Comprehensive evaluation of methods for small extracellular vesicles separation from human plasma, urine and cell culture medium. *Journal of Extracellular Vesicles, 10*(2), Article e12044.
9. Asao, T., Tobias, G. C., Lucotti, S., Jones, D. R., Matei, I., & Lyden, D. (2023). Extracellular vesicles and particles as mediators of long-range communication in cancer: Connecting biological function to clinical applications. *Extracellular Vesicles and Circulating Nucleic Acids, 4*(3), 461–485.
10. Mukherjee, A., Bisht, B., Dutta, S., & Paul, M. K. (2022). Current advances in the use of exosomes, liposomes, and bioengineered hybrid nanovesicles in cancer detection and therapy. *Acta Pharmacologica Sinica, 43*(11), 2759–2776.
11. Popović, M., de Marco, A. (2017). Canonical and selective approaches in exosome purification and their implications for diagnostic accuracy. *Translational Cancer Research*, S209-S225.
12. Chavda, V. P., Pandya, A., Kumar, L., Raval, N., Vora, L. K., Pulakkat, S., Patravale, V., Salwa, Duo, Y., Tang, B. Z. (2023). Exosome nanovesicles: A potential carrier for therapeutic delivery. *Nano Today, 49*, 101771
13. Liu, H., Geng, Z., & Su, J. (2022). Engineered mammalian and bacterial extracellular vesicles as promising nanocarriers for targeted therapy. *Extracellular Vesicles and Circulating Nucleic Acids, 3*(2), 63–86.
14. Tenchov, R., Sasso, J. M., Wang, X., Liaw, W.-S., Chen, C.-A., & Zhou, Q. A. (2022). Exosomes–nature's lipid nanoparticles, a rising star in drug delivery and diagnostics. *ACS Nano, 16*(11), 17802–17846.
15. Carta, G., & Jungbauer, A. (2020). *Protein chromatography: Process development and scale-Up, 2nd* (Edition). Wiley-VCH Verlag GmbH.
16. Hearn, M. T. W., & Anspach, B. (2001). Chemical, physical, and biochemical concepts in isolation and purification of proteins. *Separation and Purification Methods, 30*(2), 221–263.
17. Torres Crigna, A., Fricke, F., Nitschke, K., Worst, T., Erb, U., Karremann, M., Buschmann, D., Elvers-Hornung, S., Tucher, C., Schiller, M., Hausser, I., Gebert, J., Bieback, K. (2021). Interlaboratory comparison of extracellular vesicle isolation based on ultracentrifugation. *Transfus Med Hemother, 48*(1), 48–59.
18. Lobb, R. J., Becker, M., Wen, S. W., Wong, C. S., Wiegmans, A. P., Leimgruber, A., & Möller, A. (2015). Optimized exosome isolation protocol for cell culture supernatant and human plasma. *Journal of Extracellular Vesicles, 4*, 27031.

19. Yuana, Y., Levels, J., Grootemaat, A., Sturk, A., Nieuwland, R. (2014). Co-isolation of extracellular vesicles and high-density lipoproteins using density gradient ultracentrifugation. *Journal of Extracellular Vesicles, 3*
20. Weng, Y., Sui, Z., Shan, Y., Hu, Y., Chen, Y., Zhang, L., & Zhang, Y. (2016). Effective isolation of exosomes with polyethylene glycol from cell culture supernatant for in-depth proteome profiling. *The Analyst, 141*(15), 4640–4646.
21. Yakubovich, E. I., Polischouk, A. G., & Evtushenko, V. I. (2022). Principles and problems of exosome isolation from biological fluids, biochemistry (Moscow). *Supplement Series A: Membrane and Cell Biology, 16*(2), 115–126.
22. Rider, M. A., Hurwitz, S. N., & Meckes, D. G. (2016). ExtraPEG: A polyethylene glycol-based method for enrichment of extracellular vesicles. *Scientific Reports, 6*(1), 23978.
23. Sidhom, K., Obi, P. O., Saleem, A. (2020). A review of exosomal isolation methods: Is size exclusion chromatography the best option? *International Journal of Molecular Sciences, 21*(18).
24. Xu, R., Fitts, A., Li, X., Fernandes, J., Pochampally, R., Mao, J., & Liu, Y.-M. (2016). Quantification of small extracellular vesicles by size exclusion chromatography with fluorescence detection. *Analytical Chemistry, 88*(21), 10390–10394.
25. An, M., Wu, J., Zhu, J., & Lubman, D. M. (2018). Comparison of an optimized ultracentrifugation method versus size-exclusion chromatography for isolation of exosomes from human serum. *Journal of Proteome Research, 17*(10), 3599–3605.
26. Huang, K., Garimella, S., Clay-Gilmour, A., Vojtech, L., Armstrong, B., Bessonny, M., & Stamatikos, A. (2022). Comparison of human urinary exosomes isolated via ultracentrifugation alone versus ultracentrifugation followed by SEC column-purification. *Journal of Personalized Medicine, 12*(3), 340.
27. Marsh, S. R., Pridham, K. J., Jourdan, J., & Gourdie, R. G. (2021). Novel protocols for scalable production of high quality purified small extracellular vesicles from bovine milk. *Nanotheranostics, 5*(4), 488–498.
28. Visan, K. S., Lobb, R. J., Ham, S., Lima, L. G., Palma, C., Edna, C. P. Z., Wu, L. Y., Gowda, H., Datta, K. K., Hartel, G., Salomon, C., & Möller, A. (2022). Comparative analysis of tangential flow filtration and ultracentrifugation, both combined with subsequent size exclusion chromatography, for the isolation of small extracellular vesicles. *Journal of Extracellular Vesicles, 11*(9), Article e12266.
29. Marcus, R. K. (2008). Use of polymer fiber stationary phases for liquid chromatography separations: Part I - Physical and chemical rationale. *Journal of Separation Science, 31*, 1923–1935.
30. Marcus, R. K. (2009). Use of polymer fiber stationary phases for liquid chromatography separations: Part II - applications. *Journal of Separation Science, 32*(5–6), 695–705.
31. Jungbauer, A. (2005). Chromatographic media for bioseparations. *Journal of Chromatography A, 1065*, 3–12.
32. Cunico, R. L., Gooding, K. M., Wehr, T. (1998). *Basic HPLC and CE of biomolecules, bay bioanalytical laboratory*, Richmond, CA
33. Giddings, J. C. (1965). *Dynamics of chromatography*. Marcel Dekker.
34. Neue, U. D., & Columns, H. P. L. C. (1997). *Theory, technology, and practice*. Wiley-VCH.
35. Kirkland, J. J., Truszkowski, F. A., Dilks, C. H., & Engel, G. S. (2000). Superficially porous silica microspheres for fast high-performance liquid chromatography of macromolecules. *Journal of Chromatography A, 890*(1), 3–13.
36. Schuster, S. A., Wagner, B. M., Boyes, B. E., & Kirkland, J. J. (2013). Optimized superficially porous particles for protein separations. *Journal of Chromatography A, 1315*, 118–126.
37. Nelson, D. K., & Marcus, R. K. (2003). A novel stationary phase: Capillary-channeled polymer (C-CP) fibers for HPLC separations of proteins. *Journal of Chromatographic Science, 41*(9), 475–479.
38. Randunu, J. M., & Marcus, R. K. (2012). Microbore polypropylene capillary channeled polymer (C-CP) fiber columns for rapid reversed phase HPLC of proteins. *Analytical and Bioanalytical Chemistry, 404*, 721–729.

39. Stanelle, R. D., & Marcus, R. K. (2009). Nylon-6 capillary-channeled polymer (C-CP) fibers as a hydrophobic interaction chromatography stationary phase for the separation of proteins. *Analytical and Bioanalytical Chemistry, 393*(1), 273–281.
40. Wang, L., & Marcus, R. K. (2019). Evaluation of protein separations based on hydrophobic interaction chromatography using polyethylene terephthalate capillary-channeled polymer (C-CP) fiber phases. *Journal of Chromatography A, 1585*, 161–171.
41. Jackson, K. K., & Marcus, R. K. (2023). Rapid isolation and quantification of extracellular vesicles from suspension-adapted human embryonic kidney cells using capillary-channeled polymer fiber spin-down tips. *Electrophoresis, 44*, 190–202.
42. Schadock-Hewitt, A. J., & Marcus, R. K. (2014). Initial evaluation of protein A modified capillary-channeled polymer fibers for the capture and recovery of immunoglobulin G. *Journal of Separation Science, 37*(5), 495–504.
43. Bruce, T. F., Slonecki, T. J., Wang, L., Huang, S., Powell, R. R., & Marcus, R. K. (2019). Exosome isolation and purification via hydrophobic interaction chromatography using a polyester, capillary-channeled polymer fiber Phase. *Electrophoresis, 40*, 571–581.
44. Huang, S., Wang, L., Bruce, T. F., & Marcus, R. K. (2019). Isolation and quantification of human urinary exosomes by hydrophobic interaction chromatography on a polyester capillary-channeled polymer fiber stationary phase. *Analytical and Bioanalytical Chemistry, 411*, 6591–6601.
45. Jackson, K. K., Powell, R. R., Bruce, T. F., Marcus, R. K. (2021). Rapid isolation of extracellular vesicles from diverse biofluid matrices via capillary-channeled polymer fiber solid-phase extraction micropipette tips analyst *146*, 4314–4325. https://doi.org/10.1039/D1AN00373A.
46. Huang, S., Bruce, T. F., Ding, H., Wei, Y., & Marcus, R. K. (2021). Rapid isolation of lentivirus particles from cell culture media via a hydrophobic interaction chromatography method on a polyester, capillary-channeled polymer fiber stationary phase. *Analytical and Bioanalytical Chemistry, 413*, 2985–2994.
47. Brown, P. J., M. M., Sinclair, K., Tucker, E., Inam, A. (2004). Production, properties and potential applications of deep groove fiber. *Southeast Regional Meeting of the American Chemical Society*.
48. Haile, W. A., & Phillips, B. M. (1995). Deep grooved polyester fiber for wet lay applications. *Tappi Journal, 78*(9), 139–142.
49. Marcus, R. K., Davis, W. C., Knippel, B. C., LaMotte, L., Hill, T. A., Perahia, D., & Jenkins, J. D. (2003). Capillary-channeled polymer fibers as stationary phases in liquid chromatography separations. *Journal of Chromatography A, 986*(1), 17–31.
50. Stanelle, R. D., Sander, L. C., & Marcus, R. K. (2005). Hydrodynamic flow in capillary-channel fiber columns for liquid chromatography. *Journal of Chromatography A, 1100*(1), 68–75.
51. Randunu, J. M., Dimartino, S., & Marcus, R. K. (2012). Dynamic evaluation of polypropylene capillary-channeled fibers as a stationary phase in high performance liquid chromatography. *Journal of Separation Science, 35*, 3270–3280.
52. Wang, Z., & Marcus, R. K. (2014). Determination of pore size distributions in capillary-channeled polymer fiber stationary phases by inverse size exclusion chromatography (iSEC) and implications for fast protein separations. *Journal of Chromatography A, 1351*, 82–89.
53. Jiang, L., Jin, Y., & Marcus, R. K. (2015). Polyethylenimine modified Poly(ethylene terephthalate) capillary channeled-polymer (C-CP) fibers for anion exchange chromatography of proteins. *Journal of Chromatography A, 1410*, 200–209.
54. Schadock-Hewitt, A. J., Pittman, J. J., Stevens, K. A., & Marcus, R. K. (2013). Comparison of two types of nylon 6 capillary-channeled polymer (C-CP) fiber stationary phases towards the separation of proteins via ion exchange chromatography. *Journal of Applied Polymer Science, 128*, 1257–1265.
55. Stanelle, R. D., Straut, C. A., & Marcus, R. K. (2007). Nylon-6 capillary-channeled polymer fibers as a stationary phase for the mixed-mode ion exchange/reversed-phase chromatography separation of proteins. *Journal of Chromatographic Science, 45*(7), 415–421.
56. Trang, H. K., Marcus, R. K. (2020). Application of polydopamine-coated nylon capillary-channeled polymer (C-CP) fibers as a stationary phase for mass spectrometric phosphopeptide analysis electrophoresis, *41*, 215–224.

57. Huang, S., Ji, X., Jackson, K. K., Bruce, T. F., Lubman, D. M., & Marcus, R. K. (2021). Rapid separation of blood plasma exosomes from low-density lipoproteins via a hydrophobic interaction chromatography method on a polyester capillary-channeled polymer fiber phase. *Analytica Chimica Acta, 1167*, Article 338578.
58. Mata, C., Pimental, J., Marcus, R. K. Isolation of bovine milk-derived exosomes via a capillary channeled polymer (C-CP) fiber stationary
59. Phase, E. (2024). *Vesicles Circ.* Acids submitted for publication.
60. Wang, L., Bruce, T. F., Huang, S., & Marcus, R. K. (2019). Isolation and quantitation of exosomes isolated from human plasma via hydrophobic interaction chromatography using a polyester, capillary-channeled polymer fiber phase Anal. *Chimica Acta, 1082*, 186–193.
61. Wysor, S. K., & Marcus, R. K. (2023). Quantitative recoveries of exosomes and monoclonal antibodies from chinese hamster ovary cell cultures by use of a single, integrated two-dimensional liquid chromatography method. *Analytical Chemistry, 95*, 17886–17893.
62. Wysor, S. K., & Marcus, R. K. (2024). In-line coupling of capillary-channeled polymer fiber columns with optical absorbance and multi-angle light scattering detection for the isolation and characterization of exosomes. *Analytical and Bioanalytical Chemistry, 416*, 3325–3333.
63. Ji, X., Huang, S., Bruce, T. F., Tan, Z., Wang, D., Zhu, Z., Marcus, R. K., & Lubman, D. M. (2021). A novel method of high-purity extracellular vesicle enrichment from microliter-scale human serum for proteomic analysis. *Electrophoresis, 42*, 245–256.
64. Billotto, L. S., Jackson, K. K., & Marcus, R. K. (2022). Determination of the loading capacity and recovery of extracellular vesicles derived from human embryonic kidney cells and urine matrices on capillary-channeled polymer (C-CP) fiber columns. *Separations, 9*(9), 251–264.
65. Jackson, K. K., Mata, C., & Marcus, R. K. (2023). A rapid capillary-channeled polymer (C-CP) fiber spin-down tip approach for the isolation of plant-derived extracellular vesicles (PDEVs) from 20 common fruit and vegetable sources. *Talanta, 252*, Article 123779.
66. Jackson, K. K., Powell, R. R., Marcus, R. K., & Bruce, T. F. (2022). Comparison of the capillary-channeled polymer (C-CP) fiber spin-down tip approach to traditional methods for the isolation of extracellular vesicles from human urine. *Analytical and Bioanalytical Chemistry, 414*(13), 3813–3825.
67. Jackson, K. K., Powell, R. R., Bruce, T. F., & Marcus, R. K. (2020). Solid phase extraction of exosomes from diverse matrices via a polyester capillary-channeled polymer (C-CP) fiber stationary phase in a spin-down tip format. *Analytical and Bioanalytical Chemistry, 412*, 4713–4724.
68. Jackson, K. K., Powell, R. R., Bruce, T. F., Marcus, R. K. (2022). Facile, generic capture and on-fiber differentiation of exosomes via confocal immunofluorescence microscopy using a capillary-channeled polymer fiber solid-phase extraction tip. *Sensors Diagnost, 1*, 525–533.
69. Wang, Z., & Marcus, R. K. (2015). Roles of interstitial fraction and load linear velocity on the dynamic biding capacity of proteins on capillary channeled polymer fiber columns. *Biotechnology Progress, 15*, 97–109.
70. Burdette, C. Q., & Marcus, R. K. (2013). Solid phase extraction of proteins from buffer solutions employing capillary channeled polymer (C-CP) fibers as the stationary phase. *The Analyst, 138*, 1098–1106.
71. Bin Islam, M. K., Marcus, R. K. (2024). Isolation and quantification of human urinary exosomes using a tween-20 elution solvent from polyester, capillary-channeled polymer fiber columns. Anal. Chim. Acta submitted for publication.
72. Huang, S., Wang, L., Bruce, T. F., & Marcus, R. K. (2020). Evaluation of exosome loading characteristics in their purification via a glycerol-assisted hydrophobic interaction chromatography method on a polyester, capillary-channeled polymer fiber phase. *Biotechnology Progress, 36*, Article e2998.

73. Chosed, R. J., Polley, E., Pike, J. F. W., Huang, S., Zimmerman, S., Bruce, T. F., Marcus, R. K., & Roudebush, W. E. (2020). Isolation, purification and initial RNA sequence analysis of seminal fluid exosomes between pregnant and non-pregnant intrauterine insemination pregnancies. *Reproductive Biomedicine Online, 41,* Article e1.
74. Webber, J., & Clayton, A. (2013). How pure are your vesicles? *Journal of Extracellular Vesicles, 2*(1), 19861.

Immunoaffinity Chromatography and Field-Flow Fractionation Methods of EV Isolation

IAC and FFF Methods of EV Isolation

Thanaporn Liangsupree, Susanne K. Wiedmer, and Marja-Liisa Riekkola

Abstract Extracellular vesicles (EVs) and especially nanosized exosomes need time-effective, easy-to-operate, and easily automated isolation methods/techniques that are not expensive, provide a controlled yield of high purity, and preserve the integrity of the fragile isolates. In addition, the EV isolates should not be contaminated with lipoproteins or other proteins. Reliable isolation methods can help to understand further the function and behavior of EVs at a cellular level, and this information can potentially be beneficial for other applications, such as drug delivery, antibody-based therapeutics, and diagnostic tools. This chapter gives a comprehensive overview of the potential of immunochromatography (IAC) and immunocapture techniques that exploit antibodies with a specific affinity toward biomolecules on the EV membrane surface, and versatile and gentle field-flow fractionation (FFF) techniques with possibilities for on-line coupling systems for the isolation of EV subpopulations with further experimental analysis. Special attention is given to methodological, and technical challenges, and the achievements are demonstrated with reference to applications. In addition, the principle of the IAC and FFF techniques are introduced to facilitate their methodological understanding.

T. Liangsupree · S. K. Wiedmer · M.-L. Riekkola (✉)
Department of Chemistry, University of Helsinki, Helsinki, Finland
e-mail: marja-liisa.riekkola@helsinki.fi

T. Liangsupree
e-mail: t.liangsupree@gmail.com

S. K. Wiedmer
e-mail: susanne.wiedmer@helsinki.fi

R. K. Marcus and D. Lubman (eds.), *Exosomes and Extracellular Vesicles*,
https://doi.org/10.1007/978-3-031-89270-7_4

1 Introduction

Extracellular vesicles (EVs) are naturally occurring heterogeneous populations of membrane-enclosed nanoparticles (30–5000 nm) released by all eucaryotic and procaryotic cells. In the human body, they are found in cellular microenvironment and in the blood circulation [1] (plasma or serum), as well as in amniotic fluid, breast milk, cerebrospinal fluid, feces, saliva, semen, and urine [2]. Ectosomes, such as micro- and large vesicles and microparticles, are EVs of 50–1000 nm that bud out from the plasma membrane, while exosomes (50–150 nm) are EVs of endosomal origin [3]. Exosomes alone contain a large amount of information about their cell origin, which can be exploited to track the progress of different diseases. Exomeres (30–50 nm) are recently discovered particle types lacking a lipid bilayer [4]. Studies on exomeres suggest that biogenesis is most probably involved in the cell's metabolism since their inner cargo has been discovered to contain proteins from endoplasmic reticulum, microtubules, and mitochondria that are involved in glycolysis and mTORC1 metabolic pathways [5].

EVs have attracted tremendous attention in the last decade as key mediators of intercellular communication. Due to their safety, intrinsic bioactivity, and ability to reach specific cells by passing, e.g., inter-endothelial junctions or the blood–brain-barrier, they have also become evident as versatile therapeutic agents, biomarkers, and potential drug carriers [6]. In addition to cancer diagnosis, EVs have diagnostic potential for angiogenesis, immune response, cardiovascular and neurodegeneration-related diseases [3]. The potential of harnessing EVs in the diagnosis and treatment of diseases—including cancer and neurological and cardiovascular disorders—has been recognized [7].

Even though there is a growing interest in EVs and a high number of techniques and methods are available for isolation, separation, and fractionation of subpopulations of EVs from prokaryotes and eukaryotes [8, 9], especially conventional methods are slow, laborious, and inefficient and may result in low yield and purity [10]. In addition, often contaminants of similar size and density, such as lipoproteins, are co-eluted and the heterogeneous structure and composition as well as overlapping size ranges of EVs from different sources (e.g., ectosomes and endosomes) make their isolation and application in medical diagnosis and therapies challenging. The sample matrix produces additional challenges as well since the isolation of pure EV subpopulations from cell culture media or urine is easier compared to more complex blood plasma and serum matrices. For diagnostic purposes, the target is just to get a high yield and EVs do not need to be of high purity, but for therapeutic purposes, such as cellular programming, drug delivery, or immunotherapy, it is very important to have intact, pure, and well-defined subpopulations. Since the size ranges of exosomes and microvesicles overlap in the size range of 50–150 nm, their differentiation with only density- and size-based techniques is not possible. As an alternative, immunoaffinity chromatography (IAC) offers possibilities for selective and specific isolation of EVs of the exosomal size range, and the technique often targets common EV surface markers, such as tetraspanins (e.g., CD9, CD63, and CD81). In addition, field-flow fractionation (FFF) is a flexible separation technique where EV subpopulations can be isolated and purified in quite high yields inside an open confined channel with a continuous flow under the influence of an external field. Due to the lack of shear forces from the stationary phase, FFF is gentle and does not deteriorate EV subpopulations. Versatility can be enhanced by various couplings with other techniques. This chapter will introduce the great potential of both immunoaffinity chromatography and field-flow fractionation for the successful isolation of EV subpopulations.

2 Immunoaffinity Isolation Techniques

2.1 General Overview

Immunoaffinity isolation utilizes the property of antibodies to bind the antigen of interest (i.e., target molecules) specifically and reversibly. Antibodies are immunoglobulins (Ig) naturally produced by the B-cell lineage lymphocytes in mammalian systems as part of the immune response and are further categorized into five classes (IgA, IgD, IgE, IgG, and IgM) based on their heavy chain types [11, 12]. Among the subclasses, the classical Y-shaped IgG-class antibody is the most frequently used antibody for immunoaffinity purification applications due to its higher specificity, affinity, and stability [12]. Binding between antibody and antigen is primarily governed by non-covalent interaction and occurs at the antigen-binding sites located on the variable regions of the antibody molecules and at the specific region on the antigen (i.e., epitope). In some applications, antibody-antigen binding is reversed under specific conditions to recover the target molecules for further analysis. In general, an immunoaffinity isolation platform requires an insoluble solid support onto which antibodies are immobilized. When a chromatographic column is employed as the solid support, the isolation process is termed immunoaffinity chromatography (IAC) (Fig. 1a). With other types of supports, such as magnetic beads and microfluidic devices, the isolation is commonly known as immunoaffinity capture or immunocapture (Fig. 1b) [8]. Immunoaffinity procedures generally employed for both IAC and immunocapture consist of three main steps including sample introduction, washing, and desorption.

Like other chromatographic systems, an IAC platform uses a chromatographic column as the solid support. For EV isolation, a column is immobilized with antibodies having specificity toward EVs. The isolation procedure begins with column conditioning with an appropriate solution, i.e., an application or binding buffer having the optimal pH and ionic strength for ligand binding [13]. Moreover, consideration should be given to maintain biological activity and stability of antibodies as well as EVs throughout the binding process, e.g., by utilizing physiological buffer solutions and physiological pH, such as phosphate-buffer saline at pH 7.4, for binding. For EV isolation, the sample containing EVs is introduced to the IAC system manually or by a pump at a specified flow rate. After this, the system is washed with the application

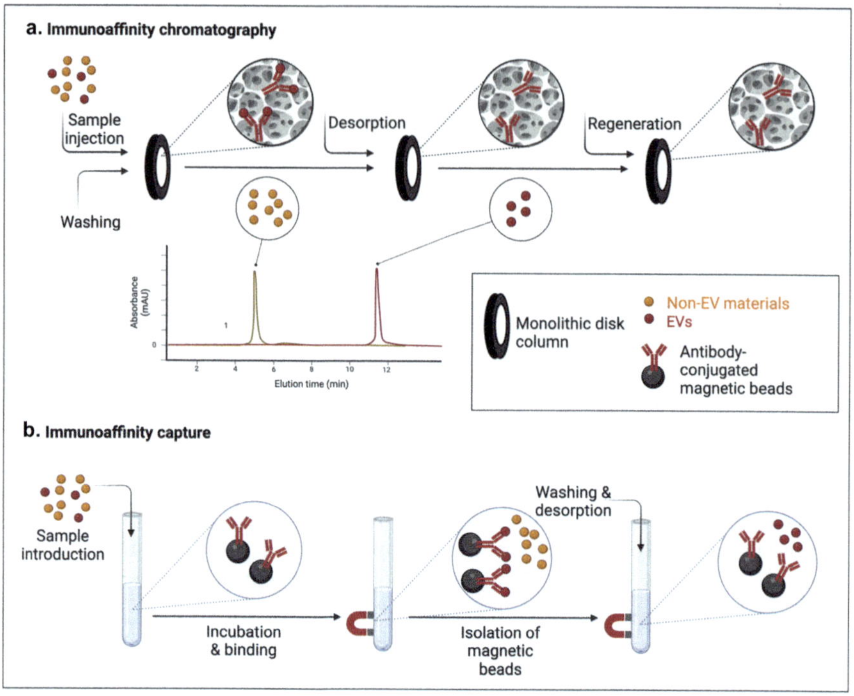

Fig. 1 Simplified EV isolation workflow using **a** immunoaffinity chromatography with a polymeric monolithic disk column as the stationary support; and **b** immunoaffinity capture using antibody-conjugated magnetic beads. Created with Biorender.com

buffer to remove unbound material. The bound EVs are eventually desorbed from the antibodies with an appropriate elution solution. The IAC column can be readily on-line coupled to other separation systems and/or detectors for real-time information on the eluting components, as well as to a fraction collector for further use of the isolated EV fractions. In addition to real-time monitoring and compatibility with other systems, other benefits of using an IAC system include potential automation, column regenerations allowing for multiple uses, and scalability.

As opposed to IAC, the protocol for immunocapture is highly dependent on the type of solid supports used. For instance, for EV isolation using magnetic beads, the beads are conjugated with EV-specific antibodies. After the sample containing EVs is introduced to the system, sufficient time is allowed for incubation and binding between EVs and the antibodies to occur. Thereafter, the antibody-conjugated magnetic bead and EV complexes are separated from the sample matrix by an external magnetic field [14], and the EVs are desorbed from the beads with an elution solution. In some applications, antibody-magnetic beads and EV complexes are directly further analyzed with other bioanalytical techniques or approaches, such as flow cytometry, Western blotting, electron microscopy, or real-time quantitative reverse transcription polymerase chain reaction (qRT-PCR) [15]. The EV-antibody-magnetic bead conjugates are not compatible with other traditional EV characterization techniques, such as nanoparticle tracking analysis (NTA) or mass spectrometry-based methods [16]. While immunocapture is considered cost-effective and user-friendly, it is laborious due to several steps, such as washing which cannot be automated [17].

2.2 *Immunoaffinity Ligand for Extracellular Vesicle Isolation*

Immunoaffinity isolation techniques are among the most utilized approaches for EV isolation [8, 18, 19]. Both IAC and immunocapture exploit antibodies having a specific affinity toward biomolecules on the EV membrane surface, such as tetraspanins and adhesion molecules, as illustrated in Fig. 2. Tetraspanin proteins, a family of integral proteins including CD9, CD63, CD81, and CD82, are particularly enriched on the surface membrane of EVs from endocytic origins, i.e., exosomes [20–22]. In addition, as tetraspanins are widely distributed across different tissues, and are present in higher concentrations compared to their originating cells [23, 24], they have been used as traditional EV protein markers. Additionally, kinetics and interaction studies have shown that tetraspanin antibodies against CD9, CD63, and CD81 proteins demonstrate relatively high affinity for EVs [25, 26], proving their suitability for selective and specific EV isolation. Recent studies have reported the use of anti-tetraspanins, particularly anti-CD9, anti-CD63, and anti-CD81, in both IAC and immunocapture settings for EV isolations from various biological fluids (See Table 1). While in most applications, a single type of anti-tetraspanin is used, it is also feasible to employ a cocktail of antibodies (e.g., anti-CD9, anti-CD63, and anti-CD81) for EV isolation EVs [27, 28].

While tetraspanin antibodies target general populations of EVs, alternative antibodies have also been employed for the isolation of EVs secreted from specific origins. For instance, a monoclonal antibody 763.74, specific for chondroitin sulphate peptidoglycan 4 epitope has been used for isolating melanoma cell-derived exosomes from patient blood plasma [48, 49]. Similarly, T-cell-derived exosomes have been separated from plasma of head and neck squamous cell carcinoma patients using anti-CD3-conjugated magnetic beads, as CD3 is expressed exclusively on T lymphocytes [42, 50]. In another application, antibodies targeting L1CAM or CD171, a transmembrane L1 cell adhesion molecule, were used to isolate EVs derived from neuronal origin [38, 45, 51]. Other antibodies have also been exploited in specific EV isolations from biological fluids, such as anti-A33 for colon epithelial cell-specific exosomes [39], anti-EpCAM for tumor-derived EVs [52, 53], anti-epidermal growth factor receptor (EGFR) for cancer-specific exosomes [47], anti-CD61 for platelet-derived EVs [31, 30], as well as anti-prostate specific membrane antigen (PSMA) for prostate derived EVs [43] (see Table 1). Immunocapture by utilizing an antibody specific for TET8 protein present in plant EVs has also been reported [40, 41].

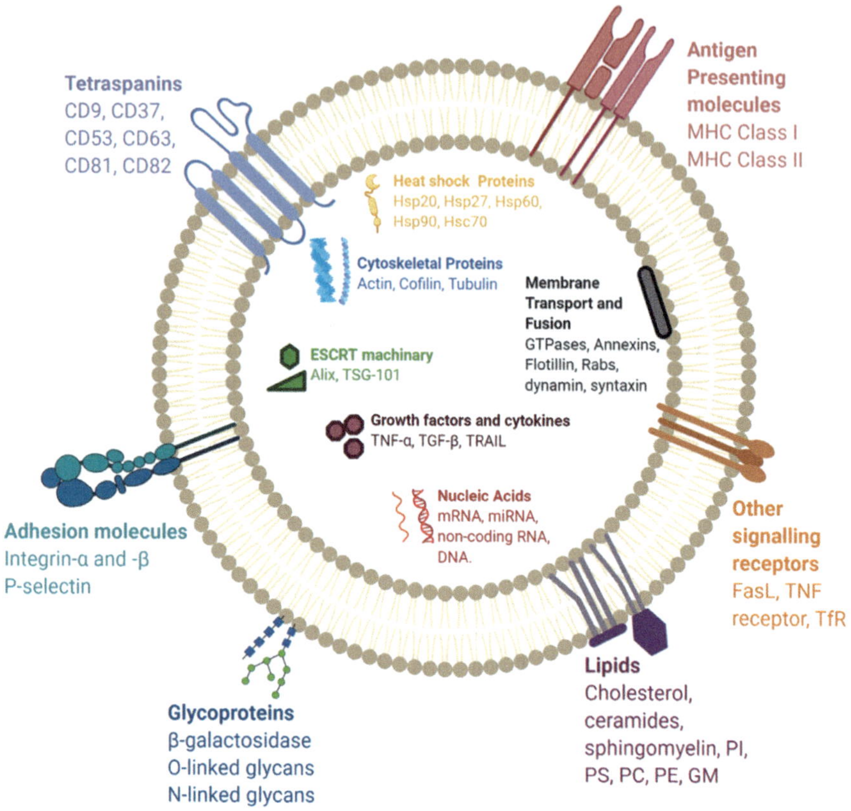

Fig. 2 Composition of exosomes. Reprinted with permission from [21]. Copyright 2021 springer nature under creative commons CC BY license

2.3 *Solid Supports*

Another important aspect of immunoaffinity isolation is the choice of solid support. In IAC settings, chromatographic columns with different materials and functionalities can be either commercially obtained, such as aldehyde-hydrazide resins [16] and carboxy imidazole poly(methyl methacrylate) monolithic disk columns [25, 30–32, 54] (see Fig. 3), or prepared in-house, e.g., methacrylate-based copolymer columns [35] and epoxy-based monoliths [29]. Ligand immobilization techniques are comprehensively reviewed and described in detail in [55]. Particularly for isolation and separation of large biomolecules and nanoparticles, monolithic columns have been extensively employed due to their superiority compared to traditional particle-based columns (Table 1) [54, 56, 57]. Monolithic columns are characterized by large interconnected through-pores, enabling the separation of large biomolecules via convective mass transfer instead of diffusive transport at high flow rates with low back pressure [58, 59]. Among different materials, polymer-based stationary phases can

Table 1 Selected applications on the separation of extracellular vesicles by immunoaffinity techniques

Isolation system	Solid support	Immunoaffinity ligand	Sample matrix	References
Immunoaffinity chromatography	Monolith filter	Anti-CD63	Urine	[29]
	Poly-methacrylate monolithic disk column	Anti-CD61	Human plasma	[30]
		Anti-CD9 and anti-CD61	Human plasma	[31]
		Anti-CD9, anti-CD63, and anti-CD81	Human plasma	[32, 33]
	Hydrazide resin	Anti-CD9	Human serum	[16]
	Sepharose column	Anti-CD81	Human placenta	[34]
	Methacrylate-based copolymer	Single domain (VHH) antibody	Cell culture media and human plasma	[35]
Immunoaffinity capture	Latex beads	Anti-CD63	Human milk	[36]
	Gold nanoparticle	Anti-CD63	Human serum	[37]
	Agarose resin	Anti-CD171	Human plasma	[38]
	Magnetic beads	Anti-A33	Cell culture medium	[39]
		Anti-CD9, anti-CD63, anti and CD81 (cocktail)	Human saliva	[27]
			Cell culture	[28]
		Anti-TET8	*Arabidopsis thaliana* leaves	[40, 41]
		Anti-CD3	Human plasma	[42]
		Anti-PSMA	Human plasma	[43]
		Anti-EpCAM	Cell culture	[44]
		L1CAM	Human serum	[45]
		Anti-MHC class II	Human milk	[36]
		Anti-CD9	Human plasma	[46]
	Magnetic beads and microfluidic platform	Anti-EGFRvIII	GLI36vIII conditioned medium	[47]

withstand a wide range of pH values necessary for EV desorption from the affinity ligand [8].

In immunocapture settings, one of the most frequently reported solid support is the magnetic bead platform. Magnetic beads, ranging from nm to µm in size, are superparamagnetic particles consisting of a magnetic core (e.g., iron oxides) and

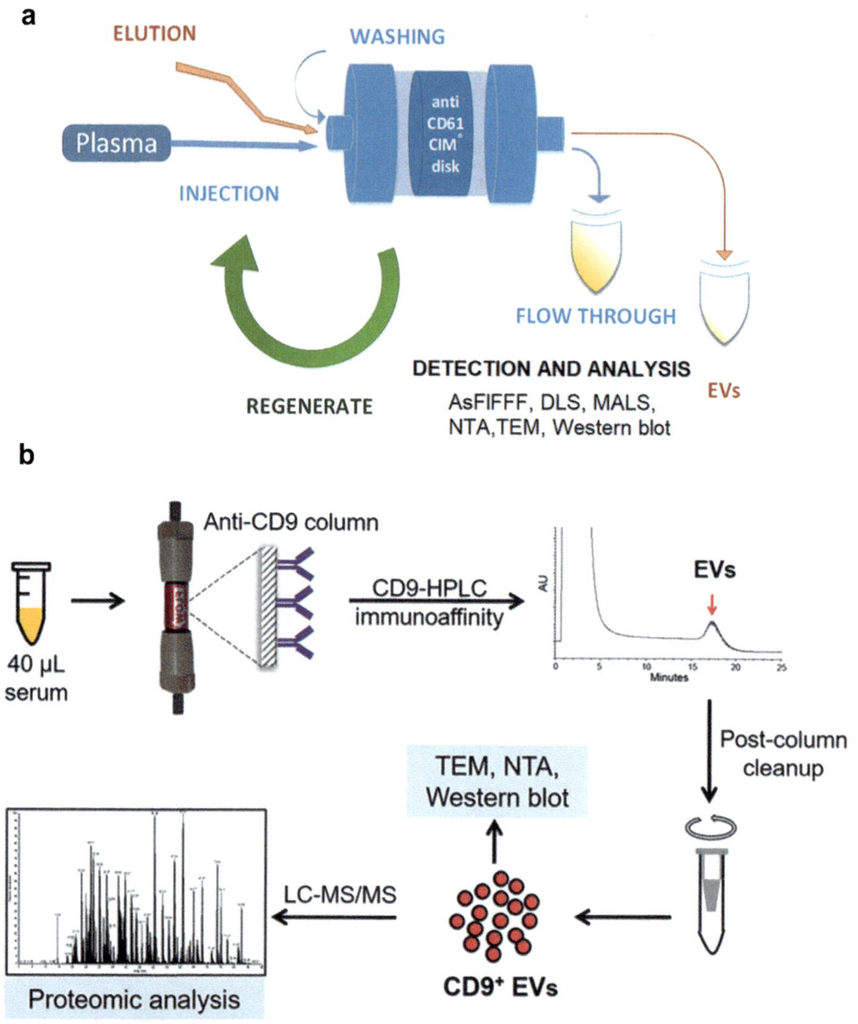

Fig. 3 Immunoaffinity chromatographic set-ups for EV isolation: **a** Immunoaffinity monolithic disk column immobilized with anti-CD61 for plasma-derived EV isolation from human plasma. Reprinted with permission from [30]. Copyright 2019 Elsevier; and **b** CD9-HPLC-IAC workflow for the isolation of EVs from human serum, followed by proteomic analysis. Reprinted with permission from [16]. Copyright 2021 American Chemical Society

a non-magnetic coating [60]. Commercially available magnetic beads are generally functionalized with biological ligands, such as streptavidin, ready for further coupling with immunoaffinity ligands for EV isolation (Fig. 4) [36, 61, 62]. In addition to bead-based methods, an elongated biotinylated magnetic nanowire conjugated with anti-CD9, anti-CD63, and anti-CD81 antibodies for the capture of exosomes from blood plasma of cancer patients has been reported [63] (Fig. 4). Other bead-based solid supports, such as latex beads [36] and gold nanoparticles, have also been utilized for EV isolation [37]. In addition to the bead-based platform, microfluidic devices have also been used for EV isolation and analysis. For instance, an immuno-magnetic exosome RNA (iMER) platform utilized magnetic beads immobilized with anti-EGFRvIII for cancer-derived EV isolation. The chip contains chambers at which RNA extraction and qPCR detection occur (Fig. 5a) [47]. In another application, a microfluidic device (ExoSearch chip) was used for exosome isolation and *in-situ* multiplexed detection (Fig. 5b) [64].

3 Field-Flow Fractionation Isolation Technique

3.1 General Overview of Field-Flow Fractionation

Field-flow fractionation (FFF) is an excellent technique for fractionation, isolation, and separation of macromolecules, colloids, and nanoparticles in an open thin ribbon-like channel, where an external force field affects the fractionation of sample components, dissolved or suspended in a flowing carrier liquid (Fig. 6). Typically, channels with dimensions of 20–50 cm in length, 2–3 cm in width, and 0.01–0.05 cm in thickness are used. The technique was pioneered in 1996 by J. Calvin Giddings who applied FFF to the separation of polystyrenes, proteins, and viruses [65]. Since then, several different FFF subtechniques have been developed, reflecting the variety of applicable external force fields needed for the fractionation of a large spectrum of applications. Among the several FFF techniques, centrifugal/sedimentation, asymmetrical flow FFF (AF4), which is one type of flow FFF (FlFFF), hollow-fiber flow (HF5), split flow thin cell (SPLITT), electrical, and thermal (ThFFF) fractionation devices are nowadays commercially available, whereas the other ones, such as dielectrophoretic, magnetic, and acoustic FFF are accessible, but still under further development in research laboratories. In addition, a wide range of different detectors, among which ultraviolet–visible (UV–Vis), multiangle light scattering (MALS), dynamic light scattering (DLS), and refractive index (RI) detectors have been the most widely used for biomacromolecules, increase the versatility of FFF techniques even further. By connecting MALS-RI detection with FFF, simultaneous determination of the weight-average molar mass, the molar mass distribution, and the z-average root-mean-square radii of the separated molecules is possible. Depending on the FFF techniques and methods, they can be useful for characterization of molar mass, particle size, and polydispersity of different samples and are feasible for preparative purposes, quality control of different kinds of industrial processes, and for fractionation of certain sample components for further analysis. FFF is applicable to samples ranging from 10^3 g/mol to *ca.* 10^{18} g/mol and to particles from 1 nm up to *ca.* 100 μm [66]. FFF techniques have been applied to a wide range of complex macromolecular and micro- and nano-scale materials, such as engineered nanoparticles, environmental colloids and particles, drug and gene therapy delivery vehicles, natural

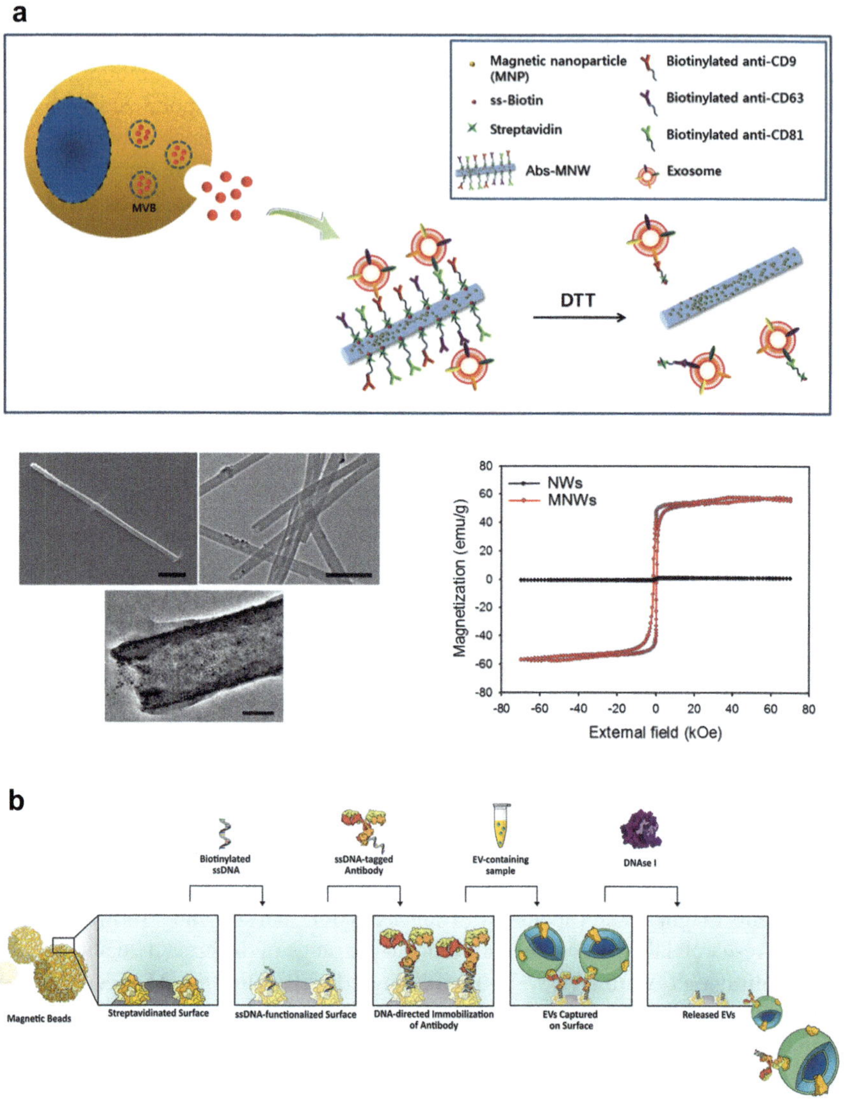

Fig. 4 Immunoaffinity capture for EV isolation: **a** Magnetic nanowires immobilized with anti-CD9, anti-CD63, and anti-CD81 for isolation of circulating exosome. Copyright Springer Nature 2019 under Creative Commons CC BY license [63], and **b** Capture and release platform for small EV isolation based on DNA-directed immobilization of anti-CD63 antibody. Reprinted with permission from [61]. Copyright 2021 American Chemical Society

and synthetic macromolecules, proteins, subcellular particles, as well as virus-like particles [67].

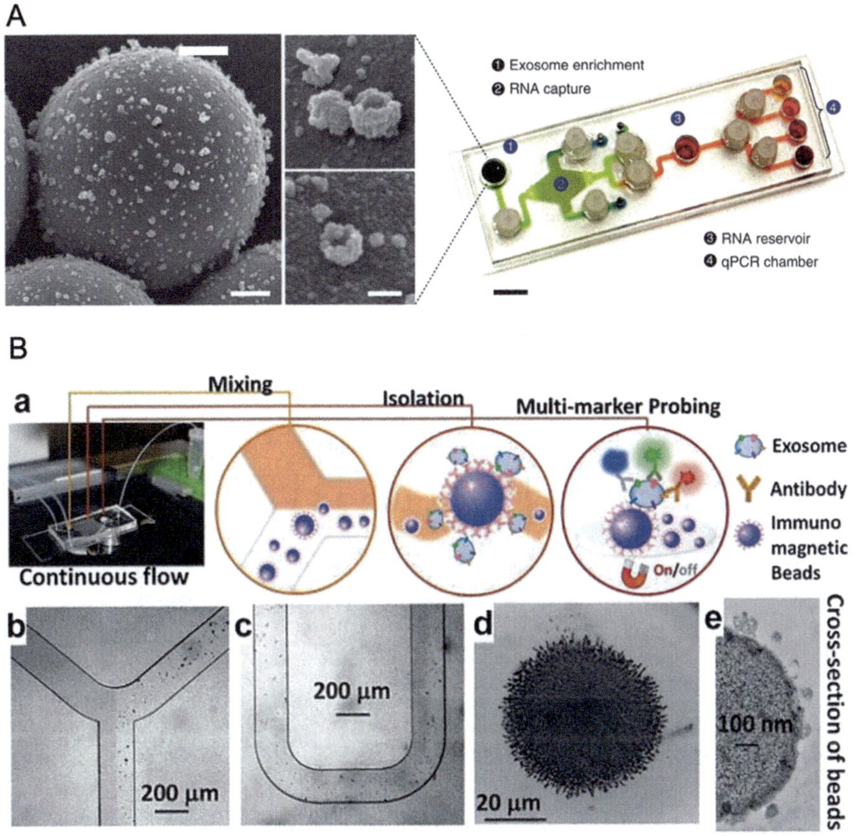

Fig. 5 Microfluidic platforms for immunocapture isolation of EVs: **a** An iMER platform consisting of the exosome enrichment compartment utilizing immunocapture (1), an RNA extraction site (2), a main reservoir for reverse transcription (3), and qPCR chamber (4). Reprinted with permission from [47]. Copyright 2015 Springer Nature under Creative Commons CC BY license; and **b** Continuous-flow ExoSearch microfluidic chip workflow **a–c** showing immunomagnetic beads bound with EVs that can be released and collected via an on/off switchable magnet **d** Reprinted with permission from [64]. Copyright 2016 Royal Society of Chemistry under Creative Commons CC BY license

FFF is sometimes called a one-phase chromatography technique because it resembles chromatography where the separation of distinct sample zones is achieved by differential elution along an axial column, whereas in FFF the separation takes place in a single moving fluid without adhesive surfaces where an external field is required. Even though the samples can be studied with more than one FFF technique, each FFF technique has its optimal target applications. The technique selected depends mainly on the purpose of the study, and on the type and physicochemical characteristics of the sample. However, it is clear that asymmetrical flow FFF with only one wall permeable to the flow is the dominant of all FFF techniques in biomacromolecules studies, as is the case of EV studies.

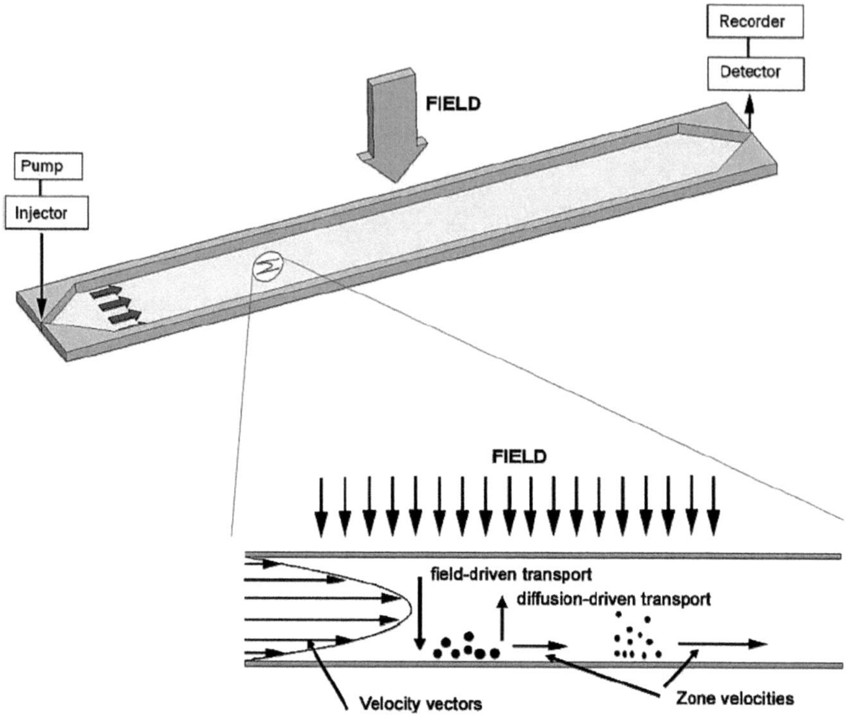

Fig. 6 Basic principle of field-flow fractionation

3.2 Recent Development of Field-Flow Fractionation Methods

Despite ongoing intense research on EVs, particularly diagnostics and therapeutics studies of their subpopulations have been hampered by the limitations of conventional isolation techniques [68], such as the required time, need for specialized laboratory equipment, and limited yield and purity. The most common methods for purification of exosomes are based on differential centrifugation, filtration/size exclusion, and immunoprecipitation [8, 69–71]. In addition, more recent techniques, e.g., asymmetrical flow field-flow fractionation (AF4 and AsFlFFF) have been successfully used for the isolation of EV subpopulations, see Fig. 7, Table 2, and [72]. Low quality purification of EVs is particularly problematic with biological matrices such as blood plasma, containing, e.g., lipoproteins of similar size to exosomes. Although isolation of subfractions is not necessarily always reliable, exosomes have been used as diagnostic biomarkers. Among different FFF techniques AF4 has by far been the most widely used technique in EV studies, but cyclical electrical field-flow fractionation (CyElFFF) and hollow-fiber flow field-flow fractionation (HF5) have also recently been used [73–75]. Commonly 10 kDa regenerated cellulose membranes have been utilized in AF4, in combination with spacers with thicknesses between 190 μm and 490 μm. The most common detection techniques have been UV and MALS, although fluorescence detection has also been demonstrated, as exemplified in quite recent work on the separation and characterization of EVs from a C2C12 cell line [73]. The miniaturized HF5 device used for the separation utilized protein intrinsic fluorescence for detailed characterization of the vesicles, suggesting different DNA/protein contents in small and large EVs.

After isolation, EVs should preferably be further characterized to verify the isolation process. Biophysical characterization, imaging, and protein and nucleic acid characterizations are some of the most used methods for exosome characterization. The collected fractions have usually undergone analysis by cryogenic transmission electron microscopy (cryo-TEM) [87, 88], transmission electron microscopy (TEM) [4, 30, 76, 78, 79, 83, 84], scanning electron microscopy (SEM) [31], dynamic light scattering (DLS), or by nanoparticle tracking analysis (NTA) [77, 85]. TEM is still the method of choice for fast characterization of particle sizes. In particular, extracellular vesicles isolated from cell lines have been investigated, but some studies on vesicles from urine [79, 80, 82], serum [83, 84], and plasma [30, 31, 84] samples have also been reported. Recently platelet-derived extracellular vesicles (PEVs) extracted

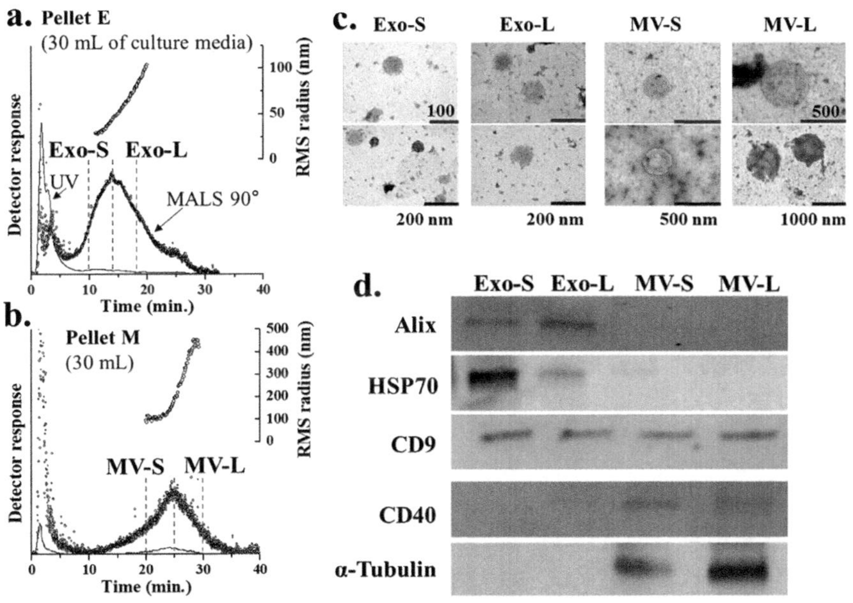

Fig. 7 AF4 fractograms of **a** pellet E (exosome fraction) and **b** pellet M (microvesicle fraction) along with the plots of root-mean-square radius values by MALS. **c** TEM images of the small and large size fractions (Exo-S and Exo-L for exosomes and MV-S and MV-L for microvesicles) and **d** Western blot results with the exosome markers (ALIX, HSP70, and CD9) and microvesicle markers (CD40 and α-tubulin). Reprinted with permission from [76]. Copyright 2022 American Chemical Society

Table 2 Selected applications on the fractionation and isolation of extracellular vesicles by FFF

Year	Type of sample; description	Type of FFF	Detection	Further analysis of collected fractions	Comments	References
2015	Lyophilized EV standard HBM-BLCL21-30 purified from culture supernatant of EBV transformed lymphoblastoid B cell line	AF4	UV, MALS	TEM, DLS, NTA	The cross-flow velocity and the channel thickness had the biggest influence, whereas the focusing time had the least influence on the AF4 separation	[77]
2015	Isolation of human thyroid carcinoma BCPAP, TPC1, and FTC133 cell line cells by sequential/ ultracentrifugation (UC)	AF4	MALS	Cryo-TEM	TPCI and BCPAP cells released greater numbers of exosomes per cell in response to mitogen-activated protein kinase inhibition	[78]
2017	EVs isolated from human urine samples using differential centrifugation	AF4	UV–Vis,	TEM, Western blotting, lipidomic analysis by nUPLC-ESI–MS/MS	Lipidomic changes in exosomes from control and prostate cancer patients	[79]
2018	EVs isolated from human A375 melanoma cell	CyElFFF	UV, MALS	No	The effect of various carrier fluids was investigated	[74]
2018	EVs from human urine samples purified by UC and size exclusion chromatography	AF4	UV-MALS	No	AF4 was suggested to be a good quality control method for urinary EVs, giving information about abundance, particle sizes, and purity	[80]
2018	EVs isolated from cultured murine melanoma cell line B16-F10 using UC	AF4	UV, QELS (quasi-elastic (dynamic) light scattering)	TEM, zeta potential measurements, Western blotting, AFM, protein analysis	The concept of exomeres was introduced	[4]

(continued)

Table 2 (continued)

Year	Type of sample; description	Type of FFF	Detection	Further analysis of collected fractions	Comments	References
2019	EVs isolated from human embryonic kidney cell line 293 cells by centrifugation	AF4	UV–Vis	nUHPLC-ESI–MS/MS of lipids	Lipidomic changes in subcellular fractions and exosomes upon oxidative stress were observed	[81]
2019	EVs isolated from human plasma using IAC	AF4	UV, MALS, DLS	Protein assay, Western blotting, TEM	First study demonstrating the existence of platelet-derived EVs with diameters of 30–130 nm	[30]
2019	EVs isolated from cultured murine melanoma cell line B16-F10 using UC	AF4	UV, QELS (quasi-elastic (dynamic) light scattering)	TEM	Small EV subpopulations of exomeres, Exo-S and Exo-L were identified	[72]
2020	Lyophilized exosome standards extracted from human urine	EAF4	MALS and NTA	No	EAF4-MALS-NTA hyphenation was demonstrated; exosomes spiked into rabbit serum; no zeta potential was obtained due to lack of exosome stability	[82]
2020	EVs isolated from human serum using a commercial exosome isolation kit, followed by UC and ultrafiltration (UF)	Frit inlet AF4	MALS	TEM; Western blotting; DLS; nUHPLC-ESI–MS/MS of proteins	Lipoprotein removal was more efficient using UC than UF; VLDLs were not completely removed from the similar sized exosomes	[83]

(continued)

Table 2 (continued)

Year	Type of sample; description	Type of FFF	Detection	Further analysis of collected fractions	Comments	References
2020	Human plasma and serum samples	AF4	UV, MALS	SDS-PAGE, Western blotting, TEM, LC–MS/MS of peptides and proteins, gene ontology enrichment, and KEGG (Kyoto Encyclopedia of Genes and Genomes) pathway analysis	Most proteins identified in human plasma EVs were involved in extracellular matrix (ECM) structural constituent and associated with the ECM-receptor interaction pathway	[84]
2020	EVs isolated by differential centrifugation from U87glioma cell line cells	CyElFFF	UV, MALS, fluorescence	No	The electrophoretic mobility and conductivity of small EVs and medium sized EVs depend on buffer ionic strength but not on EV concentration	[75]
2020	EVs isolated from human plasma using on-line IAC-AsFlFFF system	AF4	UV, MALS, DLS, DAD	Field-emission SEM; zeta potential measurements; HILIC-MS/MS of amino acids and glucose; total cholesterol analysis; total protein analysis	Online coupling of IAC with AF4; the isolated subpopulations have different chemical compositions	[31]
2021	Isolation of EVs from cultured C2C12 cells by ultracentrifugation and density gradient centrifugation	HF FlFFF	UV/Vis, fluorescence, MALS	No	On-line fluorescence and DAD detection were useful for achieving information on the protein content of the EV subpopulations	[73]
2022	Isolation of EVs from DU145 HRPC cell line culture using sequential centrifugation and UF	AF4	UV and MALS	TEM; Western blotting; nUHPLC-ESI–MS/MS of lipids	UF was more efficient than UC for initial EV isolation from cell cultures; lipids were more enriched in large exosomes than in small exosomes	[76]

(continued)

Table 2 (continued)

Year	Type of sample; description	Type of FFF	Detection	Further analysis of collected fractions	Comments	References
2022	Lyophilized EVs from human colon carcinoma cell line COLO-1	AF4	UV	NTA, SEM, ELISA, Western blotting, Capillary electrophoresis	Off-line AF4-CE was demonstrated to be an effective methodology for separating and quantifying EVs in complex biological samples	[85]
2024	Platelet-derived extracellular vesicles (PEVs) were isolated from whole blood from Wistar rats	AF4	UV and MALS	TEM, ATR-FTIR, lipidomics (UHPLC-HRMS), miRNA expression	Isolation and characterization of different subpopulation of EVs showed differences in the lipid, nucleic acid, and protein content	[86]

from whole blood from Wistar rats were fractionated and isolated by AF4-UV/MALS [86]. The subpopulations were further studied by TEM, FTIR, and UHPLC-HRMS (for lipidomics). The lipidomics data showed distinct variations in the lipid content of the different subpopulations. In addition, clear differences in the cargoes of the EVs were observed upon exposure of the EVs to particulate matter. Gene expression analysis was conducted, and the study demonstrated alterations in miRNAs upon exposure to the particulate matter.

Various couplings between the FlFFF instrumentation and other separation techniques have also been demonstrated, such as electrical FFF (EAF4) on-line coupled with nanoparticle tracking [82], IAC isolation on-line coupled with AF4 fractionation [31], as well as AF4 offline coupled with capillary electrophoresis (CE) [32]. The recently developed system where IAC with polymeric monolithic disks, immobilized with antibodies was on-line coupled to AF4 equipped with multiple detectors, such as UV, MALS, DLS, and DAD, was successfully used for the isolation and fractionation of plasma-derived EV subpopulations, including $CD9^+$, $CD61^+$ $CD63^+$, and $CD81^+$ EVs under 120 nm (exosome-sized 50–80 nm and 80–120 nm and exomere-sized <50 nm) with minimal losses and contamination and without any lipoproteins that were simultaneously separated and fractionated (see Figs. 8 and 9). The established automated methodology was fast, reliable, and reproducible and gave a high purity and high yields of subpopulations. The composition of these subpopulations was subjected to additional characterization by different techniques including NTA, electron microscopy, CE coupled with laser-induced fluorescent detection, Raman spectroscopy, comprehensive two-dimensional gas chromatography (GCXGC)- mass spectrometry (MS) and hydrophilic interaction chromatography (HILIC)-MS. Fatty acid composition of EV subpopulations isolated by on-line IAC-AF4, and that of lipoproteins and plasma, studied after derivatization by GCXGC-TOFMS, proved that lipids contributed to the differences, not only between the platelet-derived EVs and other EVs, but also between all EVs analyzed and emphasized the highly diverse nature of EVs found in the circulation (Fig. 9a) [33]. In addition, amino acid and glucose content of EV subpopulations, analyzed by HILIC-MS, differed in amino acid compositions (Fig. 9b).

4 Conclusions

Immunoaffinity chromatography and immunocapture belong to the most commonly used EV isolation approaches. They exploit antibodies with a specific affinity toward biomolecules on the EV membrane surface, such as tetraspanins and adhesion molecules, enabling highly selective and specific capture of high-purity EVs from biological matrices. Due to the availability of different immunoaffinity ligands, EVs released from diseased cells can be effectively isolated, facilitating diagnostics and therapeutic applications. The size-based fractionation and isolation of EVs by FFF and especially by asymmetrical field-flow fractionation (AF4) give flexibility by allowing the fractionation parameters to be modified even between runs. In addition,

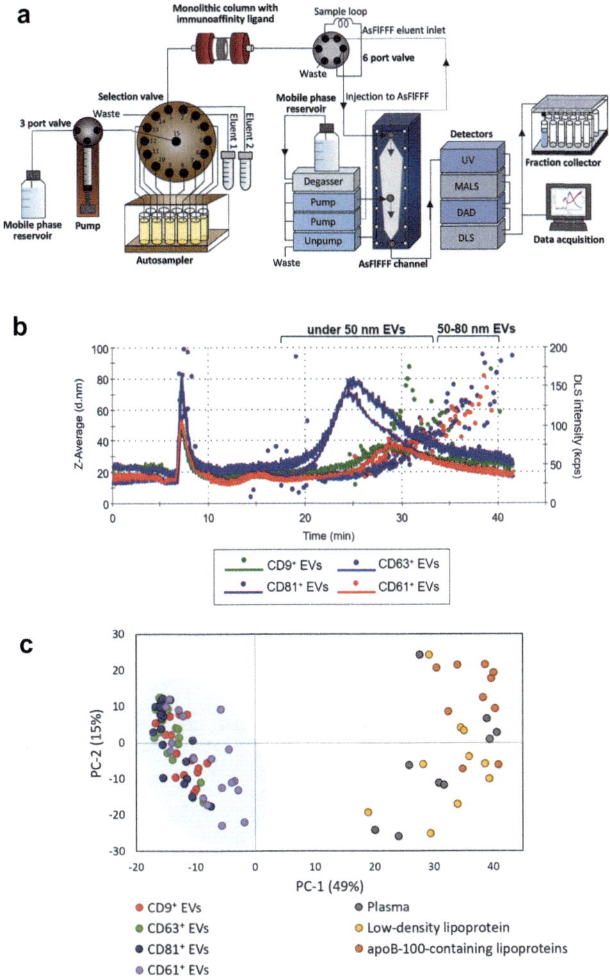

Fig. 8 **a** An on-line immunoaffinity isolation system coupled with AF4 for EV isolation and fractionation. Reprinted with permission from [31]. Copyright 2020 American Chemical Society under Creative Commons CC BY license; and **b** an example of AF4 fractograms of isolated EVs subpopulations. Reprinted with permission from [33]. Copyright 2022 Elsevier under Creative Commons CC BY license. **c** Score plot based on multivariate analysis of human plasma, UC-LDL, apoB-100-containing lipoproteins, and plasma-derived $CD9^+$, $CD63^+$, $CD81^+$, and $CD61^+$ of exomere- and exosome-sized isolated and fractionated by on-lined coupled IAC—AF4. Reprinted with permission from [33]. Copyright 2022 Elsevier under Creative Commons CC BY license

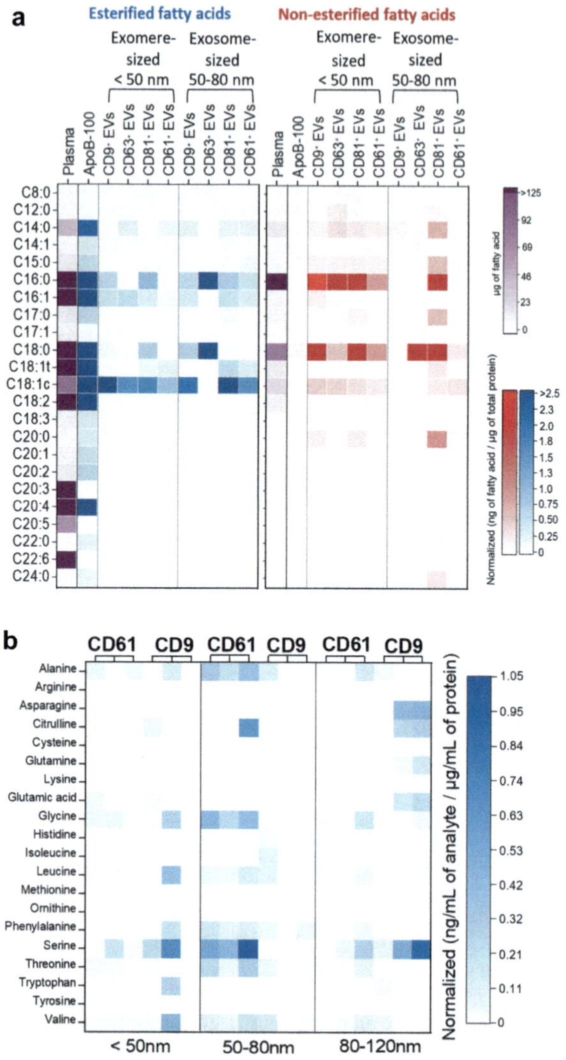

Fig. 9 **a** Heat map showing esterified (left panel) and non-esterified (right panel) fatty acid levels in human plasma (µg), apoB-100-containing-lipoproteins, and by on-line IAC-AF4 isolated EV subpopulations (ng of fatty acid per µg of total protein) analyzed by GCxGC—TOFMS. Reprinted with permission from [33]. Copyright 2022 Elsevier under Creative Commons CC BY license. **b** Heat map showing levels of amino acids (ng/mL per µg/mL of total protein) in $CD61^+$ and $CD9^+$ EV subpopulations ($n = 3$), isolated by on-line IAC-AF4. Reprinted with permission from [31]. Copyright 2020 American Chemical Society under Creative Commons CC BY license

due to the lack of shear forces from the stationary phase, FFF is a gentle technique, and a wide range of different detectors available increase its versatility even further. In addition, on-line coupled IAC-AF4 systems offer possibilities for fully automated reliable and fast isolation and fractionation of nanosized EV subpopulations under 120 nm from human plasma resulting in high purity subpopulations with high yields.

Acknowledgements Financial support was provided by the Jane and Aatos Erkko Foundation and the Research Council for Natural Sciences and Engineering, Academy of Finland (grant number 1311369).

Competing Interests The authors declare no competing interests.

References

1. Colombo, M., Raposo, G., & Théry, C. (2014). Biogenesis, secretion, and intercellular interactions of exosomes and other extracellular vesicles. *Annual Review of Cell and Developmental Biology, 30*, 255–289.
2. Yáñez-Mó, M., Siljander, P. R. M., Andreu, Z., et al. (2015). Biological properties of extracellular vesicles and their physiological functions. *Journal of Extracellular Vesicles, 4*, 27066. https://doi.org/10.3402/JEV.V4.27066
3. Kalluri, R., LeBleu, V. S. (2020). The biology, function, and biomedical applications of exosomes. *Science* (1979) 367. https://doi.org/10.1126/SCIENCE.AAU6977/ASSET/F50 50DD7-9124-4A75-9FBE-2BCDF3F27CD8/ASSETS/GRAPHIC/367_AAU6977_F6.JPEG.
4. Zhang, H., Freitas, D., Kim, H. S., et al. (2018). Identification of distinct nanoparticles and subsets of extracellular vesicles by asymmetric flow field-flow fractionation. *Nature Cell Biology, 20*, 332–343. https://doi.org/10.1038/s41556-018-0040-4
5. Zijlstra, A., Di Vizio, D. (2018). Size matters in nanoscale communication. *Nature Cell Biology, 20*, 228–230. https://doi.org/10.1038/s41556-018-0049-8
6. Villa, F., Quarto, R., Tasso, R. (2019). Extracellular vesicles as natural, safe and efficient drug delivery systems. *Pharmaceutics, 11*, 557. https://doi.org/10.3390/PHARMACEUTIC S11110557
7. Cheng, L., Hill, A. F. (2022). Therapeutically harnessing extracellular vesicles. *Nature Reviews Drug Discovery, 21*(5), 379–399. https://doi.org/10.1038/s41573-022-00410-w
8. Liangsupree, T., Multia, E., & Riekkola, M. L. (2021). Modern isolation and separation techniques for extracellular vesicles. *Journal of Chromatography A, 1636*, 461773. https://doi.org/10.1016/j.chroma.2020.461773
9. Wysor, S. K., & Marcus, R. K. (2024). In-line coupling of capillary-channeled polymer fiber columns with optical absorbance and multi-angle light scattering detection for the isolation and characterization of exosomes. *Analytical and Bioanalytical Chemistry, 416*, 3325–3333. https://doi.org/10.1007/S00216-024-05283-Z/FIGURES/6
10. Balbi, C., Parisse, P., Vondracek, H., et al. (2024). Impact of isolation methods on extracellular vesicle functionality in vitro and in vivo. *Adv Biol, 8*, 2300185. https://doi.org/10.1002/ADBI.202300185
11. Alberts, B., Johnson, A., Lewis, J., et al. (2002). B cells and antibodies
12. Thompson, N. E., Burgess, R. R. (2015). Immunoaffinity chromatography: Advantages and limitations. *Analytical Separation Science*, 483–502. https://doi.org/10.1002/9783527678129.ASSEP020.
13. Rodriguez, E. L., Poddar, S., Iftekhar, S., et al. (2020). Affinity chromatography: A review of trends and developments over the past 50 years. *Journal of Chromatography B, 1157*, 122332. https://doi.org/10.1016/J.JCHROMB.2020.122332
14. Kozhevnikova, D., Chernyshev, V., Yashchenok, A. (2023). Progress in isolation and molecular profiling of small extracellular vesicles via bead-assisted platforms. *Biosensors (Basel), 13*. https://doi.org/10.3390/BIOS13070688.
15. Oksvold, M. P., Neurauter, A., & Pedersen, K. W. (2015). Magnetic bead-based isolation of exosomes. *Methods in Molecular Biology, 1218*, 465–481. https://doi.org/10.1007/978-1-4939-1538-5_27
16. Zhu, J., Zhang, J., Ji, X., et al. (2021). Column-based technology for CD9-HPLC immunoaffinity isolation of serum extracellular vesicles. *Journal of Proteome Research, 20*, 4901–4911. https://doi.org/10.1021/ACS.JPROTEOME.1C00549/ASSET/IMA GES/LARGE/PR1C00549_0004.JPEG
17. Jiawei, S., Zhi, C., Kewei, T., & Xiaoping, L. (2022). Magnetic bead-based adsorption strategy for exosome isolation. *Frontiers in Bioengineering and Biotechnology, 10*, 942077. https://doi.org/10.3389/FBIOE.2022.942077/BIBTEX
18. Zhang, H., Zhang, Q., Deng, Y., et al. (2021). Improving Isolation of Extracellular Vesicles by Utilizing Nanomaterials. *Membranes, 12*, 55. https://doi.org/10.3390/MEMBRANES120 10055

19. Gao, J., Li, A., Hu, J., et al. (2022a). Recent developments in isolating methods for exosomes. *Frontiers in Bioengineering and Biotechnology, 10*. https://doi.org/10.3389/FBIOE.2022.1100892.
20. Berditchevski, F., & Odintsova, E. (2007). Tetraspanins as regulators of protein trafficking. *Traffic, 8*, 89–96. https://doi.org/10.1111/J.1600-0854.2006.00515.X
21. Gurung, S., Perocheau, D., Touramanidou, L., Baruteau, J. (2021). The exosome journey: from biogenesis to uptake and intracellular signalling. *Cell Communication and Signaling, 19*(1), 1–19. https://doi.org/10.1186/S12964-021-00730-1.
22. Théry, C., Zitvogel, L., & Amigorena, S. (2002). Exosomes: Composition, biogenesis and function. *Nature Reviews Immunology, 2*, 569–579.
23. Andreu, Z., & Yáñez-Mó, M. (2014). Tetraspanins in extracellular vesicle formation and function. *Frontiers in Immunology, 5*, 442. https://doi.org/10.3389/fimmu.2014.00442
24. Rydland, A., Heinicke, F., Flåm, S. T., et al. (2023). Small extracellular vesicles have distinct CD81 and CD9 tetraspanin expression profiles in plasma from rheumatoid arthritis patients. *Clinical and Experimental Medicine, 23*, 2867–2875. https://doi.org/10.1007/S10238-023-01024-1/FIGURES/4
25. Liangsupree, T., Multia, E., Forssén, P., et al. (2022). Kinetics and interaction studies of anti-tetraspanin antibodies and ICAM-1 with extracellular vesicle subpopulations using continuous flow quartz crystal microbalance biosensor. *Biosensors & Bioelectronics, 206*, 114151. https://doi.org/10.1016/J.BIOS.2022.114151
26. Kowalczyk, A., Gajda-Walczak, A., Ruzycka-Ayoush, M., et al. (2023). Parallel SPR and QCM-D quantitative analysis of CD9, CD63, and CD81 Tetraspanins: A simple and sensitive way to determine the concentration of extracellular vesicles isolated from human lung cancer cells. *Analytical Chemistry, 95*, 9520–9530. https://doi.org/10.1021/ACS.ANALCHEM.3C00772/ASSET/IMAGES/LARGE/AC3C00772_0006.JPEG
27. Reseco, L., Molina-Crespo, A., Atienza, M., et al. (2024). Characterization of extracellular vesicles from human saliva: Effects of age and isolation techniques. *Cells, 13*, 95. https://doi.org/10.3390/CELLS13010095/S1
28. Sharma, S., LeClaire, M., Wohlschlegel, J., Gimzewski, J. (2020b). Impact of isolation methods on the biophysical heterogeneity of single extracellular vesicles. *Scientific Reports, 10*(1), 1–11. https://doi.org/10.1038/s41598-020-70245-1
29. Neumair, J., D'Ercole, C., De March, M., et al. (2023). Macroporous epoxy-based monoliths functionalized with anti-CD63 nanobodies for effective isolation of extracellular vesicles in urine. *International Journal of Molecular Sciences, 24*, 6131. https://doi.org/10.3390/IJMS24076131/S1
30. Multia, E., Tear, C. J. Y., Palviainen, M., et al. (2019). Fast isolation of highly specific population of platelet-derived extracellular vesicles from blood plasma by affinity monolithic column, immobilized with anti-human CD61 antibody. *Analytica Chimica Acta, 1091*, 160–168. https://doi.org/10.1016/J.ACA.2019.09.022
31. Multia, E., Liangsupree, T., Jussila, M., et al. (2020). Automated on-line isolation and fractionation system for nanosized biomacromolecules from human plasma. *Analytical Chemistry, 92*, 13058–13065. https://doi.org/10.1021/ACS.ANALCHEM.0C01986/ASSET/IMAGES/LARGE/AC0C01986_0005.JPEG
32. Morani, M., Mai, T. D., Krupova, Z., et al. (2020). Electrokinetic characterization of extracellular vesicles with capillary electrophoresis: A new tool for their identification and quantification. *Analytica Chimica Acta, 1128*, 42–51. https://doi.org/10.1016/J.ACA.2020.06.073
33. Liangsupree, T., Multia, E., Saarinen, J., et al. (2022). Raman spectroscopy combined with comprehensive gas chromatography for label-free characterization of plasma-derived extracellular vesicle subpopulations. *Analytical Biochemistry, 647*, 114672. https://doi.org/10.1016/J.AB.2022.114672
34. Burkova, E. E., Dmitrenok, P. S., Bulgakov, D. V., et al. (2018). Exosomes from human placenta purified by affinity chromatography on sepharose bearing immobilized antibodies against CD81 tetraspanin contain many peptides and small proteins. *IUBMB Life, 70*, 1144–1155. https://doi.org/10.1002/IUB.1928

35. Filipović, L., Spasojević, M., Prodanović, R., et al. (2022). Affinity-based isolation of extracellular vesicles by means of single-domain antibodies bound to macroporous methacrylate-based copolymer. *New Biotechnology, 69*, 36–48. https://doi.org/10.1016/J.NBT.2022.03.001
36. Torregrosa Paredes, P., Gutzeit, C., Johansson, S., et al. (2014). Differences in exosome populations in human breast milk in relation to allergic sensitization and lifestyle. *Allergy, 69*, 463–471. https://doi.org/10.1111/ALL.12357
37. Panwar, D., Shrivastava, D., Kumar, A., et al. (2023). Efficient strategy to isolate exosomes using anti-CD63 antibodies conjugated to gold nanoparticles. *AMB Express, 13*, 1–14. https://doi.org/10.1186/S13568-023-01592-1/FIGURES/5
38. Dunlop, R. A., Banack, S. A., & Cox, P. A. (2023). L1CAM immunocapture generates a unique extracellular vesicle population with a reproducible miRNA fingerprint. *RNA Biology, 20*, 140–148. https://doi.org/10.1080/15476286.2023.2198805
39. Mathivanan, S., Lim, J. W. E., Tauro, B. J., et al. (2010). Proteomics analysis of A33 immunoaffinity-purified exosomes released from the human colon tumor cell line LIM1215 reveals a tissue-specific protein signature. *Molecular and Cellular Proteomics, 9*, 197–208. https://doi.org/10.1074/mcp.M900152-MCP200
40. He, B., Cai, Q., Qiao, L., et al. (2021). RNA-binding proteins contribute to small RNA loading in plant extracellular vesicles. *Nat Plants, 7*, 342. https://doi.org/10.1038/S41477-021-00863-8
41. Huang, Y., Wang, S., Cai, Q., & Jin, H. (2021). Effective methods for isolation and purification of extracellular vesicles from plants. *Journal of Integrative Plant Biology, 63*, 2020. https://doi.org/10.1111/JIPB.13181
42. Theodoraki, M. N., Hoffmann, T. K., & Whiteside, T. L. (2018). Separation of plasma-derived exosomes into CD3(+) and CD3(−) fractions allows for association of immune cell and tumour cell markers with disease activity in HNSCC patients. *Clinical and Experimental Immunology, 192*, 271–283. https://doi.org/10.1111/CEI.13113
43. Brett, S. I., Lucien, F., Guo, C., et al. (2017). Immunoaffinity based methods are superior to kits for purification of prostate derived extracellular vesicles from plasma samples. *Prostate, 77*, 1335–1343. https://doi.org/10.1002/PROS.23393
44. Černe, K., Kelhar, N., Resnik, N., et al. (2023). Characteristics of extracellular vesicles from a high-grade serous ovarian cancer cell line derived from a platinum-resistant patient as a potential tool for aiding the prediction of responses to chemotherapy. *Pharmaceuticals, 16*. https://doi.org/10.3390/PH16060907.
45. Jiang, C., Hopfner, F., Berg, D., et al. (2021). Validation of α-Synuclein in L1CAM-immunocaptured exosomes as a biomarker for the stratification of parkinsonian syndromes. *Movement Disorders, 36*, 2663–2669. https://doi.org/10.1002/MDS.28591
46. Zarovni, N., Corrado, A., Guazzi, P., et al. (2015). Integrated isolation and quantitative analysis of exosome shuttled proteins and nucleic acids using immunocapture approaches. *Methods, 87*, 46–58. https://doi.org/10.1016/J.YMETH.2015.05.028
47. Shao, H., Chung, J., Lee, K., et al. (2015). Chip-based analysis of exosomal mRNA mediating drug resistance in glioblastoma. *Nature Communications, 6*(1), 1–9. https://doi.org/10.1038/ncomms7999.
48. Sharma, P., Diergaarde, B., Ferrone, S., et al. (2020a). Melanoma cell-derived exosomes in plasma of melanoma patients suppress functions of immune effector cells. *Scientific Reports*, 10(1), 1–11. https://doi.org/10.1038/s41598-019-56542-4.
49. Sharma, P., Ludwig, S., Muller, L., et al. (2018). Immunoaffinity-based isolation of melanoma cell-derived exosomes from plasma of patients with melanoma. *Journal of Extracellular Vesicles, 7*, 1435138. https://doi.org/10.1080/20013078.2018.1435138
50. Theodoraki, M. N., Yerneni, S., Gooding, W. E., et al. (2019). Circulating exosomes measure responses to therapy in head and neck cancer patients treated with cetuximab, ipilimumab, and IMRT. *Oncoimmunology, 8*. https://doi.org/10.1080/2162402X.2019.1593805.
51. Gomes, D. E., & Witwer, K. W. (2022). L1CAM-associated extracellular vesicles: A systematic review of nomenclature, sources, separation, and characterization. *Journal of Extracellular Biology, 1*, e35. https://doi.org/10.1002/JEX2.35

52. Königsberg, R., Obermayr, E., Bises, G., et al. (2011). Detection of EpCAM positive and negative circulating tumor cells in metastatic breast cancer patients. *Acta Oncol (Madr), 50*, 700–710. https://doi.org/10.3109/0284186X.2010.549151
53. Tauro, B. J., Greening, D. W., Mathias, R. A., et al. (2013). Two distinct populations of exosomes are released from LIM1863 colon carcinoma cell-derived organoids. *Molecular and Cellular Proteomics, 12*, 587–598. https://doi.org/10.1074/mcp.M112.021303
54. Liangsupree, T., Multia, E., Metso, J., et al. (2019). Rapid affinity chromatographic isolation method for LDL in human plasma by immobilized chondroitin-6-sulfate and anti-apoB-100 antibody monolithic disks in tandem. *Scientific Reports, 9*. https://doi.org/10.1038/s41598-019-47750-z.
55. Hermanson, G. T. (2013). Immobilization of ligands on chromatography supports. *Bioconjugate Techniques*, 589–740. https://doi.org/10.1016/B978-0-12-382239-0.00015-7.
56. Ott, S., Niessner, R., & Seidel, M. (2011). Preparation of epoxy-based macroporous monolithic columns for the fast and efficient immunofiltration of Staphylococcus aureus. *Journal of Separation Science, 34*, 2181–2192. https://doi.org/10.1002/JSSC.201100208
57. Rajamanickam, V., Herwig, C., Spadiut, O., Zajickov, Z. (2015). Monoliths in bioprocess technology. chromatography, *2*, 195–212. https://doi.org/10.3390/CHROMATOGRAPHY2020195.
58. Pfaunmiller, E. L., Paulemond, M. L., Dupper, C. M., & Hage, D. S. (2013). Affinity monolith chromatography: A review of principles and recent analytical applications. *Analytical and Bioanalytical Chemistry, 405*, 2133–2145.
59. Woonton, B. W., Smithers, G. W. (2010). Novel adsorbents and approaches for nutraceutical separation. In *Separation, Extraction and Concentration Processes in the Food, Beverage and Nutraceutical Industries* (pp. 148–179). Elsevier Ltd.
60. Ruffert, C. (2015). Magnetic beads—basics and applications. *ECS Transactions, 64*, 49. https://doi.org/10.1149/06431.0049ECST
61. Brambilla, D., Sola, L., Ferretti, A. M., et al. (2021). EV separation: release of intact extracellular vesicles immunocaptured on magnetic particles. *Analytical Chemistry, 93*, 5476–5483. https://doi.org/10.1021/ACS.ANALCHEM.0C05194/ASSET/IMAGES/LARGE/AC0C05194_0005.JPEG
62. Garcia, N. A., González-King, H., Grueso, E., et al. (2019). Circulating exosomes deliver free fatty acids from the bloodstream to cardiac cells: Possible role of CD36. *PLoS ONE, 14*, e0217546. https://doi.org/10.1371/JOURNAL.PONE.0217546
63. Lim, J., Choi, M., Lee, H., et al. (2019). Direct isolation and characterization of circulating exosomes from biological samples using magnetic nanowires. *Journal of Nanobiotechnology, 17*, 1–12. https://doi.org/10.1186/S12951-018-0433-3/TABLES/1
64. Zhao, Z., Yang, Y., Zeng, Y., & He, M. (2016). A microfluidic ExoSearch chip for multiplexed exosome detection towards blood-based ovarian cancer diagnosis. *Lab on a Chip, 16*, 489–496. https://doi.org/10.1039/c5lc01117e
65. Giddings, J. C. (1966). A new separation concept based on a coupling of concentration and flow nonuniformities. *Separation Science, 1*, 123–125. https://doi.org/10.1080/01496396608049439
66. Ratanathanawongs, S. K., Lee, I., Giddings, J. C. (1991). Separation and characterization of 0.01—50-µm particles using flow field-flow fractionation, 229–246. https://doi.org/10.1021/BK-1991-0472.CH015.
67. Wiedmer, S. K., & Riekkola, M. L. (2023). Field-flow fractionation—an excellent tool for fractionation, isolation and/or purification of biomacromolecules. *Journal of Chromatography A, 1712*, 464492. https://doi.org/10.1016/J.CHROMA.2023.464492
68. De Sousa, K. P., Rossi, I., Abdullahi, M., et al. (2023). Isolation and characterization of extracellular vesicles and future directions in diagnosis and therapy. *Wiley Interdiscip Rev Nanomed Nanobiotechnol, 15*, e1835. https://doi.org/10.1002/WNAN.1835
69. Akbar, A., Malekian, F., Baghban, N., et al. (2022). Methodologies to isolate and purify clinical grade extracellular vesicles for medical applications. *Cells, 11*. https://doi.org/10.3390/CELLS11020186.

70. Martins, T. S., Vaz, M., Henriques, A. G. (2022). A review on comparative studies addressing exosome isolation methods from body fluids. *Analytical and Bioanalytical Chemistry, 415*(7), 1239–1263. https://doi.org/10.1007/S00216-022-04174-5.
71. Xu, K., Jin, Y., Li, Y., et al. (2022). Recent progress of exosome isolation and peptide recognition-guided strategies for exosome research. *Frontiers in Chemistry, 10*, 844124. https://doi.org/10.3389/FCHEM.2022.844124/BIBTEX
72. Zhang, H., Lyden, D. (2019). Asymmetric-flow field-flow fractionation technology for exomere and small extracellular vesicle separation and characterization. *Nature Protocols, 14*(4), 1027–1053. https://doi.org/10.1038/s41596-019-0126-x
73. Marassi, V., Maggio, S., Battistelli, M., et al. (2021). An ultracentrifugation—hollow-fiber flow field-flow fractionation orthogonal approach for the purification and mapping of extracellular vesicle subtypes. *Journal of Chromatography A, 1638*, 461861. https://doi.org/10.1016/J.CHROMA.2020.461861
74. Petersen, K. E., Shiri, F., White, T., et al. (2018). Exosome isolation: cyclical electrical field flow fractionation in low-ionic-strength fluids. *Analytical Chemistry, 90*, 12783–12790. https://doi.org/10.1021/ACS.ANALCHEM.8B03146/SUPPL_FILE/AC8B03146_SI_001.PDF
75. Shiri, F., Petersen, K. E., Romanov, V., et al. (2020). Characterization and differential retention of Q beta bacteriophage virus-like particles using cyclical electrical field–flow fractionation and asymmetrical flow field–flow fractionation. *Analytical and Bioanalytical Chemistry, 412*, 1563–1572. https://doi.org/10.1007/S00216-019-02383-Z/FIGURES/5
76. Kim, Y. B., Bin, L. G., & Moon, M. H. (2022). Size separation of exosomes and microvesicles using flow field-flow fractionation/multiangle light scattering and lipidomic comparison. *Analytical Chemistry, 94*, 8958–8965. https://doi.org/10.1021/ACS.ANALCHEM.2C00806/SUPPL_FILE/AC2C00806_SI_001.PDF
77. Sitar, S., Kejžar, A., Pahovnik, D., et al. (2015). Size characterization and quantification of exosomes by asymmetrical-flow field-flow fractionation. *Analytical Chemistry, 87*, 9225–9233. https://doi.org/10.1021/ACS.ANALCHEM.5B01636/ASSET/IMAGES/MEDIUM/AC-2015-01636F_0012.GIF
78. Agarwal, K., Saji, M., Lazaroff, S. M., et al. (2015). Analysis of exosome release as a cellular response to MAPK pathway inhibition. *Langmuir, 31*, 5440–5448. https://doi.org/10.1021/ACS.LANGMUIR.5B00095/ASSET/IMAGES/MEDIUM/LA-2015-000958_0003.GIF
79. Yang, J. S., Lee, J. C., Byeon, S. K., et al. (2017). Size dependent lipidomic analysis of urinary exosomes from patients with prostate cancer by flow field-flow fractionation and nanoflow liquid chromatography-tandem mass spectrometry. *Analytical Chemistry, 89*, 2488–2496. https://doi.org/10.1021/ACS.ANALCHEM.6B04634/SUPPL_FILE/AC6B04634_SI_001.PDF
80. Oeyen, E., Van Mol, K., Baggerman, G., et al. (2018). Ultrafiltration and size exclusion chromatography combined with asymmetrical-flow field-flow fractionation for the isolation and characterisation of extracellular vesicles from urine. *Journal of Extracellular Vesicles, 7*, 1490143. https://doi.org/10.1080/20013078.2018.1490143
81. Yang, J. S., Kim, J. Y., Lee, J. C., & Moon, M. H. (2019). Investigation of lipidomic perturbations in oxidatively stressed subcellular organelles and exosomes by asymmetrical flow field–flow fractionation and nanoflow ultrahigh performance liquid chromatography–tandem mass spectrometry. *Analytica Chimica Acta, 1073*, 79–89. https://doi.org/10.1016/J.ACA.2019.04.069
82. Drexel, R., Siupa, A., Carnell-Morris, P., et al. (2020). Fast and purification-free characterization of bio-nanoparticles in biological media by electrical asymmetrical flow field-flow fractionation hyphenated with multi-angle light scattering and nanoparticle tracking analysis detection. *Molecules, 25*, 4703. https://doi.org/10.3390/MOLECULES25204703.
83. Kim, Y. B., Yang, J. S., Bin, L. G., & Moon, M. H. (2020). Evaluation of exosome separation from human serum by frit-inlet asymmetrical flow field-flow fractionation and multiangle light scattering. *Analytica Chimica Acta, 1124*, 137–145. https://doi.org/10.1016/J.ACA.2020.05.031

84. Wu, B., Chen, X., Wang, J., et al. (2020). Separation and characterization of extracellular vesicles from human plasma by asymmetrical flow field-flow fractionation. *Analytica Chimica Acta, 1127*, 234–245. https://doi.org/10.1016/J.ACA.2020.06.071
85. Gao, Z., Hutchins, Z., Li, Z., & Zhong, W. (2022). Offline coupling of asymmetrical flow field-flow fractionation and capillary electrophoresis for separation of extracellular vesicles. *Analytical Chemistry, 94*, 14083–14091. https://doi.org/10.1021/ACS.ANALCHEM.2C03550/SUPPL_FILE/AC2C03550_SI_001.PDF
86. Mao, Z., Wu, Y., Kong, L., et al. (2024). Changes in cargoes of platelet derived extracellular vesicles heterogeneous subpopulations induced by PM0.1—Undisclosed cardiovascular injury communication mechanism. *Environmental Pollution, 348*, 123845. https://doi.org/10.1016/J.ENVPOL.2024.123845.
87. Emelyanov, A., Shtam, T., Kamyshinsky, R., et al. (2020). Cryo-electron microscopy of extracellular vesicles from cerebrospinal fluid. *PLoS ONE, 15*, e0227949. https://doi.org/10.1371/JOURNAL.PONE.0227949
88. Kapoor, K. S., Kong, S., Sugimoto, H., et al. (2024). Single extracellular vesicle imaging and computational analysis identifies inherent architectural heterogeneity. *ACS Nano, 18*, 11717–11731. https://doi.org/10.1021/ACSNANO.3C12556/SUPPL_FILE/NN3C12556_SI_002.XLSX

Novel Methods for Exosome Isolation Based on Precipitation and Chromatography

Alan A. Doucette

Abstract Following explosive growth in mass spectrometry and associated technology for comprehensive proteome characterization, the post-genome era presented new opportunities to discover predictive or diagnostic molecular disease biomarkers. To simplify sample collection, enhance early-stage detection of low abundance proteins, extracellular vesicles (EVs) present the opportune "liquid biopsy", whose protein cargo reflects the molecular dynamics inherent to the source "diseased" tissue. Despite availability of multiple standardized protocols for EV isolation, new strategies continue to evolve to permit rapid and robust collection of EVs at high recovery and purity. This chapter focuses on approaches to isolate exosomes, based on precipitation or chromatographic separation, offering the benefits of being easily implemented with minimal specialized equipment while also lending scalability and high throughput. Several biologically derived or inorganic materials-based approaches have been exploited to probe selective exosome binding and capture, enabling their isolate from contaminant proteins present in biological fluids. These techniques provide promising opportunities for proteomic examination of highly enriched samples and address the needs of improved detection technologies for disease biomarker discovery.

Keywords Disease biomarkers · Proteomics · Exosome isolation · Chromatography · Precipitation

---

A. A. Doucette (✉)
Department of Chemistry, Dalhousie University, Halifax, NS, Canada
e-mail: alan.doucette@dal.ca

R. K. Marcus and D. Lubman (eds.), *Exosomes and Extracellular Vesicles*,
https://doi.org/10.1007/978-3-031-89270-7_5

1 Background

1.1 A Long History of Biomarker Discovery

An accurate medical diagnosis is an essential first step in disease treatment. The process has evolved to rely on a sophisticated array of quantitative analytical tests. While the tests themselves have certainly increased in complexity, the search for reliable disease biomarkers is a defining characteristic in the history of modern medicine. Arguably, one can trace the foundations of modern medicine to Egypt, over 5000 years ago [1]. Records indicate that the Egyptian civilization demonstrated a detailed understanding of anatomy, while documenting numerous diseases and corresponding treatments [1]. However, their physicians relied only on patient symptoms to diagnose a disorder and did not perform medical tests. To assist their diagnosis, ancient Indian *vaydia* or physicians (~1500 BCE) employed all 5 senses when conducting physical exams, in tandem with detailed questioning of the patient before attempting treatments [2]. In Babylon (~1000 BCE), disease was still seen as a burden imposed on the ill by magic and demons [3]. Yet Babylonian scholars demonstrated rigorous scientific thinking, including lengthy observation of multiple disease symptoms [3]. They applied logic and reasoning to try and explain an ailment. They also prescribed "effective and tested" pharmaceutical treatments following a diagnosis [3].

By ~ 400 BCE, disease theory had evolved through Greek scholars, including the written works of Hippocrates, who prescribed the proper balance of the 4 humours (blood, phlegm, black and yellow bile) as being vital "chemical" components of a healthy body. It would take another 2 millennia for germ theory to displace these ancient ideas of disease origins. Nonetheless, the use of "molecular" disease diagnostics was already shaping in Greece; Hippocrates frequently wrote of making careful observations of urine (uroscopy) to differentiate a particular ailment [4]. As an example, from Aphorisms, Hippocrates wrote "*When bubbles settle on the surface of the urine, they indicate disease of the kidneys, and that the complaint will be protracted*" [4]. Indeed, the formation of bubbles in the urine is likely an indicator of elevated protein levels, in turn reflecting damage or malfunction of the kidneys. Uncovering chemical markers of a given biological state has been a goal of medical practitioners for nearly 2500 years [5].

A biomarker is any objective and measurable indicator of an individual's current physiological state. Often thought of as a physical (molecular) substance, biomarkers may also extend to physiological, radiological, or histological characteristics [6]. Thus, clinically relevant biomarkers include any measurable trait, from body temperature, X-ray images, to genetic, metabolic, or proteomic entities. In the post-genome era and the emergence of nanomedicine, tools for molecular screening have become widely dispersed and are often indispensable to proper diagnostic [7]. The list of potential diagnostic indicators has therefore grown exponentially. To harmonize their description, a working group of the US FDA and NIH harmonized the classification of biomarkers by publishing the "BEST" Resource (Biomarkers, EndpointS, and other Tools) in 2016 [8]. Within this resource, biomarkers provide utility through 7

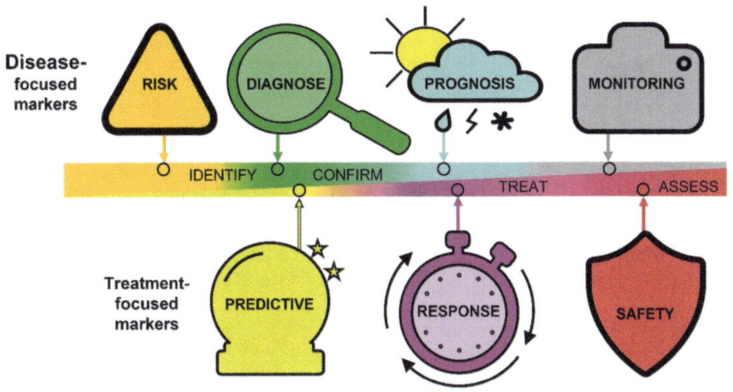

Fig. 5.1 Classifications of disease biomarkers, which serve various purposes over to the progression of a disease to assist patient treatment and outcome. Biomarkers to the left of the flow chart are generally employed early during treatment or disease diagnosis, while those to the right are employed later. The panels in the upper half of the figure are disease-focused markers (risk, diagnose, prognosis, monitoring), while those in the bottom portion of the image (predictive, response, safety) are employed to assess disease treatment

different classifications: diagnostic, monitoring, prognostic, predictive, risk susceptibility, responsive, and safety (Fig. 5.1). Table 5.1 provides an overview of the evolving technologies used to evaluate biomarkers in modern medicine.

1.2 Modern Practices for Protein Biomarker Discovery

As illustrated in Table 5.2, diagnostic biomarkers have derived from both large (proteins, genes) and small (metabolic) biomolecules. This chapter focuses primarily on protein biomarkers; the promise of uncovering new diagnostic protein biomarkers has been a driving force motivating rapid developments in proteomic technology since its earliest beginnings in the 1990s [9]. Completion of the human genome project ushered in a new era in molecular medicine, wherein high-throughput protein analysis allowed the direct characterization of the translational products of genetic material. Proteins offer a greater potential to manifest changes in the physiology following the onset of a disease. Considering that the genome is essentially unchanged over the lifetime of the patient, it cannot reflect the rapidly changing diseased state at a molecular level. By contrast, the proteome is dynamic and changes in expression to reveal or conceal a subset of proteins that correlates to the diseased state. Alternatively, protein abundance changes (up or down) may also mark a physiological state. More subtle to detect are changes in protein function, though they may manifest through post-translational modifications, protein folding, or protein–protein interactions [10]. While any such change can disclose insightful biomarkers,

Table 5.1 Evolution of technologies employed to evaluate clinical biomarkers

Date	Diagnostic tool	Application	Noted names	Further reading
5000 BCE	Natural senses (sight, touch, hearing, smell taste)	Physical exam	Charak; Imhotep; Shen Nung; Hippocrates…	www.ncbi.nlm.nih.gov/books/NBK458/
3000 BCE	Ants: to test sweetness of urine	Gluocose in urine	Ancient Hindu writings	https://doi.org/10.12816/0003257
1800	Microscope	Histology	Marie Francois Bichat	https://doi.org/10.4172/2157-7099.1000377
1816	Stethescope	Lung exam	Rene Laënnec	https://doi.org/10.3121/cmr.4.3.230
1867	Clinical thermometer	Body temperature	Sir Thomas Allbut	https://doi.org/10.1007/978-3-030-21,886-7_3
1882	Gram staining	Identify bacteria	Hans Christian Gram	https://doi.org/10.1128/br.16.1.1-29.1952
1896	Sphygmomanometer	Blood pressure	Scipione Riva-Rocci	https://doi.org/10.1111/j.1742-1241.2005.00548.x
1896	X-ray	Skeletal imaging	John Hall-Edwards	https://doi.org/10.1038/053274b0
1901	Echocardiogram	Irregular heart rhythm	Willem Einthoven	https://doi.org/10.3402/jchimp.v2i1.14383
1907	Clinical laboratory medicine	Hematology/urinalysis	Otto Folin	https://doi.org/10.1309/LMTOH5WLUQX26AIC
1913	Mammography	Breast cancer screen	Albert Solomon	https://doi.org/10.3390/healthcare11111596
1927	Angiography	Blood flow	Egas Moniz	https://doi.org/10.1053/sa.2000.17786
1942	Immunofluorescence	Histology	Albert Coons	https://doi.org/10.3181/00379727-47-13084P
1956	Auto Analyzer	Blood chemistry	Leonard Skeggs	PMID: 25,336,760
1956	Ultrasound	Abdominal/obstetrics	Ian Donald	PMID: 24,753,947
1965	Dextrostix	Glucose test strip	Helen Murray Free	US Patent 3,092,465

(continued)

Table 5.1 (continued)

Date	Diagnostic tool	Application	Noted names	Further reading
1967	At-home test kits	Pregnancy	Margaret Crane	US Patent 3,579,306
1971	Computer tamography (CT scan)	Brain/whole body	G. N. Hounsfield; J. Ambrose	https://doi.org/10.1148/radiol.230803
1974	Positron emission tomography (PET scan)	Molecular imaging	Hounsfield; Cormack; Hoffman	https://doi.org/10.1117/1.JMI.4.1.011013
1977	Magnetic resonance imaging (MRI)	Tumour detection	R. Damadian; P. C. Lauterbur	https://doi.org/10.1126/science.171.3976.1151
1985	ELISA	Protein antigens	Peter Perlman; Eva Engvall	https://doi.org/10.1128/CVI.00053-16
1985	PCR	HIV, genetic testing	Kary Mullis	https://doi.org/10.2144/btn-2020-0057
1989	LC–MS/MS	Newborn screen	David S. Millington	https://doi.org/10.1016/j.jmsacl.2024.01.006

identifying the relevant proteins is far from trivial. Protein biomarker discovery has been compared to finding a needle in a stack of other needles.

The sensitivity and selectivity of a given protein biomarker will vary considerably depending on the degree of change and correlation to the specific disease. Equally important is the capacity to obtain a clear quantitative measure of the protein biomarker in question. Consider that the human proteome comprises far more discrete components than its genome counterpart. When post-translationally modified products are considered, the number of distinct chemical entities, or proteoforms, is estimated to be upwards of 10^6 [11]. As such, even the most modern methods of proteome analysis are unable to fully capture the complete depth of such a complex biological system.

Mass spectrometry forms the cornerstone of modern proteome analysis [12]. Today's mass spectrometers boast higher resolution, sensitivity, and scan speed. By coupling state-of-the-art, tandem mass spectrometry systems with high throughput separation (online liquid chromatography), together with appropriate statistical routines to analyze the generated data, it is now possible to detect, and even quantify tens of thousands of discrete protein components in a mixture in a routine fashion [13]. Yet despite this impressive capacity, MS analysis of a proteomic system inevitably leaves many protein components unaccounted for. Most often, those proteins considered most relevant to mark a specific disease state are the most difficult proteins to detect. Specifically, lower abundance proteins are "masked" by high abundance components of the mixture [14]. Considering plasma, albumin alone makes up over

Table 5.2 Examples of commonly employed chemical diagnostic markers

Abbreviation	Biomarker description	Marker type	Condition	Classification
ALP	Alkaline phosphotase	Enzyme	Liver function	Safety
BMI	Body-mass-index	Body measurement	Nutritional status	Risk
CA-125	Cancer antigen 125	Protein	Ovarian cancer	Monitoring
CA 19–9	Carbohydrate antigen 19–9	Protein	Pancreatic cancer	Monitoring
CRP	C-reactive protein	Protein	Inflamation or infection	Risk
EGFR	Epidermal growth factor receptor	Gene	Non-small cell lung carcinoma	Predictive
Electrolytes	Na^+, K^+, Ca^{2+}, Mg^{2+}, Cl^-, $CO_3{}^{2-}$, $PO_4{}^{3-}$, etc....	Ions	Broad (infection, kidney, heart...)	Monitoring
EpCaM	Epithelial cell adhesion molecule	Protein	Cancer progression	Predictive
ERBB2	Erythroblastic oncogene B	Gene	Breast cancer	Predictive
HBa1C	Hemoglobin A1C	Protein	Diabetes	Monitoring
HGC	Human chorionic gonadotropin	Protein	Pregnancy	Diagnostic
KRAS	Kirsten rat sarcoma virus	Gene	Colorectal cancer	Predictive
LDL	Low density lipoprotein	Protein	Coronary disease	Responsive
P53	Tumor protein p53	Gene	Cancer assessment	Risk
Phe/Tyr	Phenylalanine to tyrosine ratio	Amino acids	Phenylketonuria (PKU)	Diagnostic
PSA	Prostate specific antigen	Protein	Prostate cancer	Prognostic
Rapid test	Antibodies to SARS CoV-2 spike proteins	Antibody	COVID infection	Diagnostic
Sugars	Blood glucose concentration	Small molecule	Diabetes	Monitoring
TnC	Cardiac troponin C	Protein	Cardiac injury	Diagnostic

50% of the total protein content by mass, while the top 22 most abundant proteins represent over 99% of the entire plasma proteome by mass [15]. The dynamic range of proteins in plasma from a healthy patient has been reported to be over 10 orders of magnitude [16]. Consider now the emergence of a diseased tissue, which can release altered proteins into the background plasma proteome mixture. With a desire to detect proteins originating from the diseased tissue at the earliest possible stage, it can be expected that the proteins released from this small host sample (an emerging tumor, circulating cells) will be of minimal abundance relative to the background.

A direct biopsy of a diseased tissue can be employed for proteome analysis [17], though may not be possible to obtain. Even if so, biopsies are often invasive to the patient and may not yield sufficient material to diagnose an early stage of the disease. Plasma, urine, or other biological fluids are favored as a possible source of protein biomarkers. Their collection represents a minimally invasive form of sampling, and it can be expected that these fluids contain proteins released from all areas of the body. Proximal sampling of diseased cells, particularly a tissue biopsy, would enhance the likelihood of detecting proteomic changes as they are most concentrated at the source. Ideally, the optimal sample would represent a purified set of proteins released directly by the diseased tissue but being as easy and direct to collect as plasma or urine. In fact, this is possible, and the reason why *exosomes* offer such high promise as a source of protein biomarkers.

2 A Liquid Tissue Biopsy

Extracellular vesicles (EVs) have been found to originate from practically all cell types. Exosomes are a subclassification of EVs, generally categorized according to size (40–150 nm) and density (1.08–1.19 mg/mL), though both characteristic parameters exhibit variability depending on the reporting source [18, 19]. Moreover, exosomes are decorated on their surrounding lipid membrane by a characteristic set of proteins, including tetraspanins (CD9, CD81, CD63), heat shock proteins (HSP 70, 90), and other marker proteins within their cargo (Alix, TSG101) [20]. These proteins not only provide unambiguous confirmation of exosome identity by way of immunoassays [21], but also present unique "handles" to capture exosomes through various immunoaffinity processes. Beyond proteins, exosomes also carry genetic material, amino acids, and other small metabolites.

Of relevance as a source of protein biomarkers is the understanding of the functional role of exosomes in cancer [22], inflammation [23], and other diseases [24]. Considering the mechanism of exosome formation, their cargo is directly derived from the parent diseased or tumorous cells and therefore acts as a proxy to sample proteins from the source tissue [19]. This would include characteristic protein biomarkers to differentiate the sample from a normal (healthy) state. The role of exosomes for clinical diagnostics and as therapeutic agents is described in detail in Chap. 1 of this text.

The total content of exosome-derived proteins in plasma was reported as 166 $\pm$ 50 μg/mL, which represents only 0.1–0.3% relative to other plasma proteins [25]. Fortunately, exosomes are highly stable in circulation within plasma [26], as well as during storage [27]. Therefore, robust approaches to isolate exosomes can be employed to deplete the high concentration of background proteins present in the host fluid. Special attention should be devoted to remove contaminant proteins, such as highly abundant albumin, which can adhere to the surface of the exosomes. However, this can be incorporated into the exosome purification protocol. The resulting sample can be characterized to confirm and exosome recovery, most typically by microscopy and/or nanoparticle tracking analysis (NTA) [28]. When employed successfully, the purified exosome proteome represents an ideal starting point to survey for disease biomarker [29]. Mass spectrometry-based approaches are ideally suited to survey exosomal proteins, including quantitative profiling through targeted and untargeted methods [30].

The remaining portions of this chapter are devoted to describing novel methods of isolating exosomes, and associated strategies for proteomic characterization of the EV cargo.

3 Exosome Isolation Methods

3.1 Conventional Approaches

With such a high opportunity to obtain an enriched source of biomarkers, technological advances devoted to the isolation of exosome proteins have been a topic of accelerated research. Poupardin et al. provided a high-level overview of exosome-related research over the period 2013–2022, employing machine learning to categorize various themes of research [31]. A primary conclusion of this systematic analysis was that the approach to isolate exosomes influenced the resulting proteomic data, indicating that not all isolation methods achieve equivalent results. They noted that the classic approach to exosome isolation, i.e., density centrifugation, remained the most common means of exosome capture, with 50.4% of all studies in the past decade attributed to density-based methods. Isolation by precipitation (20.7%) or by size (13.4%) rounded out the other common purification strategies.

Conventional approaches to isolate exosomes have been described in multiple reviews [18, 32–35], and so a detailed accounting will not be repeated here. Figure 5.2 summarizes several of the conventional exosome isolation strategies. These approaches leverage the characteristic features of exosome density, size, or canonical marker proteins indicative of exosome material. Naturally, each approach presents certain advantages, together with competing challenges and shortcomings. The ideal technology for exosome isolation would offer exceptional purity of sample, together with high yield. However, there are other important parameters to consider, such as the ability to scale the approach, the throughput of the preparation, as well as the cost. Differential ultracentrifugation, for example, is considered the gold standard for isolating "pure" exosomes [36]. However, it requires significant preparation time, has limited scalability, modest yield, and demands sophisticated equipment that is not applicable to the clinic. By contrast, polymer precipitation of exosomes is quick, requires no special equipment, is easy to scale, and provides extremely high yield.

However, the sample purity of polymer-precipitated exosomes can be a significant concern [37].

A standardized, optimal approach to recover purified exosomes remains to be defined. For these reasons, novel approaches to isolate exosomes continue to be developed. The reader is referred to other chapters of this text for detailed descriptions of some of these approaches. Herein, some promising, though perhaps unconventional methods of exosome isolation ahead of proteomics/ mass spectrometry are described. Focus is placed on precipitation and chromatographic approaches, which are both highly scalable, simple to implement, and require minimal specialized equipment.

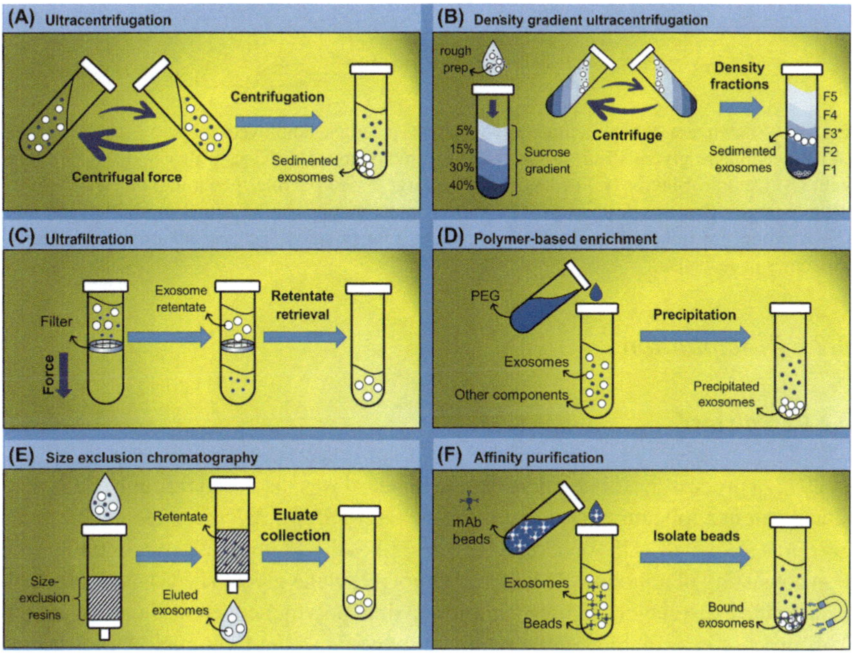

Fig. 5.2 Examples of commonly employed exosome enrichment protocols. (**A**) Ultracentrifugation: Centrifugal force is applied to separate sedimented exosomes. (**B**) Density gradient ultracentrifugation: A sucrose gradient is used to separate density fractions, resulting in sedimented exosomes. (**C**) Ultrafiltration: A filter is used to retrieve exosome retentate. (**D**) Polymer-based enrichment: Polyethylene glycol (PEG) can be added to precipitate exosomes. (**E**) Size exclusion chromatography: the resin separate slows the passage of smaller material while larger exosomes flow through quickly. (**F**) Affinity purification: Monoclonal antibodies (mAb) bound to magnetic beads selectively bind to exosome markers, which are then recovered with a magnetic force. Image adapted from [30]

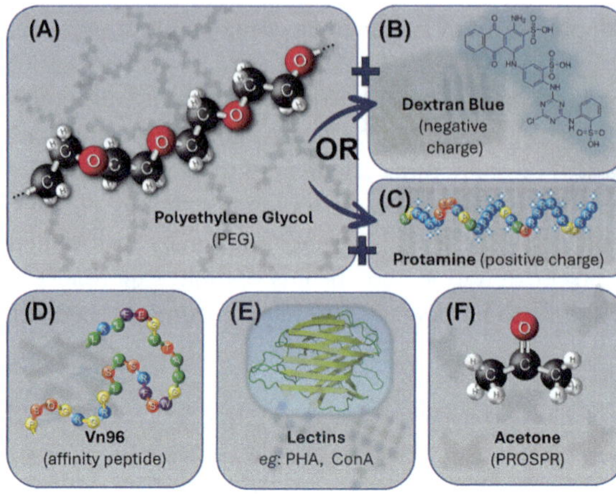

Fig. 5.3 Shown are the structures of chemical reagents employed to precipitate exosomes. (**A**) Polyethylene glycol PEG (polymer-based precipitation). PEG can also be used in tandem with (**B**) Dextran Blue or (**C**) Protamine, for charge-based polymer precipitation. (**D**) The Vn96 peptide targets heat shock proteins, (**E**) Various plant lectins can be added to bind glycan units. (**F**) Acetone precipitation works by aggregating free proteins such as albumin, while leaving exosomes in suspension

3.2 *Precipitation*

3.2.1 Polymer

In 2012, a process to isolate microvesicle material was disclosed through a US provisional patent application. Granted in 2015, "*Methods for Microvesicle Isolation and Selective Removal*" (US 9,005,888 B2), describes the addition of a precipitating solution consisting of aqueous, PBS-buffered polyethylene glycol (PEG), typically 8–10 kDa, to the exosome-containing sample at a final polymer concentration from 5–8%. This process is available commercially and known as the ExoQuick™ preparation. Advantages of this approach include high exosomal protein recovery, ease of scaling, and minimal infrastructure needed, suggesting a potential for clinical applications. Yoo et al. employed polymer precipitation to isolate serum EVs from 60 patients inflicted by rheumatoid arthritis (RA) [38]. Proteome analysis with tandem mass tagging led to the identification of 6 candidate protein markers, including elevated levels of lymphatic vessel endothelial hyaluronicacid receptor-1 (LYVE-1) in the exosome pellet, and a corresponding reduced expression of amyloid A (AA) in serum EVs. These prognostic markers have the potential to improve patient outcomes by monitoring disease activity in response to methotrexate therapy.

A concern with polymer precipitation, however, is the potential for high level of contaminant proteins in the recovered pellet, which can challenge the analysis of protein biomarkers. Zubiri et al. compared ultracentrifugation and ExoQuick™ for isolation of urinary exosomes from patients with diabetic nephropathy (DN),

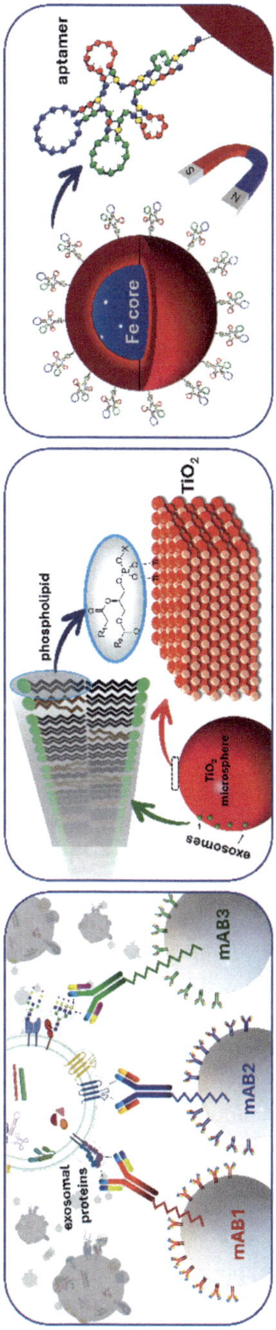

Fig. 5.4 Chemical ligands, bound to microbeads, which bind to specific components of the exosome. The first panel illustrates protein-based antibodies, targeting proteins expressed on exosome surface. The middle panel shows titanium dioxide microspheres, which targets the phospholipid membrane. The panel at right shows DNA-based aptamers, affixed to an iron core–shell particle allowing magnetic bead isolation

a congenital kidney disease correlated to elevated glucose levels. [39]. Owing to the elevated protein content of urine from DN patients, LC–MS/MS analysis was unable to detect characteristic exosome marker proteins in the polymer pellets, while high levels of albumin were recorded. Immunoaffinity depletion of albumin and IgG, in tandem ultracentrifugation, successfully enabled MS identification of several marker proteins, including MLL3, a histone N-methyltransferase (found exclusively in the diseased patient exosomes), AMBP, alpha-1-microglobulin/bikunin precursor (elevated in DM patients), as well as VDAC1, voltage-dependent anion-selective channel protein 1, seen to be decreased in DN patients. This protein panel could facilitate early diagnosis of DN, as well as monitoring disease progression.

Prior to the recent patent disclosure, PEG solutions were readily employed as a precipitating medium. It had previously been shown that PEG will precipitate particulate materials including viruses and bacteriophages [40]. The work of Yamamoto was in turn inspired by prior research on the use of PEG or other polymers as a precipitating agent, including for "free" proteins in solution [41]. The mechanism of protein precipitation by PEG is not clearly definitive but is suggested to be based on the excluded volume concept [42]. While the polymer is not assumed to directly interact with protein (nor exosome), it rather occupies a large portion of the solvent volume and thus "concentrates" solute until it exceeds its solubility limit. In essence, the addition of PEG to a solution lowers the solubility of all components, be they exosomes, lipoprotein complexes, or free proteins such as albumin, implying high yield through lower purity of the recovered samples. Nonetheless, inducing exosomes to precipitate from solution is a desirable approach, if purity is not a critical endpoint.

3.2.2 Charge Precipitation

Deregibus et al. extended the concept of polymer-based exosome precipitation by exploiting the inherent negative charge properties of exosomes [43]. They employed protamine, a highly cationic peptide containing multiple basic arginine residues, in tandem with PEG (35 kDa),this strategy was also patented by the same authors (WO 2017/178472 A1: *Isolation of extracellular vesicles (EVs) from biological sample fluids*). They showed protamine alone will readily precipitate exosomes, though the addition of PEG with protamine improved the resuspension of exosomes from the precipitation pellet. Though promising, no further applications of this precipitation strategy have been reported to date.

Konoshenko et al. employed the anionic polymer reagent, Dextran Blue (2000 kDa), in tandem with PEG (1.5%) to isolate exosomes from plasma and urine [44]. Dextran Blue consists of the synthetic dye, Cibacron Blue F3GA, covalently bound to Dextran, an anhydroglucose polymer. Cibacron has a long history of use as a non-selective ligand that binds proteins such as albumin, through electrostatic as well as hydrophobic interactions. As applied to exosome purification, the addition of Dextran Blue binds and aggregates membrane-bound particles from biofluids more effectively than it does membrane-free proteins. Thus, a lower concentration of PEG could be used, resulting in a higher purity pellet [44].

Titanium dioxide has been widely used for the selective binding and enrichment of phosphorylated peptides through electrostatic interactions. Under acidic conditions, the TiO_2 surface carries a positive charge, while phosphate would remain negatively charged. Gao et al. developed an exosome isolation strategy based on the attraction of TiO_2 to the phospholipid membrane [45] By adding TiO_2 microspheres to bulk solutions, exosomes could be recovered in high yield following only 5 min incubation and with low-speed centrifugation of the TiO_2 beads. The authors applied their strategy to isolate exosomes from the serum of pancreatic cancer patients compared to healthy donors. Quantitative MS analysis identified 59 proteins with increased expression in the exosome fraction of cancerous patients, and a further 6 proteins were downregulated. Several of these proteins were previously implicated in pancreatic cancer. The combined panel of proteins contributed to a clear separation of tumor vs non-tumor samples by principal components analysis [45].

3.2.3 Immunoaffinity

Immunoaffinity capture offers a more selective alternative to a non-specific exosome recovery approach by polymer precipitation. The approach employs antibodies against common exosome marker proteins (such as CD63, or CD81). While a sufficiently high concentration of antibodies could induce crosslinking (and subsequent precipitation) of exosomes, the most typical use of immunoaffinity precipitation is to covalently link the antibody to a bead support. These beads could be contained within a column for flow-based (chromatographic) capture, or alternatively, they may be added in bulk to the solution whereby bound exosomes are isolated by centrifugation, by magnetic bead capture [46], or recovered in microfluidic devices [47]. Tauro et al. compared ultracentrifugation, density gradient ultrafiltration and immunoaffinity capture using commercial CD326 (EpCAM) magnetic microbeads to recover exosomes from human colon carcinoma cells. They showed the immunoaffinity capture to enrich exosome-associated marker proteins, with 2 to over 50-fold higher MS peptide spectral counts compared to the centrifugation enrichment protocols. The recovered exosomes showed a tighter size distribution (40–60 nm) and displayed several protein markers previously implicated in cancer.

Regardless of the format to recover the affinity beads, the concept of employing antibodies to capture exosomes presents an inherently higher cost, particularly when performed at high scale, given the cost of the protein antibody [48]. To alleviate these costs, alternative ligands have been presented to selectively bind exosomes, such as the use of lectins by Choi et al. [49]. In their study, three different lectins with affinity to various sugar moieties were immobilized on nanoparticles held in a microfluidic chip, and compared to CA19-9 for affinity-based capture of exosomes from pancreatic cancer cell lines, as well as clinical samples of plasma obtained from metastatic vs nonmetastatic pancreatic cancer patients. A distinct glycan profile enabled selective isolation of pancreatic cancer exosomes. Shtam et al. added lectins (phytohemagglutinin M, PHA, or concanavalin A, Con-A) directly to cell culture media, urine, or plasma, to recover lectin-aggregated exosomes after low-speed centrifugation.

Exosomes can be released from the aggregated pellet by the addition of excess sugar [50]. Mass spectrometry analysis successfully identified several exosome-associated proteins. However, the co-precipitation of glycoproteins by lectin affinity purification necessitates additional preparatory steps to achieve higher purity exosomes.

Aptamers have also been employed to selectively capture exosomes through affinity binding. These ligands are typically employed for sensing applications [51]. A multiplexed array of protein-specific aptamers has been assembled into a single microsensing chip, to enable proteome-wide analysis at high throughput scales. This SOMAscan technology (slow off-rate modified DNA aptamers) was applied by Webber et al. to quantify over 1000 unique proteins from an ultracentrifuge-purified preparation of exosomes derived from prostate cancer cells [52].

Zhang et al. covalently linked a DNA aptamer against CD63 to Fe_3O_4@TiO_2 magnetic nanobeads, to isolate urinary exosomes from healthy donors [53]. The exosomes were released through the addition of DNase I to digest the DNA aptamer. Bottom-up LC–MS/MS analysis revealed 999 unique proteins, with many overlapping those recovered from ultracentrifugation.

3.2.4 Peptide Affinity Precipitation

An approach that combines the selectivity of immunoaffinity capture with the speed and simplicity of polymer precipitation was described by Ghosh in 2014 and employs peptide-based immunoaffinity precipitation of exosomes [54]. Synthetically assembled peptide chains can be produced at scale using established protocols. A series of short-chained synthetic peptide was therefore screened, identifying a set of peptides with high affinity to heat shock proteins (HSP60, HSP70, and HSP90) [54]. Like tetraspanins, HSPs are also known to decorate the surface of exosomes and are further seen with elevated levels of expression in cancerous tissues [55]. These cationic affinity peptides, termed "venceremins" (Vn) showed extremely high affinity toward anionic HSPs. At 27 residues in length, the Vn96 peptide (sequence *PSQGKGR-GLSLSRFSWGALTLGEFLKL*) displayed the highest affinity for HSPs [54]. When added to plasma, urine, or cell cultures, the Vn96 peptide effectively crosslinks exosomes, allowing their collection by low-speed centrifugation [56, 57]. Askeland et al. compared peptide affinity precipitation of plasma with ultracentrifugation and size exclusion chromatography, and determined the Vn96 preparation showed the lowest level of impurity proteins (albumin, lipoproteins) [58]. Precipitation also yielded the highest reproducibility of identified proteins. It was further shown that the precipitated exosome preparation resists denaturation or resuspension in a mild detergent solution, while complete dissolution is only observed in a strongly denaturing buffer containing SDS [59]. Therefore, a successive washing protocol with surfactants can be employed to eliminate all but the most strongly associated proteins of the HSP-peptide complex—i.e. those contained within the exosomes. The result is a high purity, high yield preparation of exosomal proteins. Proteomics analysis of the resulting pellet by mass spectrometry reveals a family of proteins associated with exosomes and indicative of the cancerous cell lines from which they were derived.

Interestingly, the proteins closely associated with HSPs and isolated through peptide affinity precipitation may not originate as exosomal material. Tai et al. determined that cancer cells form stable high affinity HSP complexes called the epichaperome, comprising a tightly interacting complex of chaperones, co-chaperones, adaptors, and folding enzyme [60]. Considering the "stressed" environment caused by the accelerated growth of a cancerous tumour, it is not surprising to find unique HSP chaperome complexes in these systems [61]. The Vn96 can therefore serve as an important tool to isolate this form of liquid biopsy for biomarker analysis [62]. In their analysis of Vn96-precipitated complexes from cancerous cell lines (U87, HELA, HEK293T), not only did the abundance of HSP proteins correlate with the invasiveness of the cancer cell, but they also observed high levels of histones, chaperone proteins, and metabolic enzymes in the extracellular complexes. Vn96 is sold commercially as the Microvesicle Extraction (ME™) kit.

3.2.5 Solvent-Based Precipitation

Still exploring precipitation-based approaches, an alternative strategy to purify exosomal proteins was described by Gallart-Palau in [63]. Termed PRotein Organic Solvent Precipitation (PROSPR), the approach follows the opposing concept of eliminating all non-exosomal proteins from the sample by organic solvent precipitation. The authors showed that EVs would retain their structural stability and corresponding cargo solubility in cold 80% acetone for extended periods, while all other "free" proteins will crash out of solution. What remains in the organic solvent supernatant therefore constitutes "pure" EV proteins, which can be examined by SDS extraction of the solvent-dried pellet. These rapid, low-cost, highly scalable processes therefore present viable strategies to enrich exosomes for the purpose of protein biomarker discovery. The same group later applied PROSPR to recover exosomes from human central nervous tissues, identifying 2901 unique proteins [64]. They reported an approximate sevenfold enrichment of low abundance EV proteins relative to purification by ultracentrifugation. The method is also applicable to low quantities of starting material, therefore expanding its utility for disease characterization.

3.3 Chromatography

A flow-through design can be designed to selectively retain exosomes in column format and offers the flexibility of fully automated, scalable sample handling, particularly if the column surface can be regenerated. The immunoaffinity capture approach described above is difficult to preserve, as the antibodies cannot be exposed to extremes of temperature, pH, or surfactants and therefore do not withstand robust column regenerations. Fortunately, multiple alternative chromatographic approaches have been proposed to selectively isolate exosomes. Chromatographically purified exosomes are often prepared in bulk for therapeutic evaluation.

The most widespread column approach is size exclusion chromatography (SEC), with several commercial platforms available to recover exosomes based on their characteristic size range. Care must be taken with SEC to avoid collecting similar sized particulates in the sample media, such as high-density lipoprotein (HDL) in plasma or aggregated Tamm Horsfall protein in urine [65]. The pores of the gel filtration media used in SEC may also become clogged, particularly when working with plasma or other concentrated media [66]. Lattmman et al. employed a simple freeze–thaw cycle, with low-speed centrifugation of plasma and serum samples prior to SEC isolation of EVs. They quantified 2896 unique proteins in the EV fraction and revealed several known melanoma marker proteins (MCAM, TNC, and TGFBI) to be elevated in the EV fraction, even though they were not distinguishable in the undeleted plasma relative to healthy donor samples. SEC has frequently been coupled with other purification strategies, such as ultracentrifugation or dual mode chromatography, resulting in higher purity samples, though at the expense of increased sample preparation complexity [67]. Thus, alternative chromatographic approaches have also been proposed.

3.3.1 Anion Exchange

Under physiological conditions, exosomes universally carry a net negative charge at their surface, given the presence of basic proteins (HSPs) as well as negatively charged phospholipids. Ion exchange separation can be employed by taking advantage of their lower negative charge relative to other protein impurities, including lipoproteins and albumin. Therefore, EVs can be retained on a positively charged support (i.e. anion exchange media), and then released by gradually increasing the ionic strength of the eluting buffer. The process requires significantly less time than ultracentrifugation and the buffer systems are considered biologically compatible [68]. Anion exchange chromatography has also been shown to separate exosome subpopulations based on size and charge density [68, 69], with total recovery over 70% (loss due to adsorption to chromatographic system) [69]. Anion exchange has most commonly been employed to enrich exosomes for therapeutic evaluation, though proteome profiling of the resulting purification has also been reported to provide further insight into the therapeutic benefits of EV preparation [70].

3.3.2 Cation Exchange

The LEAP platform (Ligand-based Exosome Affinity Purification) is a patented technology, (*Methods and Compositions for Purification or Isolation of Microvesicles and Exosomes* WO2018112557A1) and based on cation exchange chromatography to isolate exosomes [71]. Recall that EVs predominantly display a *net* negative charge. However, they are still known to carry some positive charges through surface proteins. Thus, the LEAP technology is designed to optimize electrostatic attraction of these positive charges through a proprietary backbone that spaces negative charges by ~5 Å

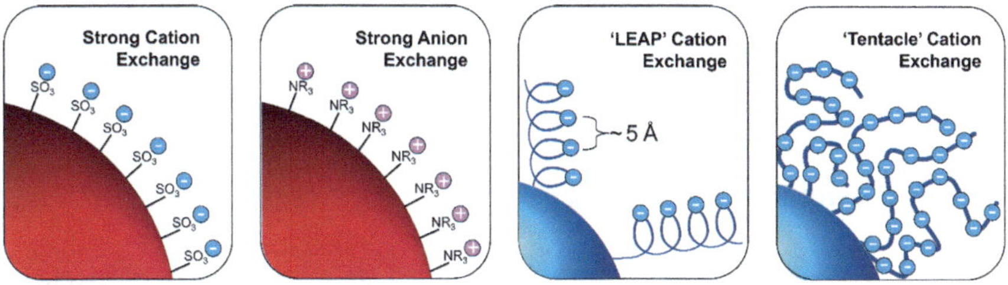

Fig. 5.5 The diagram illustrates four types of charge-based capture materials used for chromatographic isolation of exosomes. The first panel at left shows conventional strong cation exchange resin with a fixed negative charge. The next panel shows conventional anion exchange resin with a fixed positive charge. The LEAP technology employs cation exchange whereby negative charges are affixed to a proprietary polymer background and spaced by 5 Å. In the last panel tentacle cation exchangers extend the charge from the surface of the resin through flexible polymer chains, with varied charge density

(Fig. 5.5). The LEAP resin can be regenerated to allow repeated use. Platelet-derived exosomes (pEV) were purified by the LEAP technology and characterized by mass spectrometry to reveal several proteins indicative of the parent cell, including platelet membrane proteins associated with wound healing [71].

Saari et al. examined various cation exchange supports to capture exosomes, including HiTrap Sepharose FF SP columns (a conventional strong cation exchange support) as well as Fractogel EMD SO3-. The latter carries a similar negatively charged sulfoisobutyl residue, though is arranged onto flexible polymethacrylate chains which extend from the resin support [72]. This flexibility allows the negatively charged residue to bind to select positive groups on the EV surface, increasing recovery. They further demonstrated the highest purity through a mixed mode conventional/tentacle cation exchange column, yielding superior results to anion exchange isolation of platelet-derived exosomes.

3.3.3 Charge Reversible Capture System

Recently, Kawano et al. reported a novel catch and release isolation system (EV-CaRiS), which took advantage of a charge invertible curvature sensing peptide [73]. The authors designed a peptide whose net charge switched from positive to negative from pH 6 (loading buffer) to 9 (eluting buffer). Over this pH range, exosomes remain negatively charged, allowing selective binding and elution over a narrow pH ramp. Moreover, the binding characteristics of the peptide (*EZBLLZXLNZBTEBLSZXGSGSC*, where X and B indicate norleucine and 2-aminoisobutyric acid, respectively) are based on a conformational change as the peptide adopts a flexible random coil in solution, but converting to a rigid α-helix as the peptide inserts into the hydrophobic EV membrane [74]. This binding is unique toward highly curved particles and is thus independent of the surface proteins decorating the exosomes. The total time for capture and release from a column, covalently loaded with the curvature sensing peptides, was 90 min and with a recovery threefold higher than by ultracentrifugation.

4 Conclusions and Outlook

As the current "gold standard" for exosomal preparation, differential ultracentrifugation results in a highly purified sample, but lacks the robustness and throughput needed to afford routine exosome sample preparation ahead of proteome processing. Precipitation strategies are extremely simple to implement and can capitalize on a variety of precipitating agents to selectively isolate exosomes. However, the potential for automation afforded by chromatographic separation suggests this as the most desired alternative to ultracentrifugation. Size exclusion chromatography is a commonly employed strategy, though alternative modes of separation can enhance selectivity, particularly when combined in multi-mode formats. Thus, charge-based separation of affinity selection are important orthogonal mode. Multidimensional separation platforms can be fully automated, and not only offer the potential to remove non-exosomal protein contaminants but could also further isolation of exosome sub-populations. As the field of exosome analysis matures, and clinical profiling becomes normalized as a routine practice, it will be important to establish alternative "gold standard" isolation platforms, to focus on the equally difficult challenge of characterizing the recovered exosome proteins.

Acknowledgements The author acknowledges the Natural Sciences and Engineering Research Council of Canada (NSERC) for financial support.

Declaration The author declares no conflict of interest.

References

1. Metwaly, A. M., Ghoneim, M. M., Eissa, I. H., Elsehemy, I. A., Mostafa, A. E., Hegazy, M. M., Afifi, W. M., & Dou, D. (2021). Traditional ancient Egyptian medicine: A review. *Saudi Journal of Biological Sciences, 28*(10), 5823. https://doi.org/10.1016/J.SJBS.2021.06.044
2. Saini, A. (2016). Physicians of ancient India. *Journal of Family Medicine and Primary Care, 5*(2), 254. https://doi.org/10.4103/2249-4863.192322
3. Geller, M. J. (2010). Chapter 1: Medicine as science. In *Ancient Babylonian Medicine: Theory and Practice* (pp. 11–42). Wiley-Blackwell. https://doi.org/10.1002/9781444319996.CH1.
4. Armstrong, J. A. (2007). Urinalysis in Western culture: A brief history. *Kidney International, 71*(5), 384–387. https://doi.org/10.1038/SJ.KI.5002057
5. Crnogorac-Jurcevic, T. (2024). A brief history of urine examination—From ancient uroscopy to 21st century urinomics. *Advances in Biomarker Sciences and Technology, 6*, 83–90. https://doi.org/10.1016/J.ABST.2024.06.001
6. García-Gutiérrez, M. S., Navarrete, F., Sala, F., Gasparyan, A., Austrich-Olivares, A., & Manzanares, J. (2020). Biomarkers in psychiatry: concept, definition, types and relevance to the clinical reality. *Frontiers in Psychiatry, 11*, Article 527209. https://doi.org/10.3389/FPSYT.2020.00432/BIBTEX

7. Di Bella, M. A. (2022). Overview and update on extracellular vesicles: Considerations on exosomes and their application in modern medicine. *Biology, 11*(6), 804. https://doi.org/10.3390/BIOLOGY11060804
8. Food and Drug Administration, & National Institutes of Health (2016) BEST (Biomarkers, Endpoints, and other tools) resource. Silver Spring, MD, FDA-NIH Biomarker Working Group
9. Mann, M. (2016). Origins of mass spectrometry-based proteomics. *Nature Reviews Molecular Cell Biology, 17*(11), 678–678. https://doi.org/10.1038/nrm.2016.135
10. Duan, G., & Walther, D. (2015). The roles of post-translational modifications in the context of protein interaction networks. *PLoS Computational Biology, 11*(2), 1004049. https://doi.org/10.1371/JOURNAL.PCBI.1004049
11. Aebersold, R., Agar, J. N., Amster, I. J., Baker, M. S., Bertozzi, C. R., Boja, E. S., Costello, C. E., Cravatt, B. F., Fenselau, C., Garcia, B. A., Ge, Y., Gunawardena, J., Hendrickson, R. C., Hergenrother, P. J., Huber, C. G., Ivanov, A. R., Jensen, O. N., Jewett, M. C., Kelleher, N. L., & Zhang, B. (2018). How many human proteoforms are there? *Nature Chemical Biology, 14*(3), 206. https://doi.org/10.1038/NCHEMBIO.2576
12. Jiang Y, Rex DA, Schuster D, Neely BA, Rosano GL, Volkmar N, Momenzadeh A, Peters-Clarke TM, Egbert SB, Kreimer S, Doud EH, Crook OM, Yadav AK, Vanuopadath M, Hegeman AD, Mayta ML, Duboff AG, Riley NM, Moritz RL, Meyer JG (2024) Comprehensive overview of bottom-up proteomics using mass spectrometry. ACS Meas Sci Au 4(4):338–417. https://doi.org/10.1021/acsmeasuresciau.3c00068
13. Muntel, J., Gandhi, T., Verbeke, L., Bernhardt, O. M., Treiber, T., Bruderer, R., & Reiter, L. (2019). Surpassing 10 000 identified and quantified proteins in a single run by optimizing current LC-MS instrumentation and data analysis strategy. *Molecular Omics, 15*(5), 348–360. https://doi.org/10.1039/C9MO00082H
14. Liu, T., Qian, W.-J., Mottaz, H. M., Gritsenko, M. A., Norbeck, A. D., Moore, R. J., Purvine, S. O., Camp, D. G., & Smith, R. D. (2006). Evaluation of multiprotein Immunoaffinity subtraction for plasma proteomics and candidate biomarker discovery using mass spectrometry. *Molecular & Cellular Proteomics, 5*(11), 2167–2174. https://doi.org/10.1074/mcp.T600039-MCP200
15. Nanjappa, V., Thomas, J. K., Marimuthu, A., Muthusamy, B., Radhakrishnan, A., Sharma, R., Ahmad Khan, A., Balakrishnan, L., Sahasrabuddhe, N. A., Kumar, S., Jhaveri, B. N., Sheth, K. V., Kumar Khatana, R., Shaw, P. G., Srikanth, S. M., Mathur, P. P., Shankar, S., Nagaraja, D., Christopher, R., Prasad, T. S. K. (2014). Plasma Proteome Database as a resource for proteomics research: 2014 update. *Nucleic Acids Research, 42*(Database issue), D959. https://doi.org/10.1093/NAR/GKT1251.
16. Anderson, N. L., & Anderson, N. G. (2002). The human plasma proteome: History, character, and diagnostic prospects. *Molecular & Cellular Proteomics, 1*(11), 845–867. https://doi.org/10.1074/mcp.R200007-MCP200
17. Guo, T., Kouvonen, P., Koh, C. C., Gillet, L. C., Wolski, W. E., Röst, H. L., Rosenberger, G., Collins, B. C., Blum, L. C., Gillessen, S., Joerger, M., Jochum, W., & Aebersold, R. (2015). Rapid mass spectrometric conversion of tissue biopsy samples into permanent quantitative digital proteome maps. *Nature Medicine, 21*(4), 407–413. https://doi.org/10.1038/nm.3807
18. Doyle, L. M., & Wang, M. Z. (2019). Overview of extracellular vesicles, their origin, composition, purpose, and methods for exosome isolation and analysis. *Cells, 8*(7). https://doi.org/10.3390/CELLS8070727.
19. Kalluri, R., & LeBleu, V. S. (2020a). The biology, function, and biomedical applications of exosomes. *Science (New York, N.Y.), 367*(6478). https://doi.org/10.1126/SCIENCE.AAU6977
20. Deng, F., & Miller, J. (2019). A review on protein markers of exosome from different bio-resources and the antibodies used for characterization. *Journal of Histotechnology, 42*(4), 226–239. https://doi.org/10.1080/01478885.2019.1646984
21. Théry, C., Witwer, K. W., Aikawa, E., Alcaraz, M. J., Anderson, J. D., Andriantsitohaina, R., Antoniou, A., Arab, T., Archer, F., Atkin-Smith, G. K., Ayre, D. C., Bach, J. M., Bachurski, D., Baharvand, H., Balaj, L., Baldacchino, S., Bauer, N. N., Baxter, A. A., Bebawy, M., Zuba-Surma, E. K. (2018). Minimal information for studies of extracellular vesicles 2018

(MISEV2018): a position statement of the international society for extracellular vesicles and update of the MISEV2014 guidelines. *Journal of Extracellular Vesicles*, *7*(1). https://doi.org/10.1080/20013078.2018.1535750.
22. Dai, J., Su, Y., Zhong, S., Cong, L., Liu, B., Yang, J., Tao, Y., He, Z., Chen, C., & Jiang, Y. (2020). Exosomes: key players in cancer and potential therapeutic strategy. *Signal Transduction and Targeted Therapy, 5*(1), 1–10. https://doi.org/10.1038/s41392-020-00261-0.
23. Tian, Y., Cheng, C., Wei, Y., Yang, F., & Li, G. (2022). The role of exosomes in inflammatory diseases and tumor-related inflammation. *Cells, 11*(6). https://doi.org/10.3390/CELLS11061005
24. Miao, C., Wang, X., Zhou, W., & Huang, J. (2021). The emerging roles of exosomes in autoimmune diseases, with special emphasis on microRNAs in exosomes. *Pharmacological Research, 169*, Article 105680. https://doi.org/10.1016/J.PHRS.2021.105680
25. Shtam, T., Naryzhny, S., Kopylov, A., Petrenko, E., Samsonov, R., Kamyshinsky, R., Zabrodskaya, Y., Nikitin, D., Sorokin, M., Buzdin, A., & Malek, A. (2018). Functional properties of circulating exosomes mediated by surface-attached plasma proteins. *Journal of Hematology, 7*(4), 149. https://doi.org/10.14740/JH412W
26. Sanz-Rubio, D., Martin-Burriel, I., Gil, A., Cubero, P., Forner, M., Khalyfa, A., & Marin, J. M. (2018). Stability of circulating exosomal miRNAs in healthy subjects. *Scientific Reports, 8*(1), 1–10. https://doi.org/10.1038/s41598-018-28748-5.
27. Ge, Q., Zhou, Y., Lu, J., Bai, Y., Xie, X., & Lu, Z. (2014). MiRNA in plasma exosome is stable under different storage conditions. *Molecules, 19*(2), 1568. https://doi.org/10.3390/MOLECULES19021568
28. Sokolova, V., Ludwig, A. K., Hornung, S., Rotan, O., Horn, P. A., Epple, M., & Giebel, B. (2011). Characterisation of exosomes derived from human cells by nanoparticle tracking analysis and scanning electron microscopy. *Colloids and Surfaces B: Biointerfaces, 87*(1), 146–150. https://doi.org/10.1016/J.COLSURFB.2011.05.013
29. Yu, D., Li, Y., Wang, M., Gu, J., Xu, W., Cai, H., Fang, X., & Zhang, X. (2022). Exosomes as a new frontier of cancer liquid biopsy. *Molecular Cancer, 21*(1), 1–33. https://doi.org/10.1186/S12943-022-01509-9
30. Jalaludin, I., Lubman, D. M., & Kim, J. (2023). A guide to mass spectrometric analysis of extracellular vesicle proteins for biomarker discovery. *Mass Spectrometry Reviews, 42*(2), 844–872. https://doi.org/10.1002/MAS.21749
31. Poupardin, R., Wolf, M., Maeding, N., Paniushkina, L., Geissler, S., Bergese, P., Witwer, K. W., Schallmoser, K., Fuhrmann, G., Strunk, D., Poupardin, R., Wolf, M., Maeding, N., Paniushkina, L., Strunk, D., Witwer, K. W., & Geissler, S. (2024). Advances in extracellular vesicle research over the past decade: source and isolation method are connected with cargo and function. *Advanced Healthcare Materials, 2303941*, 1–14. https://doi.org/10.1002/adhm.202303941
32. Chen, J., Li, P., Zhang, T., Xu, Z., Huang, X., Wang, R., & Du, L. (2022). Review on strategies and technologies for exosome isolation and purification. *Frontiers in Bioengineering and Biotechnology, 9*. https://doi.org/10.3389/FBIOE.2021.811971/FULL.
33. Li, P., Kaslan, M., Han Lee, S., Yao, J., & Gao, Z. (2017). Progress in exosome isolation techniques. *Theranostics, 7*(3), 789–804. https://doi.org/10.7150/thno.18133
34. Shirejini, S. Z., & Inci, F. (2022). The Yin and Yang of exosome isolation methods: Conventional practice, microfluidics, and commercial kits. *Biotechnology Advances, 54*, Article 107814. https://doi.org/10.1016/J.BIOTECHADV.2021.107814
35. Witwer, K. W., Buzás, E. I., Bemis, L. T., Bora, A., Lässer, C., Lötvall, J., Nolte-, E. N., Piper, M. G., Sivaraman, S., Skog, J., Théry, C., Wauben, M. H., & Hochberg, F. (2013). Standardization of sample collection, isolation and analysis methods in extracellular vesicle research. *Journal of Extracellular Vesicles*, *2*(1). https://doi.org/10.3402/jev.v2i0.20360
36. Greening, D. W., Xu, R., Ji, H., Tauro, B. J., & Simpson, R. J. (2015). A protocol for exosome isolation and characterization: Evaluation of ultracentrifugation, density-gradient separation, and immunoaffinity capture methods. *Methods in Molecular Biology, 1295*, 179–209. https://doi.org/10.1007/978-1-4939-2550-6_15/TABLES/2

37. Lobb, R. J., Becker, M., Wen, S. W., Wong, C. S. F., Wiegmans, A. P., Leimgruber, A., & Mö Ller, A. (2015). Optimized exosome isolation protocol for cell culture supernatant and human plasma. *Journal of Extracellular Vesicles, 4*, 27031. https://doi.org/10.3402/jev.v4.27031
38. Yoo, J., Lee, S. K., Lim, M., Sheen, D., Choi, E. H., & Kim, S. A. (2017). Exosomal amyloid A and lymphatic vessel endothelial hyaluronic acid receptor-1 proteins are associated with disease activity in rheumatoid arthritis. *Arthritis Research & Therapy, 19*(1). https://doi.org/10.1186/S13075-017-1334-9
39. Zubiri, I., Posada-Ayala, M., Sanz-Maroto, A., Calvo, E., Martin-Lorenzo, M., Gonzalez-Calero, L., de la Cuesta, F., Lopez, J. A., Fernandez-Fernandez, B., Ortiz, A., Vivanco, F., & Alvarez-Llamas, G. (2014). Diabetic nephropathy induces changes in the proteome of human urinary exosomes as revealed by label-free comparative analysis. *Journal of Proteomics, 96*, 92–102. https://doi.org/10.1016/J.JPROT.2013.10.037
40. Yamamoto, K. R., Alberts, B. M., Benzinger, R., Lawhorne, L., & Treiber, G. (1970). Rapid bacteriophage sedimentation in the presence of polyethylene glycol and its application to large-scale virus purification. *Virology, 40*(3), 734–744. https://doi.org/10.1016/0042-6822(70)90218-7
41. Polson, A., Potgieter, G. M., Largier, J. F., Mears, G. E. F., & Joubert, F. J. (1964). The fractionation of protein mixtures by linear polymers of high molecular weight. *Biochimica et Biophysica Acta, 82*(3), 463–475. https://doi.org/10.1016/0304-4165(64)90438-6
42. Atha, D. H., & Ingham, K. C. (1981). Mechanism of precipitation of proteins by polyethylene glycols. Analysis in terms of excluded volume. *Journal of Biological Chemistry, 256*(23), 12108–12117. https://doi.org/10.1016/S0021-9258(18)43240-1
43. Deregibus, M. C., Figliolini, F., D'Antico, S., Manzini, P. M., Pasquino, C., De Lena, M., Tetta, C., Brizzi, M. F., & Camussi, G. (2016). Charge-based precipitation of extracellular vesicles. *International Journal of Molecular Medicine, 38*(5), 1359–1366. https://doi.org/10.3892/IJMM.2016.2759/HTML
44. Konoshenko, M. Y., Lekchnov, E. A., Bryzgunova, O. E., Kiseleva, E., Pyshnaya, I. A., & Laktionov, P. P. (2021). Isolation of extracellular vesicles from biological fluids via the aggregation–precipitation approach for downstream miRNAs detection. *Diagnostics, 11*(3), 384. https://doi.org/10.3390/DIAGNOSTICS11030384.
45. Gao, F., Jiao, F., Xia, C., Zhao, Y., Ying, W., Xie, Y., Guan, X., Tao, M., Zhang, Y., Qin, W., & Qian, X. (2019). A novel strategy for facile serum exosome isolation based on specific interactions between phospholipid bilayers and TiO2. *Chemical Science, 10*(6), 1579–1588. https://doi.org/10.1039/C8SC04197K
46. Kaboord, B., Smith, S., Patel, B., & Meier, S. (2015). Enrichment of low-abundant protein targets by immunoprecipitation upstream of mass spectrometry. *Methods in Molecular Biology (Clifton, N.J.), 1295*, 135–151. https://doi.org/10.1007/978-1-4939-2550-6_12
47. Wang, J., Ma, P., Kim, D. H., Liu, B. F., & Demirci, U. (2021). Towards microfluidic-based exosome isolation and detection for tumor therapy. *Nano Today, 37*, Article 101066. https://doi.org/10.1016/J.NANTOD.2020.101066
48. Shtam, T., Evtushenko, V., Samsonov, R., Zabrodskaya, Y., Kamyshinsky, R., Zabegina, L., Verlov, N., Burdakov, V., Garaeva, L., Slyusarenko, M., Nikiforova, N., Konevega, A., & Malek, A. (2020). Evaluation of immune and chemical precipitation methods for plasma exosome isolation. *PLoS ONE, 15*(11). https://doi.org/10.1371/JOURNAL.PONE.0242732.
49. Choi, Y., Park, U., Koo, H. J., Park, J. seok, Lee, D. H., Kim, K., & Choi, J. (2021). Exosome-mediated diagnosis of pancreatic cancer using lectin-conjugated nanoparticles bound to selective glycans. *Biosensors and Bioelectronics, 177*. https://doi.org/10.1016/J.BIOS.2021.112980
50. Shtam, T. A., Burdakov, V. S., Landa, S. B., Naryzhny, S. N., Bairamukov, V. Y., Malek, A. V., Orlov, Y. N., & Filatov, M. V. (2017). Aggregation by lectins as an approach for exosome isolation from biological fluids: Validation for proteomic studies. *Cell and Tissue Biology, 11*(2), 172–179. https://doi.org/10.1134/S1990519X17020043
51. Zhang, K., Yue, Y., Wu, S., Liu, W., Shi, J., & Zhang, Z. (2019). Rapid capture and nondestructive release of extracellular vesicles using aptamer-based magnetic isolation. *ACS Sensors, 4*(5), 1245–1251. https://doi.org/10.1021/ACSSENSORS.9B00060

52. Webber, J., Stone, T. C., Katilius, E., Smith, B. C., Gordon, B., Mason, M. D., Tabi, Z., Brewis, I. A., & Clayton, A. (2014). Proteomics analysis of cancer exosomes using a novel modified aptamer-based array (SOMAscanTM) platform. *Molecular & Cellular Proteomics, 13*(4), 1050–1064. https://doi.org/10.1074/MCP.M113.032136
53. Zhang, N., Sun, N., & Deng, C. (2021). Rapid isolation and proteome analysis of urinary exosome based on double interactions of Fe3O4@TiO2-DNA aptamer. *Talanta, 221*, Article 121571. https://doi.org/10.1016/J.TALANTA.2020.121571
54. Ghosh, A., Davey, M., Chute, I. C., Griffiths, S. G., Lewis, S., Chacko, S., Barnett, D., Crapoulet, N., Fournier, S., Joy, A., Caissie, M. C., Ferguson, A. D., Daigle, M., Meli, M. V., Lewis, S. M., & Ouellette, R. J. (2014). Rapid isolation of extracellular vesicles from cell culture and biological fluids using a synthetic peptide with specific affinity for heat shock proteins. *PLoS ONE, 9*(10), Article e110443. https://doi.org/10.1371/JOURNAL.PONE.0110443
55. Melendez, K., Wallen, E. S., Edwards, B. S., Mobarak, C. D., Bear, D. G., & Moseley, P. L. (2006). Heat shock protein 70 and glycoprotein 96 are differentially expressed on the surface of malignant and nonmalignant breast cells. *Cell Stress & Chaperones, 11*(4), 334–342. https://doi.org/10.1379/CSC-187.1
56. Bijnsdorp, I. V., Maxouri, O., Kardar, A., Schelfhorst, T., Piersma, S. R., Pham, T. V., Vis, A., Van Moorselaar, R. J., & Jimenez, C. R. (2017). Feasibility of urinary extracellular vesicle proteome profiling using a robust and simple, clinically applicable isolation method. *Journal of Extracellular Vesicles, 6*(1). https://doi.org/10.1080/20013078.2017.1313091
57. Taylor, C., Chacko, S., Davey, M., Lacroix, J., Macpherson, A., Finn, N., Wajnberg, G., Ghosh, A., Crapoulet, N., Lewis, S. M., & Ouellette, R. J. (2020). Peptide-affinity precipitation of extracellular vesicles and cell-free dna improves sequencing performance for the detection of pathogenic mutations in lung cancer patient plasma. *International Journal of Molecular Sciences, 21*(23), 1–19. https://doi.org/10.3390/IJMS21239083
58. Askeland, A., Borup, A., Østergaard, O., Olsen, J. V., Lund, S. M., Christiansen, G., Kristensen, S. R., Heegaard, N. H. H., & Pedersen, S. (2020). Mass-spectrometry based proteome comparison of extracellular vesicle isolation methods: Comparison of ME-kit, size-exclusion chromatography, and high-speed centrifugation. *Biomedicines, 8*(8). https://doi.org/10.3390/BIOMEDICINES8080246
59. Griffiths, S. G., Cormier, M. T., Clayton, A., & Doucette, A. A. (2017). Differential proteome analysis of extracellular vesicles from breast cancer cell lines by chaperone affinity enrichment. *Proteomes, 5*(4). https://doi.org/10.3390/PROTEOMES5040025
60. Tai, W., Guzman, M. L., & Chiosis, G. (2016). The epichaperome: the power of many as the power of one. *Oncoscience, 3*(9–10), 266–267. https://doi.org/10.18632/ONCOSCIENCE.321.
61. Rodina, A., Wang, T., Yan, P., Gomes, E. D., Dunphy, M. P. S., Pillarsetty, N., Koren, J., Gerecitano, J. F., Taldone, T., Zong, H., Caldas-Lopes, E., Alpaugh, M., Corben, A., Riolo, M., Beattie, B., Pressl, C., Peter, R. I., Xu, C., Trondl, R., & Chiosis, G. (2016). The epichaperome is an integrated chaperome network that facilitates tumour survival. *Nature, 538*(7625), 397–401. https://doi.org/10.1038/NATURE19807
62. Griffiths, S. G., Ezrin, A., Jackson, E., Dewey, L., & Doucette, A. A. (2019). A robust strategy for proteomic identification of biomarkers of invasive phenotype complexed with extracellular heat shock proteins. *Cell Stress and Chaperones, 24*(6), 1197–1209. https://doi.org/10.1007/S12192-019-01041-8
63. Gallart-Palau, X., Serra, A., Wong, A. S. W., Sandin, S., Lai, M. K. P., Chen, C. P., Kon, O. L., & Sze, S. K. (2015). Extracellular vesicles are rapidly purified from human plasma by PRotein Organic Solvent PRecipitation (PROSPR). *Scientific Reports, 5*. https://doi.org/10.1038/SREP14664.
64. Gallart-Palau, X., Serra, A., & Sze, S. K. (2016). Enrichment of extracellular vesicles from tissues of the central nervous system by PROSPR. *Molecular Neurodegeneration, 11*(1), 1–13. https://doi.org/10.1186/S13024-016-0108-1/FIGURES/5
65. Lozano-Ramos, I. S., Bancu, I., Oliveira-Tercero, A., Armengol, M. P., Menezes-Neto, A., Portillo, H. A. Del, Lauzurica-Valdemoros, R., & Borràs, F. E. (2015). Size-exclusion

chromatography-based enrichment of extracellular vesicles from urine samples. *Journal of Extracellular Vesicles, 4*. https://doi.org/10.3402/jev.v4.27369
66. Bö Ing, A. N., Van Der Pol, E., Grootemaat, A. E., Coumans, F. A. W., Sturk, A., & Nieuwland, R. (2014). Single-step isolation of extracellular vesicles by size-exclusion chromatography. *Journal of Extracellular Vesicles, Sept, 8*(3). https://doi.org/10.3402/jev.v3.23430
67. Sidhom, K., Obi, P. O., & Saleem, A. (2020). A review of exosomal isolation methods: Is size exclusion chromatography the best option? *International Journal of Molecular Sciences, 21*(18), 6466. https://doi.org/10.3390/IJMS21186466
68. Heath, N., Grant, L., De Oliveira, T. M., Rowlinson, R., Osteikoetxea, X., Dekker, N., & Overman, R. (2018). Rapid isolation and enrichment of extracellular vesicle preparations using anion exchange chromatography. *Scientific Reports, 8*(1), 1–12. https://doi.org/10.1038/s41598-018-24163-y
69. Koch, L. F., Best, T., Wüstenhagen, E., Adrian, K., Rammo, O., & Saul, M. J. (2023). Novel insights into the isolation of extracellular vesicles by anion exchange chromatography. *Frontiers in Bioengineering and Biotechnology, 11*, 1298892. https://doi.org/10.3389/FBIOE.2023.1298892/BIBTEX
70. Fang, S. Bin, Zhang, H. Y., Meng, X. C., Wang, C., He, B. X., Peng, Y. Q., Xu, Z. Bin, Fan, X. L., Wu, Z. J., Wu, Z. C., Zheng, S. G., & Fu, Q. L. (2020). Small extracellular vesicles derived from human MSCs prevent allergic airway inflammation via immunomodulation on pulmonary macrophages. *Cell Death & Disease, 11*(6), 1–15. https://doi.org/10.1038/s41419-020-2606-x.
71. Johnson, J., Law, S. Q. K., Shojaee, M., Hall, A. S., Bhuiyan, S., Lim, M. B. L., Silva, A., Kong, K. J. W., Schoppet, M., Blyth, C., Ranasinghe, H. N., Sejic, N., Chuei, M. J., Tatford, O. C., Cifuentes-Rius, A., James, P. F., Tester, A., Dixon, I., & Lichtfuss, G. (2023). First-in-human clinical trial of allogeneic, platelet-derived extracellular vesicles as a potential therapeutic for delayed wound healing. *Journal of Extracellular Vesicles, 12*(7), 12332. https://doi.org/10.1002/JEV2.12332
72. Saari, H., Pusa, R., Marttila, H., Yliperttula, M., & Laitinen, S. (2023). Development of tandem cation exchange chromatography for high purity extracellular vesicle isolation: The effect of ligand steric availability. *Journal of Chromatography A, 1707*, Article 464293. https://doi.org/10.1016/J.CHROMA.2023.464293
73. Kawano, K., Kuzuma, Y., Yoshio, K., Hosokawa, K., Oosugi, Y., Fujiwara, T., Yokoyama, F., & Matsuzaki, K. (2024). Extracellular-vesicle catch-and-release isolation system using a net-charge invertible curvature-sensing peptide. *Analytical Chemistry, 96*(9), 3754–3762. https://doi.org/10.1021/ACS.ANALCHEM.3C03756
74. Kawano, K., Yokoyama, F., Kamasaka, K., Kawamoto, J., Ogawa, T., Kurihara, T., & Futaki, S. (2021). Design of the N-terminus substituted curvature-sensing peptides that exhibit highly sensitive detection ability of bacterial extracellular vesicles. *Chemical and Pharmaceutical Bulletin, 69*(11), 1075–1082. https://doi.org/10.1248/CPB.C21-00516
75. Kalluri, R., & LeBleu, V. S. (2020b). The biology, function, and biomedical applications of exosomes. *Science (New York, N.Y.), 367*(6478). https://doi.org/10.1126/SCIENCE.AAU6977
76. Silver Spring (MD): Food and Drug Administration (US); Bethesda (MD): National Institutes of Health (US). (2016). BEST (Biomarkers, EndpointS, and other Tools) Resource. In *BEST (Biomarkers , EndpointS , and other Tools Resource*. Food and Drug Administration (US). https://www.ncbi.nlm.nih.gov/books/NBK326791/.
77. Yoo, J., Kwang Lee, S., Lim, M., Sheen, D., Choi, E.-H., & Ae Kim, S. (n.d.). *Exosomal amyloid A and lymphatic vessel endothelial hyaluronic acid receptor-1 proteins are associated with disease activity in rheumatoid arthritis*. https://doi.org/10.1186/s13075-017-1334-9
78. Zhang, B., Li, H., Kong, L., Yang, N., Yang, S., Qi, L., Liu, T., Wang, X., & Qin, W. (2023). Tandem enrichment of serum exosomes and exosomal RNA with titanium dioxide. *Journal of Chromatography A, 1693*, Article 463882. https://doi.org/10.1016/J.CHROMA.2023.463882

A Column-Based CD9-HPLC Immunoaffinity Isolation Method of Extracellular Vesicles from Serum

Jie Zhang, Jianhui Zhu, Komal Abhange, and David M. Lubman

Abstract There are currently major challenges in EV isolation of small amounts, i.e. microliter-scale, of human serum while minimizing contamination of blood proteins and lipoprotein particles. One solution we have developed involves a column-based CD9 antibody-immobilized HPLC immunoaffinity (CD9-HPLC-IAC) technology for EV isolation from serum for downstream proteomic analysis. The CD9-HPLC-IAC method can achieve EV isolation from 40 μL of serum in 30 min with a yield of 8.0×10^9 EVs. Mass spec analysis yielded 482 proteins identified in EVs from patient serum using liquid chromatography tandem mass spectrometry (LC–MS/MS). Among the detected proteins were common exosomal markers such as CD63, CD81, CD82, Alix, and TSG101. In the EV protein content, the top 10 serum proteins were significantly decreased by using the CD9-HPLC-IAC method compared to ultracentrifugation (UC) and size exclusion chromatography (SEC). In addition, contamination from alipoproteins and albumin were minimized. This technology could in principle be extended to other important surface protein markers for detection of cancer or other diseases.

Keywords Extracellular vesicles · LC–MS/MS · Immunoaffinity · Proteomics

Abbreviations

EV	Extracellular Vesicles
IAC	Immunoaffinity Column
SEC	Size Exclusion Chromatography
UC	Ultracentrifugation
PBS	Phosphate Buffered Saline

J. Zhang · J. Zhu · K. Abhange · D. M. Lubman (✉)
Department of Surgery, The University of Michigan Medical Center, Ann Arbor, MI, USA
e-mail: dmlubman@umich.edu

R. K. Marcus and D. Lubman (eds.), *Exosomes and Extracellular Vesicles*,
https://doi.org/10.1007/978-3-031-89270-7_6

1 Introduction

Extracellular vesicles (EVs) are membrane-enclosed nanovesicles secreted by cells. There are different types of EVs where exosomes are generally between 30–150 nm in diameter. These EVs are released into the bloodstream [1–5] and carry a complex cargo of functional proteins, miRNA, and lipids which can mediate intercellular communication by transferring biological information between cells [6–8]. The lipid bilayer membrane of EVs protects the contents from degradation which may otherwise occur during exposure to biofluids. EVs are very stable and can be stored for extended periods of time. EVs can be found in various biofluids including serum/plasma, urine, and saliva. They may potentially serve as a source of biomarkers for detection of disease progression based on minimally invasive assays [9–15].

There are various challenges to isolating EVs from serum for further protein or metabolite analysis [15–17]. These involve eliminating a large amount of background of high abundance proteins, especially albumin and the alipoproteins that would otherwise eliminate mass spec analyses. Also, often there is a limited amount of serum available in clinical situations where it is essential to be able to isolate the EVs with high recovery to be able to effectively analyze the limited number of EVs. Conventional widely used EV separation techniques include ultracentrifugation (UC) [18–21], polymeric precipitation [21, 22], size exclusion chromatography (SEC) [23–25] and immunoaffinity [26] among others. However, major issues including the contaminating impurities co-enriched with EVs, such as large extracellular vesicles, lipid clusters, and blood proteins, remain to be solved. These methods may often be used together such as UC-SEC [18] or UC-UC [18, 19] to further eliminate blood protein contamination. In addition, the UC or SEC-based methods usually require a large amount of serum/plasma (i.e. 2–6 mL), which may not be available and are generally not very efficient.

The ability to characterize the protein content in EVs for the discovery of protein signatures and biomarker candidates for disease diagnosis and prognosis is essential. There are challenges in proteomic profiling of serum/plasma-derived EVs, which include (1) the low protein concentration in EVs, where the average mass of protein per EV is 0.5×10^{-6} ng, and (2) the contaminating impurities of blood proteins and lipoprotein particles which suppress mass spectrometry (MS) signals of low-abundance EV proteins [27–35]. We have explored multiple isolation techniques, UC, SEC, and a combination of UC&SEC and UC-UC to compare the proteomic profiles of serum-derived EVs but these are time-consuming and require the use of an ultracentrifuge and a significant amount of starting material.

In this chapter, we describe a column-based CD9 antibody-immobilized HPLC immunoaffinity (CD9-HPLC-IAC) technology coupled with a post-column cleaning procedure to isolate EVs [36]. This CD9-HPLC-IAC approach enables isolation of EVs from micro-scale amounts of serum, i.e. 40 μL, within 30 min. It also allows altering the gradient to wash the column for extended periods of time to further clean the EVs. Immunoaffinity isolation of EVs has also been performed with antibody-coated magnetic beads. This requires a significant amount of time and

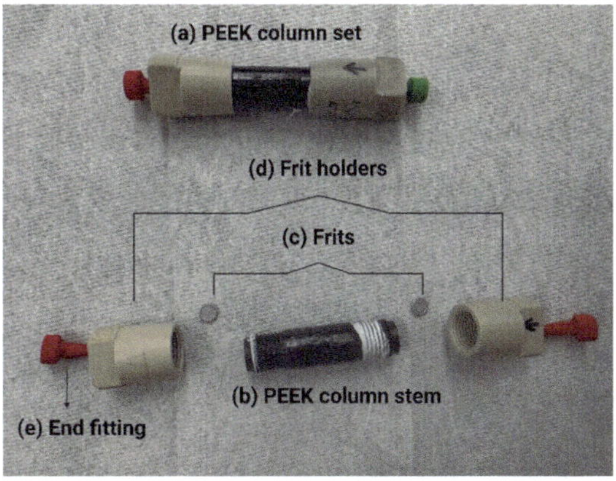

Fig. 6.1 Components of CD9 column **a** PEEK column, **b** PEEK column stem, **c** frits, **d** frit holders and **e** end fitting

cannot provide the extended on-line wash of the HPLC method. Proteomic profiling of EVs showed that this CD9-HPLC-IAC approach greatly reduced the contamination of blood proteins and apolipoproteins compared to the UC and SEC methods. The CD9-HPLC-IAC approach coupled with post-column cleaning is shown to be highly efficient for isolation of EVs which provides a means for analysis of the protein content of EVs from small volumes of serum. It may also provide a means for studying sub-populations of EVs for biomarker studies based on the antibody selected.

2 Methods

2.1 The CD9 Antibody Column

We developed a CD9-HPLC-IAC method to enable the rapid isolation of EVs from a microscale volume of serum for downstream proteomic analysis. The anti-CD9 HPLC column was developed using the UltraLink hydrazide resin conjugated with a CD9 antibody via hydrazide chemistry. The selection of CD9 antibody is based on the product information from the supplier Web site, from which the application in immunoprecipitation should be included. The protocol for making this custom column is given below and in Fig. 6.1a are shown a picture of the column and the parts used to make it.

2.2 Materials Needed for Column

- Zeba spin desalting columns

- Falcon tubes (15 mL)
- Centrifuge and mini centrifuge
- Anti-CD9 antibody
- Coupling buffer
- Sodium meta periodate
- Aniline
- Hydrazide beads
- Pierce centrifuge column
- Circular tube rotator
- 1 M NaCl solution
- 1X PBS solution
- Vaclok syringe with tube fitting
- PEEK column set

2.3 Procedure for Making Column

Desalting Column Preparation

1. Open the Zeba spin desalting column by twisting the bottom and place it in a 15 mL Falcon tube.
2. Centrifuge at 1000 g, 4 °C for 2 min. Discard the flow-through.
3. Add 1 mL coupling buffer and centrifuge at 1000 g, 4 °C for 2 min. Repeat this step for a total of 3 times.

Antibody Preparation

4. Add 400 μL coupling buffer to the CD9 antibody vial (100 μL antibody), vortex, and centrifuge briefly.
5. Transfer the antibody-buffer mixture to the equilibrated Zeba desalting column from step 3 with a new 15 mL Falcon collection tube. Centrifuge at 1000 g, 4 °C for 2 min to collect the flow-through, will be further used for antibody-bead conjugation.

Oxidization of Antibody

6. In a 1.5 mL Eppendorf tube, weigh 4 mg sodium meta periodate and dissolve it in the antibody in coupling buffer from step 5. Vortex and centrifuge briefly.
7. Cover with aluminum foil and incubate in the dark for 30 min.
8. Prepare a fresh Zeba desalting column as in steps 1–3.
9. After the incubation, load the oxidized antibody to the Zeba desalting column. Centrifuge at 1000 g, 4 °C for 2 min to collect the flow-through.
10. In a fume hood, prepare GlycoLink Coupling Catalyst by adding 18μL of aniline to 0.5 mL of GlycoLink Coupling Buffer. Vortex for 10 s and add total volume to the oxidized antibody from step 9, resulting in 0.1 M aniline final concentration.

    Note: Aniline is highly toxic. Wear gloves and handle with extreme care.

Hydrazide Bead Preparation

11. Add 2 mL hydrazide resin slurry to a Pierce centrifuge column, allow the resin to settle for 15 min. Drain the liquid from the column.
12. Wash the resin in the column three times with coupling buffer, by adding buffer to the top of the resin bed and allowing it to drain through the column.

Antibody-Hydrazide Conjugation

13. Add the oxidized antibody from step 10 to the resin (~1 mL of oxidized antibody per mL of resin). Mix well.
14. Incubate on a circular rotator for 4 h. The antibody is now conjugated to the hydrazide beads.

Wash the Conjugated Resin

15. After incubation, stand the column upright and allow the resin to settle for 15 min. Drain the liquid from the column.
16. Wash the column with 1 resin-bed volume of coupling buffer.
17. Wash the column sequentially with 5 resin-bed volumes of 1 M NaCl and PBS containing 0.05% sodium azide.

Column Packing: Assemble the antibody-conjugated hydrazide beads into a PEEK column (4.6 × 50 mm).

18. Place a frit in one end fitting and secure one end of the PEEK column. Connect this end with a Vaclok syringe for waste removal.
19. Set up the PEEK column upright. Keep the top of column open.
20. Slowly load the CD9-hydrazide beads from the top through the column, using the Vaclok syringe to collect the waste.
21. Once the CD9 beads are loaded, wash with 3 column volumes of PBS containing 0.05% sodium azide.
22. Remove any excess beads from the top. Disconnect the tube fitting and Vaclok from the bottom.
23. Secure the other end of the column with frit and end fitting.
24. Mark the inlet and outlet directions on the column.

Note: Ensure all procedures are performed at 4 °C where specified to maintain the stability and the activity of the antibodies. The prepared column should be kept at 4 °C after use for optimal performance and should be flushed with PBS for at least 5 min before the first use to ensure the removal of any unbound antibodies or reactants.

2.4 General Experimental Design

The overall workflow of the experiment is summarized in Fig. 6.2. The key component of this work is the use of the CD9-IA column. A small amount of serum, i.e. 40 μL, was loaded onto the column where the contamination fraction from serum

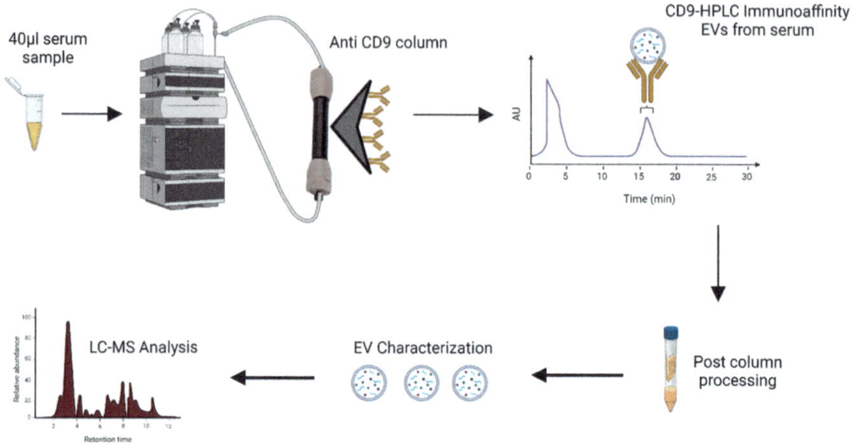

Fig. 6.2 Workflow of a CD9-HPLC-IAC for isolating EVs from 40 µL of human serum, with the EV fraction eluted between 16 and 19 min, followed by proteomic analysis

was discarded and the EV fraction was collected. A post-purification cleaning step using the 50 kDa MWCO filter was performed to desalt, concentrate, and reduce the co-eluting serum proteins of the EV fraction. After characterization with TEM, NTA analysis, and Western blotting, the EVs were processed for proteomic analysis by nano LC–MS/MS.

2.5 *CD9-HPLC-IAC Isolation of EVs from Serum*

EVs were extracted from serum using the anti-CD9 column on an HPLC system with a UV absorbance detector. The HPLC isolation of EVs was achieved within 30 min. Micro-scale volumes of serum were tested. Serum sample was diluted with PBS in a 1:5 dilution ratio and then filtered through a 0.22 µm filter to remove any particulates and large EVs prior to injecting into the HPLC system. The mobile phases were PBS (buffer A) and 1 M NaCl (buffer B), with a flow rate of 0.5 mL/min. UV absorbance was detected at 254 nm. After injecting the filtered serum sample, the column was flushed with buffer A (0–10 min), eluted with buffer B (10–19 min), and then re-equilibrated with buffer A (19–30 min). In the elution step, the binding interaction between antigen and antibody was broken with high salt concentration. The eluted fraction was collected and followed with a post-column cleaning procedure.

2.6 *Post-Column Cleaning Procedure*

A post-column cleaning procedure was performed according to the protocol established in our recent study [36]. The eluted EV fraction was transferred into a 4 mL YM-50 centrifugal device and centrifuged at 3,700 g for 10 min, followed by buffer exchange three times with PBS. The final sample volume was 100 μL. This post-column cleaning step functioned as buffer exchange, and concentration with further reduction of the contaminating serum proteins prior to LC–MS/MS.

2.7 *Protocol: Isolation of Serum EVs Using Immunoaffinity-Based CD9 HPLC Column*

Sample Preparation

1. Dilute serum samples 1:5 with 1X PBS.
2. Filter the diluted sample using 0.22 μm filter to purify them of impurities prior to HPLC analysis.

HPLC Run

3. Use 1X PBS (Buffer A) and 1 M NaCl (Buffer B) as mobile phases with a flow rate of 0.5 ml/min.
4. Set UV absorbance at 254 nm for EV detection.
5. Inject the filtered serum sample into the HPLC system.
6. The run parameters are set such that the column will be flushed with buffer A for 0–10 min, eluted with buffer B between 10–19 min and re-equilibrated with buffer A between 19–30 min.
7. Collect the elution fraction between 10–19 min as this contains the EVs.
8. The eluted EV fraction is further subjected to post column cleaning process.

Post Column Cleaning

9. Transfer eluted EV fraction to a 4 ml Amicon® Ultra Centrifugal Filter, 50 kDa MWCO
10. Centrifuge at 3700 g for 10 min. Repeat this step a total of 4 times. At the end of each step ensure 1X PBS exchange.
11. At the end of 4th spin aspirate about 100 μl of final sample containing EVs from the filter for further analysis.

2.8 *Characterization: Transmission Electron Microscopy (TEM), NanoSight Analysis and Western Blot*

The characterization of the collected EVs was performed using TEM, Nanosight, and Western Blot using procedures already described in our prior work [9]. TEM images were obtained on a JEOL 1400-plus transmission electron microscope. The concentration and size distribution of EVs were measured using a NanoSight NS300 (Malvern, UK).

2.9 *EV Protein Extraction and Tryptic Digestion*

The procedures for EV protein extraction and tryptic digestion are as given in previous work [9].

2.9.1 Nano-LC–MS/MS

EV peptides were dissolved in 0.1% formic acid (FA) and half of the sample was injected onto an Orbitrap Fusion Lumos Tribrid Mass Spectrometer (Thermo) coupled with a Dionex UltiMate 3000 UPLC system. Peptides were separated with 0.1% FA in water (A) and 80% CH_3CN containing 0.1% FA (B) on a 75 μm × 50 cm column (Thermo, P/N 164,942) under a 90 min linear gradient from 2 to 40% B at a flow rate 300 nL/min. The MS instrument was operated in positive ion mode and the data were acquired in a data-dependent mode. MS1 spectra (m/z 375−1700) were acquired in the Orbitrap MS (120 k resolution, 2e5 AGC target). Up to 20 of the most intense MS1 peaks were selected for tandem MS2 analysis by CID in the linear ion trap. The normalized collision energy was set at 34% for MS/MS. The isolation width was set at 1.6. Dynamic exclusion was enabled with an exclusion window of ±10 ppm and exclusion duration of 12 s.

All MS/MS spectra were searched against the human UniProt database using SEQUEST (Proteome Discoverer 1.4, Thermo Scientific). The search parameters were as follows: static modification, carbamidomethyl (C); dynamic modifications, oxidation (M); 2 missed cleavages allowed; MS1 mass tolerance 10 ppm; MS2 mass tolerance 0.6 Da; 1% false discovery rate (FDR). Proteins were quantified using a label-free quantitative method termed normalized spectral counting [37], where the spectral count of a given protein was normalized against the sum of the spectral counts of all proteins identified in an MS run.

3 Results and Discussion

3.1 *CD9-IAC HPLC Fraction of EVs*

Figure 6.2 shows a representative chromatogram of EV isolation from 40 μL of human serum using the CD9-HPLC-IAC method. The isolation of EVs was achieved within 30 min where the peak of the EV fraction was observed between 16 and 19 min. We found that the usable injection volume of serum in this column was between 20–60 uL before the column was saturated. The size of the PEEK column used in this study was 4.6 × 50 mm, where the column volume was about 0.8 mL. One could scale up the amount of antibody and the size of the PEEK column to accommodate larger sample volumes.

The CD9-IAC column allowed EV isolation from serum volumes as low as 20 μL, which is significantly lower than that used in the conventional methods. The column was run at the optimal flow rate, 0.5 mL/min. Since the column was packed with antibody-conjugated resin (average particle size 50–80 microns), it is suitable for low-pressure procedures. As the flow rate is increased, the pressure will increase, which will result in lowering the binding capacity. Since 40 μL of serum gave a significant yield in EVs, we chose 40 μL of serum as the injection amount for downstream characterization of EVs. A wash was applied between sample injections, which confirmed that there was no carryover between runs.

We also investigated the robustness of the CD9-IAC column by conducting the isolation experiment with 50 aliquots of the serum sample (40 μL each). The coefficient of variation (CV) of the peak areas of EVs among the 50 replicates was 9.7%, with a minor decrease of 8.4% in the yield for the 50th injection compared to the first injection. The result indicated that the CD9-IAC column can be used at least 50 times without significant degradation of performance.

3.2 *EV Characterization*

The purified EVs were characterized by transmission electron microscopy (TEM), nanoparticle tracking analysis (NTA), and Western blotting (WB). As shown in Fig. 6.3a, the TEM image shows that the isolated EVs retained morphological integrity as spherical or spheroidal, membrane-encapsulated particles of ~35–150 nm in diameter. The result indicated that the CD9-HPLC-IAC procedure did not affect the membrane structure of EVs. The size histogram of EVs showed that the mean diameters of EVs were around 70 nm.

The particle concentration was determined by NTA. Figure 6.3b shows the NTA result of 5 videos of the isolated EVs where a major peak was observed at 96 nm and two less-concentrated peaks at 71 and 140 nm, respectively. The NTA result showed that about 95% of the isolated particles were within the expected size range of exosomes (30–150 nm). Minor signals at ~200 nm accounted for <5% of the particles.

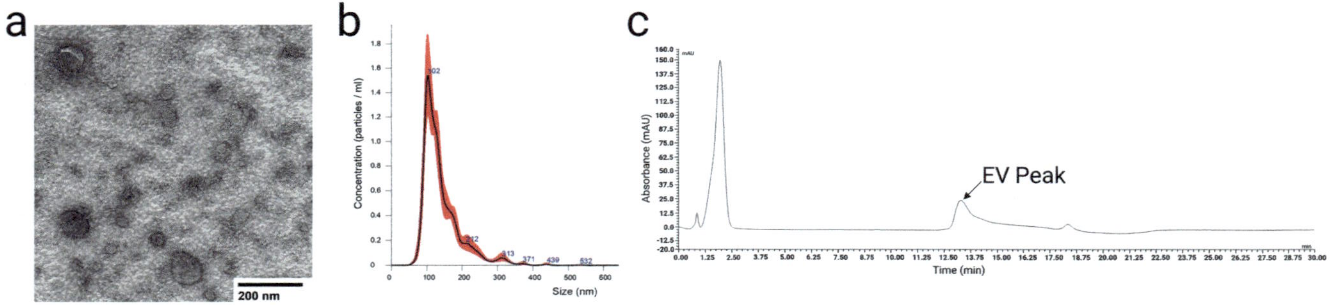

Fig. 6.3 Characterization of the EVs isolated from human serum using the anti-CD9 HPLC column **a** TEM image shows the morphology and size of the EVs; scale bar = 100 nm **b** NTA analysis of particle size distribution and concentration of the EVs **c** Chromatogram showing elution of EV peak following isolation by the anti-CD9 column

Though the number of larger EVs was low, they could contribute significantly to the protein signal. It has been reported that NTA analysis produced substantially larger sizes and broader size distributions as compared to TEM imaging, which might result from the physical principle of size determination of the NTA analysis where the size is calculated based on the motion of particles in solution and influenced by the surface protein composition of EVs [38].

The number of EVs isolated from 40 μL of serum was ~8.0×10^9, which was equivalent to ~2.0×10^{11} per mL serum. Compared to the conventional UC and SEC methods which can recover $1.0–1.5 \times 10^9$ EVs from 1 mL of serum [18], the current CD9-HPLC-IAC method achieved a 100-fold higher yield. In addition, our method was comparable to the PET C-CP fiber HIC method developed by the Marcus group [30], which can isolate ~1×10^{10} EVs from 0.1 mL of serum.

To confirm the vesicles were indeed EVs, a common EV marker, CD9, was analyzed using Western blotting (not shown). CD9 was observed in the WB result of EVs isolated from human serum by the CD9-HPLC-IAC method. Other EV markers such as CD63 and CD81 were detected in the subsequent mass spectrometry analysis. Figure 6.3c shows the chromatogram of the EV peak analyzed after elution from the anti-CD9 column.

3.3 *Proteome Profiling of EVs Isolated from Serum*

The EVs were lysed, and the proteins digested where the resulting peptides were injected for LC–MS/MS analysis. The experiment was performed at three different times with 3 aliquots of a human serum sample. In total, 482 proteins were identified among the three replicates. A Gene Ontology (GO) analysis of cellular components for molecular function, and biological processes was performed. For the cellular component, 82% of proteins were associated with exosomes, followed by cytoplasm (72%) and plasma membrane (38%), and a total of 23% of proteins were from the extracellular space. This suggests that these proteins originated from exosomes. The prevalent molecular function of the identified proteins was assigned as transporter activity, cytoskeletal protein binding, cell adhesion molecule activity, and receptor activity. GO analysis for biological processes discovered that the EV proteins were mainly involved in signal transduction (24%), cell communication (23%), protein metabolism (18%), cell growth and/or maintenance (17%), and transport (11%).

Various EV markers were identified in this study. The identified EV markers included endosome-specific tetraspanins (CD63, CD81, CD82, and CD151), the components of the endosomal sorting complex required for transport (ESCRT) complex (Alix, TSG101, HSP70, and VPS), and endosome-related membrane transport and fusion proteins (Annexins, flotillin, and Rab). Besides CD63/CD81/CD82, other tetraspanins including TSPAN4, TSPAN8, and TSPAN14 were also detected. We detected CD9 in the Western Blot analysis, which confirmed the presence of CD9 in EV samples. CD9 was not detected by MS probably due to the tryptic digestion step using the FASP method with a YM-30 kD centrifugal filter where CD9 has a MW less than 25 kD, which might be filtered out during the FASP procedure.

3.4 *Relative Quantification of Serum Contaminants Among EV Protein Content*

We then performed relative quantification of the proteomic data to assess the relative abundance of proteins in EV content and the efficiency of the CD9-HPLC-IAC method in reducing contaminating proteins in EV samples. We compared the results to that of other EV isolation methods studied in prior work where the results are shown in Fig. 6.4. The major challenge of the proteomic study of serum-derived EVs is the co-elution of the high-abundance serum proteins and lipoprotein particles [29, 35, 39, 40] especially for EVs purified from a small amount of serum [28]. The two major impurities can interfere with analysis of EV protein content by mass spectrometry [39]. The top 10 most abundant serum proteins include albumin, IgG, transferrin, fibrinogen, IgA, α-2-macroglobulin, α-1-antitrypsin, complement C3, IgM, and haptoglobin, accounting for 90% of the total serum protein mass [41], while the main components of lipoprotein particles include apolipoprotein A/B/C/E and lipoprotein(a) [35]. The relative abundance of proteins in these two groups was summed, respectively, to represent the percentage content of the top 10 serum proteins and apolipoproteins identified in EVs.

The percentage content of the top 10 serum proteins in the three EV samples was 15.0%, 11.9%, and 9.6%, respectively, and that of apolipoproteins was 8.6%, 8.8%, and 6.3%, respectively. The mean and SD value of the top 10 serum proteins and apolipoproteins were $12.2 \pm 2.7\%$ and $7.9 \pm 1.4\%$, respectively. We compared the MS result with that reported in our recent study (see Fig. 6.4) where EVs were isolated from 1 mL of human serum using an optimized UC or SEC (qEV) methods [18]. The content of the top 10 serum proteins was significantly decreased in the CD9-HPLC-IAC method as compared to the optimized UC and SEC (qEV) methods. The content of apolipoproteins was also significantly reduced in the CD9-HPLC-IAC

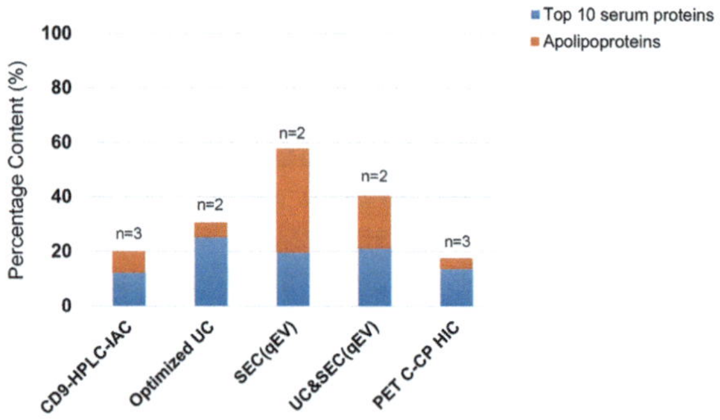

Fig. 6.4 A comparison of the relative quantitation of the contamination detected by mass spec for the top 10 serum proteins (blue) and alipoproteins (brown) for each EV isolation method compared to the total protein content

method compared to the SEC (qEV) method while comparable to the UC method. We also compared the average percentage content of the top 10 serum proteins in the EVs between the CD9-IAC method and the PET C-CP fiber HIC method [30], where we found that the IAC method is comparable to the HIC method in reducing the high-abundance serum proteins in EV isolation.

3.5 A Comparison Between the CD9-HPLC-IAC Method Versus Other EV Isolation Methods Including UC/SEC/ Magnetic Bead-Based IAC/HIC

We compared the CD9-HPLC-IAC method versus the UC and SEC methods in the isolation of EVs from human serum, including serum sample volume, processing time, morphology, size, concentration, and reducing the contamination of serum proteins and lipoprotein particles compared to our recent work. This comparison is summarized in Table 1.

In the case of morphology, the EVs isolated using the CD9-HPLC-IAC, UC or SEC methods all had intact membrane structures, indicating these approaches did not affect the morphology of EVs. Regarding the size of EVs, the TEM result showed that the mean diameters of EVs from the CD9-HPLC-IAC, UC, and SEC methods were 70 nm, 67 nm, and 77 nm [18], respectively. When comparing the yield of serum-derived EVs among the three methods, the CD9-HPLC-IAC method achieved 8.0×10^9 EVs from 40 μL of serum, or equivalent to 2.0×10^{11} per mL serum, while the UC and SEC methods can recover 1.0×10^9 and 1.3×10^9 EVs from 1 mL of serum [18]. The current CD9-HPLC-IAC method achieved a 100-fold higher yield than the UC and SEC methods. It has been reported that on average, the number of EVs from 1 mL of serum can vary between $1–3 \times 10^{12}$ particles/mL [42]. Taking into consideration the very small volume of the starting serum, i.e. 40 μL, the CD9-HPLC-IAC method achieved a high yield of EVs.

Table 1 Comparison between EV isolation methods

Parameters	CD9-HPLC-IAC	5 cycles of UC	SEC (qEV)	PET-CP-HIC
Serum volume (μl)	40	1000	500	20
Processing time (min)	30	400	60	15
For EVs				
Morphology	Intact	Intact	Intact	Intact
Mean size	70 nm	67 nm	77 nm	79 nm
Concentration	2×10^{11} per ml serum	1×10^9 per ml serum	1.3×10^9 per ml serum	1×10^{11} per ml serum

It should be noted that, compared to the UC and SEC methods, the advantage of the CD9-HPLC-IAC method for isolation of EVs from serum is that it scales down the starting volume of serum with as little as 40 μL. It enabled proteome profiling of EVs derived from a micro-scale amount of serum, which is most important when patient samples are limited in volume size. In addition, the whole procedure of the CD9-HPLC-IAC method including the post-column cleaning took 60 min, while the UC method is time-consuming which usually takes 400 min. Although the SEC method also provides as rapid EV isolation as the CD9-HPLC-IAC method, a SEC column can only be reused for up to 5 times while the antibody-based HPLC column can be reused for over 50 times. In addition, size- or density-based separation cannot distinguish subpopulations of EVs; however, antibody-based immuno-isolation, such as CD9-HPLC-IAC, can capture EV subtypes based on the specific surface proteins of EVs.

Compared to the magnetic bead-based IAC method, the column-based IAC method has the advantage of real-time monitoring of the EV fraction and minimizing contamination of blood proteins by optimizing the column wash step. The use of magnetic beads is a useful strategy for isolation of EVs from cell culture supernatant or liquid biopsy for the downstream analysis of EV surface markers using flow cytometry or imaging. However, lipoproteins in liquid biopsy often bind to magnetic beads, being one of the major contaminants in the isolated EVs for downstream proteomic analysis [34, 35]. The CD9-HPLC-IAC method enables elution of intact EVs and fast post-column cleaning, where the EV sample is obtained in PBS and ready for a full characterization including proteomic profiling. Another advantage is that the CD9-IAC-HPLC column is reusable multiple times, i.e. >50 times.

The CD9-HPLC-IAC technology developed in this study is a different technique compared to the PET C-CP HIC method reported in our previous work [32–34]. The CD9-HPLC-IAC method is based on immunoaffinity using a CD9 antibody while the PET C-CP is based on hydrophobic interaction chromatography. An important advantage of both the CD9-HPLC-IAC and PET C-CP HIC methods is that they can provide efficient isolation of EVs from small amounts of serum sample, down to 20 μL if needed. Both methods provide the equivalent yield of nearly 10^{11} EVs per mL serum. In addition, both methods can be performed rapidly on an HPLC platform with real-time monitoring, and the columns are reusable and reproducible. Most importantly, these methods provide improved performance in eliminating high abundance blood proteins and apolipoproteins, which interfere with mass spec analysis, where extensive washing on column can be achieved by these chromatographic methods. The two techniques complement each other in various studies and may be used together to purify EV populations.

3.6 Conclusion and Future Prospects

We developed a CD9-HPLC-IAC method which enables rapid and effective isolation of EVs from a micro-scale amount of serum as little as 40 μL. This approach yielded

8.0×10^9 EVs from 40 μL of serum in 30 min. Proteomic analysis of the EVs after a post-column cleaning step using the 50 kDa MWCO filter demonstrated that this method greatly reduced the contamination of serum proteins and apolipoproteins, the major impurities in serum-derived EVs. This technique enables real-time isolation of EVs from a micro-scale amount of serum for downstream proteomic analysis, which is most important when patient samples are limited in volume size. We only used 40 μL for this study to demonstrate the method, indicating that it is suitable for the development of real-time EV assay, especially for trace amount sample analysis. The column could be scaled up to accommodate larger sample volumes. The high efficiency of the CD9-HPLC-IAC method in isolation and characterization of EVs from a micro-scale volume of serum for proteomic profiling may provide a potential application for clinical analysis. In addition, the IA technology can be extended by using other EV markers and cancer markers for EV isolation [43, 44] including for metabolomics [45] and RNAs studies [42, 46–48], which is especially useful where one might be interested in various subpopulations of EVs in correlation with diseases [48–51].

Acknowledgements This work was supported by the National Cancer Institute under grants R21CA189775 (DML) and R50 CA221808 (JZ), and partially supported under grants R01 CA160254 (DML) and 1R01 CA258240 (D.M.L.). DML acknowledges support under the Maud T. Lane Professorship.

References

1. Caby, M. P., Lankar, D., Vincendeau-Scherrer, C., Raposo, G., & Bonnerot, C. (2005). Exosomal-like vesicles are present in human blood plasma. *International Immunology, 17*(7), 879–887.
2. Abhange, K., Makler, A., Wen, Y., et al. (2021). Small extracellular vesicles in cancer. *Bioactive Materials, 6*, 3705–3743.
3. Pisitkun, T., Shen, R. F., & Knepper, M. A. (2004). Identification and proteomic profiling of exosomes in human urine. *Proceedings of the National Academy of Sciences of the United States of America, 101*(36), 13368–13373.
4. Whiteside, T. L. (2016). Tumor-derived exosomes and their role in cancer progression. *Advances in Clinical Chemistry, 74*, 103–141.
5. Gurunathan, S., Kang, M. H., Jeyaraj, M., Qasim, M., Kim, J. H. (2019). Review of the isolation, characterization, biological function, and multifarious therapeutic approaches of exosomes. *Cells, 8*(4).
6. Willms, E., Cabanas, C., Mager, I., Wood, M. J. A., & Vader, P. (2018). Extracellular vesicle heterogeneity: subpopulations, isolation techniques, and diverse functions in cancer progression. *Frontiers in Immunology, 9*, 738.
7. Kowal, J., Arras, G., Colombo, M., Jouve, M., Morath, J. P., Primdal-Bengtson, B., Dingli, F., Loew, D., Tkach, M., & Thery, C. (2016). Proteomic comparison defines novel markers to characterize heterogeneous populations of extracellular vesicle subtypes. *Proceedings of the National Academy of Sciences of the United States of America, 113*(8), E968–E977.
8. Thery, C., Zitvogel, L., & Amigorena, S. (2002). Exosomes: Composition, biogenesis and function. *Nature Reviews Immunology, 2*(8), 569–579.

9. An, M., Lohse, I., Tan, Z., Zhu, J., Wu, J., Kurapati, H., Morgan, M. A., Lawrence, T. S., Cuneo, K. C., & Lubman, D. M. (2017). Quantitative proteomic analysis of serum exosomes from patients with locally advanced pancreatic cancer undergoing chemoradiotherapy. *Journal of Proteome Research, 16*(4), 1763–1772.
10. Vasconcelos, M. H., Caires, H. R., Abols, A., Xavier, C. P. R., & Line, A. (2019). Extracellular vesicles as a novel source of biomarkers in liquid biopsies for monitoring cancer progression and drug resistance. *Drug Resistance Updates, 47*, Article 100647.
11. Zhao, L., Shi, J., Chang, L., Wang, Y., Liu, S., Li, Y., Zhang, T., Zuo, T., Fu, B., Wang, G., Ruan, Y., Zhang, Y., & Xu, P. (2021). Serum-derived exosomal proteins as potential candidate biomarkers for hepatocellular carcinoma. *ACS Omega, 6*(1), 827–835.
12. Drake, R. R., & Kislinger, T. (2014). The proteomics of prostate cancer exosomes. *Expert Review of Proteomics, 11*(2), 167–177.
13. Lihon, M. V., Hadisurya, M., Wu, X., Iliuk, A., & Tao, W. A. (2023). Isolation and identification of plasma extracellular vesicles protein biomarkers. *Methods in Molecular Biology, 2660*, 207–217.
14. Whiteside, T. L. (2015). The potential of tumor-derived exosomes for noninvasive cancer monitoring. *Expert Review of Molecular Diagnostics, 15*(10), 1293–1310.
15. Jalaludin, I., Lubman, D. M., & Kim, J. (2023). A guide to mass spectrometric analysis of extracellular vesicle proteins for biomarker discovery. *Mass Spectrometry Reviews, 42*, 844–872.
16. Burton, J. B., Carruthers, N. J., & Stemmer, P. M. (2023). Enriching extracellular vesicles for mass spectrometry. *Mass Spectrometry Reviews, 42*, 779–795.
17. Yan, H., Li, Y., Cheng, S., Zeng, Y. (2021). Advances in analytical technologies for extracellular vesicles. *Analytical Chemistry, 93*, 4739−4774.
18. An, M., Wu, J., Zhu, J., & Lubman, D. M. (2018). Comparison of an optimized ultracentrifugation method versus size-exclusion chromatography for isolation of exosomes from human serum. *Journal of Proteome Research, 17*(10), 3599–3605.
19. Kim, J., Tan, Z., & Lubman, D. M. (2015). Exosome enrichment of human serum using multiple cycles of centrifugation. *Electrophoresis, 36*(17), 2017–2026.
20. Bobrie, A., Colombo, M., Krumeich, S., Raposo, G., & Thery, C. (2012). Diverse subpopulations of vesicles secreted by different intracellular mechanisms are present in exosome preparations obtained by differential ultracentrifugation. *Journal of Extracellular Vesicles, 1*, 18397.
21. Cao, F., Gao, Y., Chu, Q., Wu, Q., Zhao, L., Lan, T., & Zhao, L. (2019). Proteomics comparison of exosomes from serum and plasma between ultracentrifugation and polymer-based precipitation kit methods. *Electrophoresis, 40*(23–24), 3092–3098.
22. Rider, M. A., Hurwitz, S. N., & Meckes, D. G. (2016). ExtraPEG: A polyethylene glycol-based method for enrichment of extracellular vesicles. *Science and Reports, 6*, 23978.
23. de Menezes-Neto, A., Saez, M. J., Lozano-Ramos, I., Segui-Barber, J., Martin-Jaular, L., Ullate, J. M., Fernandez-Becerra, C., Borras, F. E., & Del Portillo, H. A. (2015). Size-exclusion chromatography as a stand-alone methodology identifies novel markers in mass spectrometry analyses of plasma-derived vesicles from healthy individuals. *Journal of Extracellular Vesicles, 4*, 27378.
24. Boing, A. N., van der Pol, E., Grootemaat, A. E., Coumans, F. A., Sturk, A., & Nieuwland, R. (2014). Single-step isolation of extracellular vesicles by size-exclusion chromatography. *Journal of Extracellular Vesicles, 3*, 23430.
25. Park, S., Jalaludin, I., Hwang, H., Ko, M., Adelipour, M., Cho, N., Kim, K. K., Lubman, D. M., Kim, J. (2023). Size-exclusion chromatography for the characterization of urinary exosomes. *Journal of Chromatography B: Analytical Technologies in the Biomedical and Life Sciences, 17*, 1228:123828.
26. Griffiths, S. G., Cormier, M. T., Clayton, A., & Doucette, A. A. (2017). Differential proteome analysis of extracellular vesicles from breast cancer cell lines by chaperone affinity enrichment. *Proteomes, 5*(4), 25.

27. Hong, C. S., Funk, S., Muller, L., Boyiadzis, M., & Whiteside, T. L. (2016). Isolation of biologically active and morphologically intact exosomes from plasma of patients with cancer. *Journal of Extracellular Vesicles, 5*, 29289.
28. Smolarz, M., Pietrowska, M., Matysiak, N., Mielanczyk, L., & Widlak, P. (2019). Proteome profiling of exosomes purified from a small amount of human serum: The problem of co-purified serum components. *Proteomes, 7*(2), 18.
29. Brennan, K., Martin, K., FitzGerald, S. P., O'Sullivan, J., Wu, Y., Blanco, A., Richardson, C., & Mc Gee, M. M. (2020). A comparison of methods for the isolation and separation of extracellular vesicles from protein and lipid particles in human serum. *Science and Reports, 10*(1), 1039.
30. Ji, X., Huang, S., Zhang, J., Bruce, T. F., Tan, Z., Wang, D., Zhu, J., Marcus, R. K., & Lubman, D. M. (2021). A novel method of high-purity extracellular vesicle enrichment from microliter-scale human serum for proteomic analysis. *Electrophoresis, 42*(3), 245–256.
31. Gao, F., Jiao, F., Xia, C., Zhao, Y., Ying, W., Xie, Y., Guan, X., Tao, M., Zhang, Y., Qin, W., & Qian, X. (2019). A novel strategy for facile serum exosome isolation based on specific interactions between phospholipid bilayers and TiO2. *Chemical Science, 10*(6), 1579–1588.
32. Wang, L., Bruce, T. F., Huang, S., & Marcus, R. K. (2019). Isolation and quantitation of exosomes isolated from human plasma via hydrophobic interaction chromatography using a polyester, capillary-channeled polymer fiber phase. *Analytica Chimica Acta, 1082*, 186–193.
33. Bruce, T. F., Slonecki, T. J., Wang, L., Huang, S., Powell, R. R., & Marcus, R. K. (2019). Exosome isolation and purification via hydrophobic interaction chromatography using a polyester, capillary-channeled polymer fiber phase. *Electrophoresis, 40*(4), 571–581.
34. Ji, X., Huang, S., Zhang, J., et al. (2021). A novel method of high-purity extracellular vesicle enrichment from microliter-scale human serum for proteomic analysis. *Electrophoresis, 42*, 245–256.
35. Sodar, B. W., Kittel, A., Paloczi, K., Vukman, K. V., Osteikoetxea, X., Szabo-Taylor, K., Nemeth, A., Sperlagh, B., Baranyai, T., Giricz, Z., Wiener, Z., Turiak, L., Drahos, L., Pallinger, E., Vekey, K., Ferdinandy, P., Falus, A., & Buzas, E. I. (2016). Low-density lipoprotein mimics blood plasma-derived exosomes and microvesicles during isolation and detection. *Science and Reports, 6*, 24316.
36. Zhu, J., Zhang, J., Ji, X., Tan, Z., & Lubman, D. M. (2021). Column-based technology for CD9-HPLC immunoaffinity isolation of serum extracellular vesicles. *Journal of Proteome Research, 20*, 4901–4911.
37. Liu, H., Sadygov, R. G., & Yates, J. R. (2004). A model for random sampling and estimation of relative protein abundance in shotgun proteomics. *Analytical Chemistry, 76*(14), 4193–4201.
38. Bachurski, D., Schuldner, M., Nguyen, P. H., Malz, A., Reiners, K. S., Grenzi, P. C., Babatz, F., Schauss, A. C., Hansen, H. P., Hallek, M., Pogge von Strandmann, E. (2019). Extracellular vesicle measurements with nanoparticle tracking analysis - An accuracy and repeatability comparison between NanoSight NS300 and ZetaView. *Journal of Extracellular Vesicles, 8* (1), 1596016.
39. Karimi, N., Cvjetkovic, A., Jang, S. C., Crescitelli, R., Hosseinpour Feizi, M. A., Nieuwland, R., Lotvall, J., & Lasser, C. (2018). Detailed analysis of the plasma extracellular vesicle proteome after separation from lipoproteins. *Cellular and Molecular Life Sciences, 75*(15), 2873–2886.
40. Yuana, Y., Levels, J., Grootemaat, A., Sturk, A., & Nieuwland, R. (2014). Co-isolation of extracellular vesicles and high-density lipoproteins using density gradient ultracentrifugation. *Journal of Extracellular Vesicles, 3*, 23262.
41. Anderson, N. L., & Anderson, N. G. (2002). The human plasma proteome: History, character, and diagnostic prospects. *Molecular and Cellular Proteomics, 1*(11), 845–867.
42. Li, M., Zeringer, E., Barta, T., Schageman, J., Cheng, A., & Vlassov, A. V. (2014). Analysis of the RNA content of the exosomes derived from blood serum and urine and its potential as biomarkers. *Philosophical Transactions of the Royal Society of London. Series B, Biological Sciences, 369*(1652), 20130502.
43. Castillo, J., Bernard, V., San Lucas, F. A., Allenson, K., Capello, M., Kim, D. U., Gascoyne, P., Mulu, F. C., Stephens, B. M., Huang, J., Wang, H., Momin, A. A., Jacamo, R. O., Katz,

M., Wolff, R., Javle, M., Varadhachary, G., Wistuba, I. I., Hanash, S., … Alvarez, H. (2018). Surfaceome profiling enables isolation of cancer-specific exosomal cargo in liquid biopsies from pancreatic cancer patients. *Annals of Oncology, 29*(1), 223–229.

44. Abhange, K., Kitata, R. B., Zhang, J., Wang, Y. T., Gaffrey, M. J., Liu, T., Gunchick, V., Khaykin, V., Sahai, V., Cuneo, K. C., Parikh, N. D., Shi, T., & Lubman, D. M. (2024). In-depth proteome profiling of small extracellular vesicles isolated from cancer cell lines and patient serum. *Journal of Proteome Research, 23*(1), 386–396.
45. Luo, X., An, M., Cuneo, K. C., Lubman, D. M., & Li, L. (2018). High-performance chemical isotope labeling liquid chromatography mass spectrometry for exosome metabolomics. *Analytical Chemistry, 90*(14), 8314–8319.
46. RNA content of the exosomes derived from blood serum and urine and its potential as biomarkers. *Philosophical Transactions of the Royal Society B: Biological Sciences, 369*(1652), 20130502.
47. Kumar, S. R.; Kimchi, E. T.; Manjunath, Y.; Gajagowni, S.; Stuckel, A. J. (2020). RNA cargos in extracellular vesicles derived from blood serum in pancreas associated conditions. *Scientific Reports, 10*(1), 2800. 27.
48. Zhang, Y., Liu, Y., Liu, H., & Tang, W. H. (2019). Exosomes: Biogenesis, biologic function and clinical potential. *Cell & Bioscience, 9*, 19.
49. Ciferri, M. C., Quarto, R., Tasso, R. (2021). Extracellular vesicles as biomarkers and therapeutic tools: from pre-clinical to clinical applications. *Biology, 10.*
50. Brett, S. I., Lucien, F., Guo, C., Williams, K. C., Kim, Y., Durfee, P. N., Brinker, C. J., Chin, J. I., Yang, J., & Leong, H. S. (2017). Immunoaffinity based methods are superior to kits for purification of prostate derived extracellular vesicles from plasma samples. *Prostate, 77*(13), 1335–1343.
51. Sharma, P., Ludwig, S., Muller, L., Hong, C. S., Kirkwood, J. M., Ferrone, S., & Whiteside, T. L. (2018). Immunoaffinity-based isolation of melanoma cell-derived exosomes from plasma of patients with melanoma. *Journal of Extracellular Vesicles, 7*(1), 1435138.

Isolation of Extracellular Vesicles and the Analysis of Their RNA Cargo

Favour C. Nwachukwu, Mattalyn E. Gordon, Malgorzata A. Witek, and Steven A. Soper

Abstract The study of extracellular vesicles (EVs) has evolved at a rapid rate over the past few years. The knowledge concerning EV origin, their role(s) in cell biology, and their molecular cargo have contributed to a deeper understanding of EVs' biology, which is facilitating their use as important clinical-based biomarkers. Much attention has been given to the importance of nucleic acids, especially a plethora of different ribonucleic acids (RNA), namely total RNA (TRNA), packaged into EVs. The immense EV research progress has been realized through efforts in developing novel technologies for EV isolation and the subsequent analysis of their low abundant cargo, including TRNA. In this chapter, we will discuss isolation technologies of EVs and the analysis of their RNA cargo, and highlight the specific challenges associated with conducting these studies. In addition, we will provide information on the differences in RNA content within EVs to that found in biological cells. Finally, we will present examples using EV-TRNA in managing a number of different disease states.

F. C. Nwachukwu · M. E. Gordon · M. A. Witek · S. A. Soper (✉)
Department of Chemistry, The University of Kansas, Lawrence, KS, USA
e-mail: ssoper@ku.edu

Center of BioModular Multiscale Systems for Precision Medicine, The University of Kansas, Lawrence, KS, USA

M. A. Witek · S. A. Soper
Kansas Institute for Precision Medicine, University of Kansas Medical Center, Kansas City, KS, USA

S. A. Soper
Bioengineering Program, The University of Kansas, Lawrence, KS, USA

Department of Mechanical Engineering, The University of Kansas, Lawrence, KS, USA

The University of Kansas Medical Center, Cancer Center, Kansas City, KS, USA

BioFluidica, Inc., San Diego, CA, USA

Department of Cancer Biology, University of Kansas Medical Center, Kansas City, KS, USA

R. K. Marcus and D. Lubman (eds.), *Exosomes and Extracellular Vesicles*,
https://doi.org/10.1007/978-3-031-89270-7_7

1 Introduction

Extracellular vesicles (EVs) are nanometer-sized lipid-bound vesicles released by cells into the extracellular space that can be isolated from bodily fluids such as plasma, saliva, urine, breast milk, and cerebrospinal fluid using a number of different strategies [1]. EVs were first identified by Peter Wolf and colleagues in 1967 as particles obtained from the ultracentrifugation of platelet-free plasma [2]. Initially, EVs were thought to be merely cell debris serving as a mechanism to dispose of cell waste until research showed that EVs could transfer their molecular cargo (e.g., proteins, lipids, nucleic acids) from the originating cells to recipient cells, thereby serving as a means of intercellular communication [3, 4].

EVs are typically classified according to their biogenesis and yield three main subtypes: exosomes, microvesicles (MVs), and apoptotic bodies (ABs). These subtypes vary in size and release pathway. Exosomes are formed by the inward budding of the cell plasma membrane to form early endosomes, which then undergo further inward budding to create intraluminal vesicles (ILVs). The endosomes containing ILVs then mature into multivesicular bodies (MVBs). The mature MVB can be degraded upon combination with lysosomes or fuse with the plasma membrane to form exosomes released through exocytosis. Exosomes usually range in size from 30 to 150 nm [1, 5, 6]. In contrast, MVs, also referred to as ectosomes, are formed by the outward budding of the plasma membrane and are typically between 100 and 1000 nm in diameter [7]. ABs are formed during plasma membrane blebbing of apoptotic cells and they range in size from 1 to 5 μm [8]. However, due to overlap in sizes of the three subtypes and lack of standardized subtype-specific markers, it is challenging to precisely determine the exact origin of EVs after they are released from the cell of origin. As a result, the International Society for Extracellular Vesicles (ISEV) recently recommended that EVs be classified as small EVs (<200 nm) or large EVs (>200 nm) or based on their phenotypical composition (e.g., CD63 + EVs) [9]. It is important to consider the findings of the Minimal Information for the Study of Extracellular Vesicles (MISEV), which was established by the ISEV and put forth guidelines for accepted practices in EVs research [10].

In recent years, there has been particular interest in EVs as a biomarker for liquid biopsy due to the presence of potentially highly informative cargo they contain and the ease of accessibility of EVs from body fluids. EVs' cargo includes proteins, nucleic acids, and lipids, which are protected by the EV lipid bilayer from nucleases, proteases, and fluctuations in pH [11, 12], making them an attractive source of diseases management biomarkers. It is evident now that the ribonucleic acids (RNAs) contained in EVs play an important role in cell-to-cell communication, affecting processes such as immune response [13, 14], tumor progression and metastasis [15], neurodegeneration [16], and many others. EV RNA has also been reported to be a potential biomarker for several pathological conditions, such as cancer [17, 18], Alzheimer's disease [16], stroke [19], and other cardiovascular diseases [15]. Therefore, this chapter will briefly cover EV isolation techniques, and then focus on the RNA content in EVs and their analysis.

EV Isolation Methods. To fully utilize EVs and their cargo in application scenario whether biological discovery or diagnostic, EVs must be efficiently isolated to secure high yields with high purity (i.e., EVs free of non-vesicular content such as lipoproteins and cell debris) for downstream analysis [20]. Both diseased and non-diseased cells release EVs [18], which can result in dilution of pathologically relevant EVs with "normal" EVs. Therefore, careful consideration of the choice of EV isolation method is the key for the subsequent molecular analysis of their cargo.

Ultracentrifugation. Ultracentrifugation (UC) uses increasing centrifugal forces and spin durations to isolate EVs based on their size and density. UC was the first method used for EV isolation [2] and is still considered the gold standard for EV isolation, accounting for 29% of all non-microfluidic EV isolations in the past decade (Based on Google Scholar advanced search, performed on July 8th, 2024; Keywords – ultracentrifugation, differential ultracentrifugation, exosomes, extracellular vesicles). There are two main approaches to UC, differential and density gradient centrifugation (DGUC) [5, 21]. The main difference between the two methods is that DGUC utilizes a preconstructed density gradient with increasing density from top to bottom, typically made of sucrose or iodoxanol [1, 22] while differential UC relies purely on different centrifugal speeds. Though still widely used, there has been a shift in the popularity of UC due to challenges such as forming exosome aggregates [23], long duration, labor intensiveness, and expensive equipment [24].

Ultrafiltration. Ultrafiltration is a common size-based technique for EV isolation. It uses the same concept as conventional filtration where suspended particles in a sample are separated based on the pore size or molecular weight cut off (MWCO) of the membrane used for the filtration [1, 25]. In this method, large particles such as cells and cell debris are removed using nanomembranes with pore sizes similar to the EVs allowing retention of the EVs. Different pore sizes can be used for isolation and pore size is dependent on the EV of interest [26]. For example, a nanomembrane concentrator with a pore size of 13 nm has been reported to be used for isolation of EVs from urine [27]. While this method is faster and requires less expensive equipment compared to UC, a challenge associated with ultrafiltration is the clogging of membranes leading to loss and potential alteration of EV structure [28].

Size Exclusion Chromatography. Size-exclusion chromatography is another size-based EV isolation method wherein a porous solid phase separates particles based on their retention on the column. As the sample moves through the column, smaller particles pass through the porous solid phase, but delaying their travel due to increased path length and causing larger particles to elute prior to the smaller ones [5, 29]. Several commercial products based on SEC technology have been developed, such as the quantitative (q)EV columns (Izon Science) that allow rapid isolation of EVs within 15 min from sample volumes ranging from 0.15 to 10 mL [30]. Exo-Spin columns (Cell Guidance Systems) allow EV isolation from 0.1 to 5 mL samples [31].

EV Precipitation. This method works by altering the solubility of EVs to precipitate them out of samples. The mechanism involves introducing water-excluding polymers such as polyethylene glycol (PEG) into the sample to tie up water molecules, causing less soluble components such as EVs to precipitate out of solution [25, 32]. To achieve this, samples are incubated in 5–15% PEG at 4 °C overnight to precipitate EVs followed by low-speed centrifugation 1,200–3,000 g of EV aggregates [33–36]. A major drawback of this method is the coprecipitation of proteins along with EVs. PEG precipitation are commercially available, such as ExoQuick from System Biosciences [30] and the exoEasy or miRCURY products from Qiagen. An alternative to PEG precipitation is lectin-induced precipitation [37]. Lectins are a family of proteins that bind specifically to carbohydrate moieties. Lectin simultaneously binds to saccharide across EV surfaces, inducing the formation of EV aggregates, causing them to precipitate out of solution. To ensure that lectins only bind to glycans on the EV surface, samples are first ultracentrifuged to remove cells and cell debris. The resulting samples are then incubated overnight with lectins. After incubation, the aggregated EVs are precipitated out by centrifugation [37].

Immunoaffinity capture of EVs. EVs possess protein markers (i.e., antigens) on their surface, which represent their cell-of-origin, making them excellent candidates to be specifically isolated via immunoaffinity. Immunoaffinity capture of EVs relies on antibodies interacting with EV surface antigens in order to isolate a specific phenotype of EVs [18]. Affinity isolation typically involves immobilization of antibodies specific to the antigen of interest onto a solid support, such as magnetic beads, microplates [38], resins, or microfluidic devices [39]. Though it is established that EVs are highly heterogeneous in size, surface markers, and cargo, it is generally agreed that tetraspanins such as CD9, CD63, and CD81 are suitable markers for EV isolation [10]. Tetraspanins are not exclusively enriched in exosomes, having also been identified in MVs [40–42]. This protein family is intimately involved in cellular communication, with functions in cell adhesion, signaling, and membrane fusion [43, 44]. Other proteins identified as EV markers are flotillin-1, annexins, and heat shock proteins, which are not specific to any subpopulation of EV, but can be used as selection markers against non-vesicular contaminants [41]. Antigens are targeted in immunoaffinity isolation techniques where EVs are being specifically selected out of all populations of EVs and other biological moieties [45–49].

Microfluidic-based EV Isolation Methods. Microfluidics is one of the new and advanced techniques reported for EV isolation. Microfluidic devices can be designed to rapidly isolate EVs based on either their physical or biochemical properties [25] and offer several advantages: low sample consumption, short processing time, cost-effectiveness, and the ability to be integrated with post-EV isolation steps such as enumeration and molecular cargo profiling [18]. In microfluidics, EVs can be isolated based on size [50], surface markers [51], centrifugation [51], acoustics [52], or electrophoresis [50], some of which will be discussed below.

The EV Herringbone microfluidic chip (^{EV}HB-Chip) isolates EVs based on immunoaffinity [53]. The herringbone feature in the chip (Fig. 1a i–ii) enhances binding between EVs and antibodies immobilized on the surface of the device by creating micro-vortices that increase the number of interactions between EVs and surface-bound antibodies (Ab). The device also consists of a nanostructured substrate

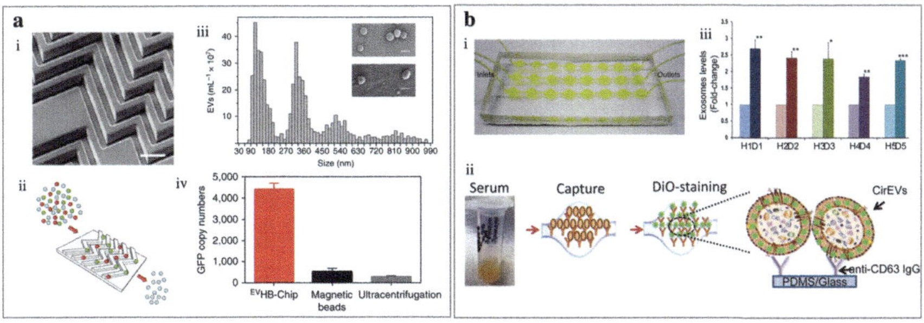

Fig. 1 Microfluidics-based EV isolation methods. **a** (i) SEM image of ^{EV}HB-Chip. (ii) Schematic of serum or plasma from healthy donors spiked with red and green fluorescent EVs processed through the device surface decorated with tumor-specific Ab. (iii) Size and concentration distribution of EVs isolated with the ^{EV}HB-Chip, analyzed by tRPS. iv) Comparison of total number of EGFRvIII copies captured with ^{EV}HB-Chip, magnetic beads, and ultracentrifugation (UC). Adapted with permission from Reátegui et al. [53]. Licensed under CC BY 4.0. **b** (i) Image of the ExoChip device. (ii) Schematic of the workflow for EV capture and analysis using ExoChip. (iii) EV levels in serum from pancreatic cancer patients ($n = 5$) and healthy donors ($n = 5$) were measured using fluorescence intensity. The fold-change in EV levels was calculated by comparing the means of the diseased (D) and healthy (H) groups. Adapted with permission from Zhang et al. [55]. **c** (i) Image of the EV-MAP chip showing the channels and diamond-shaped pillars used for immunoaffinity isolation of EVs. (ii) Workflow of sample processing and release of EVs using the EV-MAP device. (iii) NTA trace and TEM images of clinical samples isolated by PEG precipitation (upper TEM image) and affinity isolation (lower TEM image). Adapted with permission from Wijerathne et al. [19]. Licensed under CC BY 4.0. **d** (i) Schematic illustration of the AF4 working principle. In the focus stage, two opposing flows are applied to the inlet and outlet of the channel and balanced at the injection port where samples are injected, focused, and separated by elevation relative to their diffusion coefficient. In the normal stage, one flow is applied from inlet to outlet which causes particles with smaller hydrodynamic sizes and high diffusion coefficient to elute first. (ii) Fractionation profile of B16-F10 cell line sEVs obtained by applying a linear cross-flow gradient starting at a flow rate of 0.5 mL/min over 45 min. Peaks are labeled P0–P5 with UV detection shown in red, quasi-elastic light scattering in blue. Adapted from Kanwar et al. [54] with permission from the Royal Society of Chemistry. **e** (i) Image of an assembled TENPO chip (left) alongside SEM micrographs showing the TENPO magnetic nanopores (center and right), with an EV immobilized on the chip (right). (ii) NTA trace comparing size distribution of EVs isolated via TENPO vs UC. (iii) Correlation of nucleic acid cargos between TENPO and UC. Each point represents a nucleic acid marker measured in CD9/CD63/CD81 + EVs isolated using TENPO compared to the same markers measured in EVs isolated using UC. Adapted with permission from Lin et al. [56]. Licensed under CC BY 4.0. **f** (i) Represents the acoustic trapping method for EVs. A piezoelectric transducer generates a local acoustic standing wave in the channel, which induces retention of seeded particles. Sample with EVs is flown through a channel and EVs are trapped by the seeding cluster because of particle interactions. The acoustic wave is turned off and EVs are released. (ii) The trapping efficiency of the acoustic trap over six fluorescently labeled polystyrene beads. Adapted with permission from Ku et al. [57]. Copyright 2024 American Chemical Society

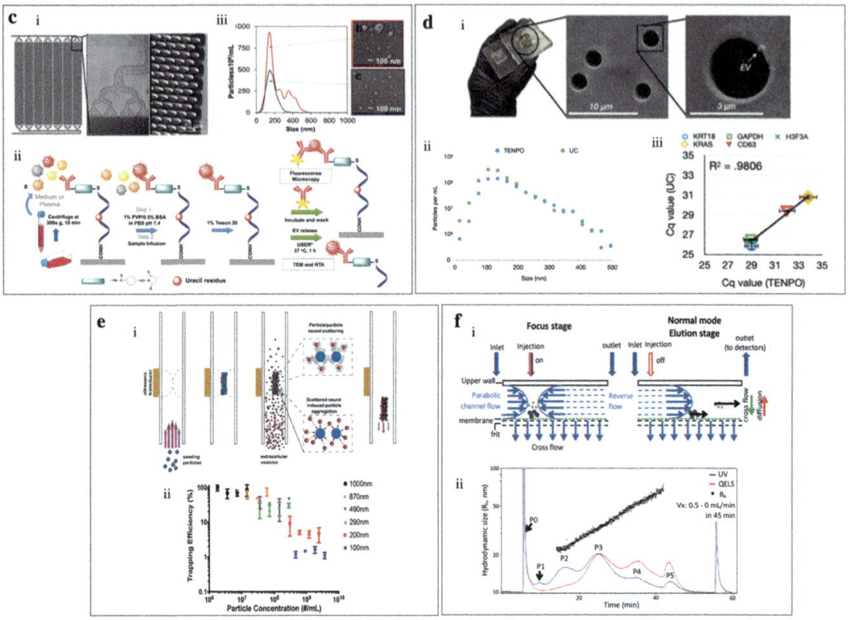

Fig. 1 (continued)

made up of an ultrathin gelatin membrane functionalized with streptavidin-coated nanoparticles that increase the surface area of the device. This combined with mixing from the herringbone structures improves EV capture. The authors reported an internal volume of 90 μL, and device capability to handle sample volumes from 0.1 to 5 mL at a flow rate of 1 mL/h. The isolated EVs were characterized in terms of their sizes to determine the device's ability to isolate tumor-specific EVs without size bias and showed that different populations of EVs were being captured (Fig. 1a, iii) and compared device performance to two standard methods for EV isolation (Fig. 1a, iv). They reported a tenfold increase in the mRNA content of EVs isolated from plasma and serum samples from glioblastoma multiforme (GBM) patients using the ^{EV}HB chip as compared to UC and magnetic beads [53].

ExoChip is another example of an immunoaffinity-based microfluidic device designed to isolate and quantify EVs directly on chip as shown in Fig. 1b, i [54]. The device was functionalized with anti-CD63 antibodies, which aided to selectively isolate CD63 + EVs. Surface-bound EVs were fluorescently labeled with DiO (3,3′-dihexadecyloxacarbocyanine) dye that stains membrane vesicles (Fig. 1b, ii), and quantification of EVs was done directly on chip by a standard plate reader. ExoChip was used to analyze serum from both healthy donors and pancreatic cancer patients. A 2.3-fold increase in EV load was observed for EVs captured from cancer patients' serum (400 μL, introduced at 8 μL min^{-1}) when compared to healthy donor samples, as shown in Fig. 1b, iii [54].

EV micro-affinity purification (EV-MAP) chip is another immunoaffinity-based device that affinity isolates subtypes of EVs using a specific surface marker [19]. For efficient immunocapture of EVs, the microfluidic device consisted of dense array of micropillars that provided high surface area for EV antigen and antibody interactions as shown in Fig. 1c i. Anti-CD8 monoclonal Abs were attached onto the micropillar surfaces of the device using a bifunctional linker that could be cleaved. After sample processing, EVs were released, Fig. 1c ii. The device was reported to have an internal volume of 30 μL and capability of processing 100 μL of plasma in 10 min. EV recovery of 85% and EV concentration of $4.3 \pm 2.1 \times 10^9$ particles/100 μL of plasma was reported. EV MAP chip performance was then compared to PEG precipitation by analyzing isolated EVs from both methods with NTA and TEM as shown in Fig. 1c iii. A narrower size distribution was observed for the EV MAP technique as compared to PEG precipitation indicating purer EVs isolated. CD8 + EV mRNA cargo was then used in molecular profiling and achieved 80% test positivity for identification of patients diagnosed with acute ischemic stroke [19].

A major challenge for researchers seeking to utilize immunoaffinity to isolate specific EV subtypes is the presence of a high amount of background EVs present in clinical samples (10^{10}–10^{12} EVs/mL in blood). Lin and colleagues [56] developed a device that sought to address this challenge by employing a track-etched magnetic nanopore (TENPO) chip as shown in Fig. 1d i, which isolated EVs through parallel immunomagnetic sorting [56]. To achieve isolation, EVs were first labeled with biotinylated antibodies specific to target markers and then conjugated with anti-biotin-modified 50 nm magnetic nanoparticles (MNPs). As the sample flows vertically through the TENPO chip's magnetic nanopores, EV-MNP complexes are captured at the edges of the nanopore due to magnetophoretic force overcoming the drag force of the fluid. They showed that the size distribution, number of EVs, and RNA cargo of TENPO-isolated EVs were seen to be consistent with that of UC-isolated EVs (see Fig. 1d ii–iii). They also reported a $>10 \times$ reduction of nucleic acids from EV background through optimization of key design parameters such as pore diameter, the number of membranes in series, and the flow rate [56].

Asymmetrical Flow Field-Flow Fractionation (AF4) is a microfluidic-based technique for isolating EVs based on size. In this method, separation occurs in a thin, empty rectangular channel where a parabolic flow is generated. Perpendicular to the parabolic flow, a secondary flow (crossflow) is applied across the channel to cause accumulation of samples at the bottom of the channel shown in Fig. 1e, i. The balance between the driving force and particle diffusion in the channel positions the samples at different layers within the channel relative to the bottom wall and causes them to elute at different times. Consequently, when parabolic flow is initiated, smaller particles with higher mean elevation elute earlier than larger ones, allowing for the separation of particles by size. AF4 was reported to separate EVs at 1 nm resolution and have a wide dynamic range capable of resolving particles from a few nm to 100 μm. Sample volumes ranging from 40 to 100 μL were processed and the fractionation process took ~2 h per sample [55].

Acoustic trapping is another microfluidic isolation technique that uses acoustic wave scattering in a cavity to achieve EV enrichment (Fig. 1f, i). In this system, acoustic standing waves are generated by a piezoelectric transducer and applied to a glass capillary. Polystyrene beads are trapped by the waves and used to facilitate the trapping of EVs as EVs are too small to be trapped by primary acoustic waves. As sample passes through the capillary, secondary acoustic forces resulting from particle–particle interactions cause EVs to be clustered and trapped by the polystyrene beads. EV particles are then released by turning off the acoustic field and washing with buffer. Trapping efficiency of the device was tested using fluorescently labeled polystyrene beads ranging between 0.1 and 1 μm in size and the result of this is shown in Fig. 1f, ii. This method was also reported to perform EV enrichment in conditioned media, urine, and plasma in as short as 30 min with a sample volume $\leq$300 μL [57].

The different isolation methods mentioned above operate on different principles, meaning that different methods will enrich different subpopulations of EVs and co-isolate contaminants to different degrees. Many researchers have investigated the effect of different isolation techniques on particle yield and purity [58–61]. One such study [62] compared four different isolation methods: precipitation-based protocols (ExoQuick® and Exo-spin™), size exclusion chromatography (SEC) using qEV columns (Izon Science), and DGUC-based OptiPrep. They utilized tunable resistive pulse sensing (tRPS) and protein analysis to analyze EVs isolated from non-small cell lung cancer (NSCLC) SK-MES-1 cell line and healthy donor plasma. EV yield was analyzed using tRPS and purity (particles per mL/protein markers) was assessed with western blotting using EV markers (HSP70, Flotillin-1, and TSG101). It was found that the precipitation-based methods (i.e., ExoQuick) produced higher EV yields compared to the other methods, such as UC, however, ExoQuick performed poorly in terms of EV purity. Size exclusion methods (i.e., qEV columns) showed the highest purity regarding the particle-to-protein ratio. These results suggest that high yield and low purity of ExoQuick are due to higher co-isolation of proteins [62].

Another important study assessed the impact of different EV isolation methods on mRNA content [63]. This was done by evaluating four methods: UC, DGUC-based OptiPrep, and 2 commercially available precipitation-based kits (ExoQuick and Total Exosome Isolation (TEI)) for yield, purity, presence of proteins, and RNA mass. EV yield and size distribution was evaluated using Nanoparticle Tracking Analysis (NTA), and it was observed that ExoQuick had the highest yield of EVs similar to the previous report [62]. However, results for UC and OptiPrep showed a narrower size distribution indicating more homogenous particle size of EVs. EV purity was evaluated with immunoelectron microscopy and Western blotting using EV markers (Alix, HSP90a, HSP70, TSG101, and CD63). Results from both methods showed that EV markers were most enriched in OptiPrep-isolated EVs, yielding the highest EV purity amongst the four methods. Isolated EV RNA using miRNeasy Micro showed the highest relative amount of RNA per 10^6 particles 40 ng while it was lowest for OptiPrep (0.7 ng). This study also noted that OptiPrep-isolated EVs showed a unique RNA profile with sizes below 50 nt as compared to the other isolation methods, which showed longer isolated RNA ranging in size between 100 and 1,500 nt. It was assumed that other methods may co-isolate RNA of non-exosomal origin.

All EV isolates analyzed for the presence of Argonaute-2 (Ago2), which is a non-vesicular RNA-binding protein, with the exception of those isolated with OptiPrep that were positive for Ago2. This finding suggested potential contamination with non-EV protein or RNA. This study showed that the choice of isolation method has a significant impact on the yield and purity of EVs and consequentially the total RNA content [63]. Methods that lead to high EV yield typically have lower purity as they tend to co-isolate contaminants along with EVs. On the other hand, methods with high purity deliver lower yields of EV-RNA, but possibly more disease-specific RNA. Therefore, it is important to follow the ISEV recommendations suggesting that the choice of EV isolation method is predicated by the research question being addressed, the type of downstream analysis in question, and specific application [10, 64].

Enumeration and Characterization of EVs. Methods used to characterize EVs aim to collect information on their size, morphology, concentration, and surface protein composition [10, 20, 65–67]. While many technologies and protocols for these analyses exist, this section provides a brief overview of the generally accepted approaches laid out by ISEV [10]. Note that these techniques each have their own strengths and limitations and correlations have not been established between EV source/isolation method and certain EV characteristics due to the high heterogeneity of EVs. Demonstrated here are methods calibrated with standards and comparisons between them are readily available [58, 59, 62, 68, 69].

Electron microscopy. A straightforward way to confirm the presence of EVs after isolation is by transmission electron microscopy (TEM). By directly visualizing EVs, information about their size and overall morphology can be confirmed (Fig. 2a). Though this method is recommended for verification of EV presence in isolated fractions, it only interrogates a very small fraction of the total sample volume and is, therefore, low throughput. As an alternative to TEM, scanning electron microscopy (SEM) [70–73] and cryo-TEM/SEM [23, 74, 75] have been demonstrated for confirmation of EV presence in samples as well (Fig. 2a). It is important to note that different sample preparation protocols in TEM and SEM can result in differences in EV size estimates and artifacts even for the same sample [69, 76, 77]. For example, typical SEM and TEM imaging require desiccation, which results in a cup-shaped morphology and EV shrinking that is not seen in methods where EVs remain hydrated (Fig. 2a, i) [77–80]. TEM can also be coupled with immunolabeling to provide more information on specific protein markers present on the isolated EV fraction [47].

Nanoparticle Tracking Analysis (NTA). Use of NTA for collecting information on the size distribution and concentration of EVs in a sample is considered standard practice. This technique determines the size of particles by capturing scattered light or dye fluorescence [81, 82] and transducing using a CCD camera. By performing this scattering measurement in real time, the Brownian motion of the particles can be assessed to determine size. This is done by utilizing the Stokes–Einstein relation, which correlates diffusion with particle size (Fig. 2b, i–iii).

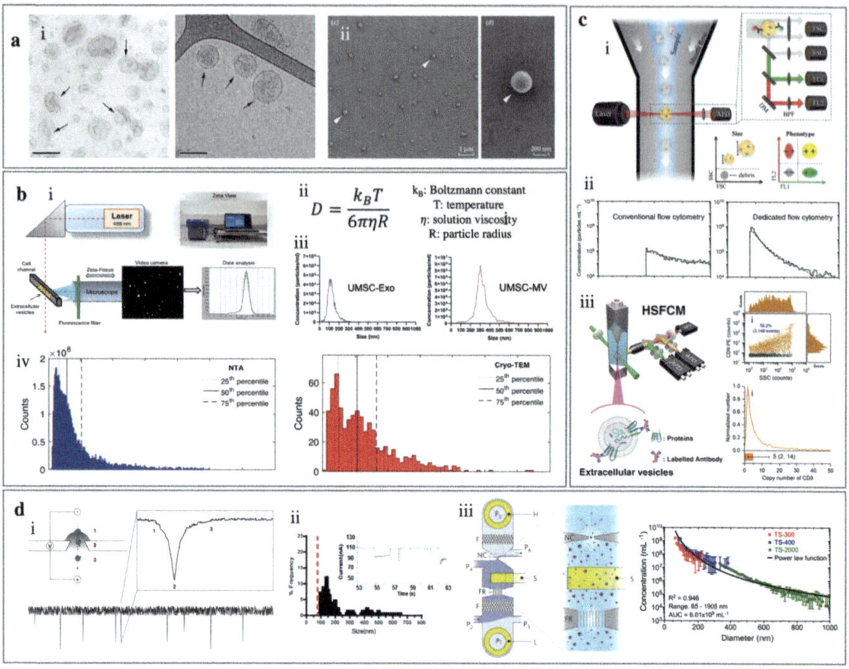

Fig. 2 EV characterization techniques. **a** Electron microscopy approaches: (i) TEM (left) using uranyl acetate staining showing cup-shaped EV morphology (Adapted with permission from Raposo et al. [80]). Licensed under CC BY NC SA 3.0), and cryo-TEM (right) showing spherically shaped EVs (Adapted with permission from Kondratov et al. [72]). EV sizes shown are 30–100 nm. (ii) SEM imaging of EVs showing round morphology and 40–300 nm in size. **b** Nanoparticle tracking analysis (NTA). (i) Experimental setup. Adapted with permission from Weber et al. [82]. (ii) Stokes–Einstein equation used to calculate particle size. Note the dependence of this measurement on sample temperature and viscosity. (iii) Comparative size distribution of exosomes and microvesicles via NTA. Adapted with permission from Yu et al. [83]. Licensed under CC BY 4.0. (iv) Comparison of size distributions obtained by NTA and TEM analysis. Adapted with permission from Noble et al. [84]. **c** (i) Flow cytometry setup and representative figures-of-merit for sizing and phenotyping of EVs. Adapted with permission from Kwon & Park [85]. Licensed under CC BY 4.0. (ii) Comparison of "conventional" FC (FACSCalibur, BD) and "dedicated" FC (A50-Micro, Apogee) for filter-isolated urine EVs. The dedicated FC was able to detect EVs down to 100 nm. Adapted with permission from Van der Pol et al. [86]. (iii) Left: Experimental setup for flow cytometry analysis of EVs reported by Tian et al. [87]. Right: Results for CD9 labeling of EVs isolated from the colorectal cancer cell line HCT15. Protein fluorescence vs. side scattering (SSC) events (top) and copy numbers (bottom). Adapted with permission from Tian et al. [87]. Copyright 2024 American Chemical Society. **d** (i) RPS working principle. Adapted with permission from Maas et al. [88]. (ii) RPS results for EVs in PBS using an Izon qNano TRPS system with a pore size of 50–330 nm. Adapted with permission from Anderson et al. [76] Copyright 2024 American Chemical Society. (iii) Left: Schematic of MRPS in the Spectradyne microfluidic. Adapted with permission from Fraikin et al. [89]. Right: Overlaid results from each Spectradyne cartridge on the same sample EVs demonstrating consistency of concentration measurements and sizing. Adapted with permission from Cimorelli et al. [90]. Licensed under CC BY 4.0. **e** Comparison of TEM, flow cytometry, NTA, and RPS measurement error for the most common EV characterization techniques. Adapted with permission from Van der Pol et al. [86]

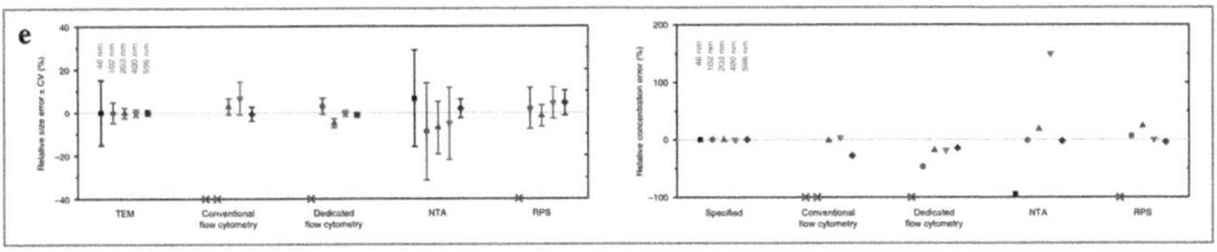

Fig. 2 (continued)

Several publications reported on the accuracy and reproducibility of NTA [86, 91–93], which generally agree that data is biased toward the detection of larger particles. This is the result of: (i) Assumptions made by the software that the sample is monodispersed and relatively dilute [94, 95]; (ii) a tendency for EVs to form aggregates [23, 78]; (iii) nonspecific measurement of non-EV particles not removed in the isolation steps; and (iv) low intensity light scatter for smaller particles. It should be noted that NTA measurements are based on hydrodynamic radius rather than the true diameter, which leads to differences in sizing when compared to TEM (Fig. 2b, iv) [77, 84]. Additionally, measurements for most NTA measurements involve averaging several separate recordings, resulting in large variability in estimating the particles' concentration [91].

Flow cytometry (FC). FC is capable of EV sizing as well as protein marker identification in single, dye-labeled EVs by measuring fluorescence of flowing EVs past a detector and collecting their side- and forward-scattered light (Fig. 2c, i) [96]. FC for EV characterization is inherently limited, as most conventional FC instruments have a lower detection limit of 200–500 nm particles and struggle to detect the low fluorescence intensities from EVs having lower numbers of fluorophore binding sites than cells due to their much smaller size [86, 97, 98]. Nano-FC approaches have been introduced to somewhat circumvent this issue [99–103]. This type of measurement is possible in commercial equipment using high-power lasers (200 mW), increasing dwell time at the detector, and employing high-performance photomultiplier tubes [104]. Tian et al. (2018) were able to profile EVs as small as 40 nm using a custom-built FC (Fig. 2c, iii) [87].

Resistive Pulse Sensing (RPS). RPS operates by measuring the current across a small electrically biased pore filled by a conductive solution. The size of this pore depends on the particle size and is typically 4–5 times the diameter of the smallest particle to be analyzed. When particles, such as EVs, translocate across the nanopore they displace ions in the conducting solution with a number proportional to the size of the target volume and result in a transient reduction in the current, which is registered as an event in the data trace (Fig. 2d, i). An advantage of this technique is that it is label-free and can provide both enumeration and sizing of nanoparticles.

The primary challenge associated with RPS is matching nanopore size with the size heterogeneity of EVs, with larger EVs relative to the pore size posing the risk of blocking the pore and smaller EVs being too small to cause significant resistance to be detected [105]. This can be addressed by using tunable RPS sensors (tRPS) such as the qNano platform (Izon) [76], or a microfluidic RPS (MRPS) from Spectradyne (Fig. 2d, ii–iii) [89, 90]. In TRPS, a hole-punched membrane can be stretched or loosened to accommodate different population sizes. Regardless of the approach, like with NTA, care must be taken to reduce false signals by similarly sized particles such as lipoproteins or protein aggregates, which cannot be differentiated based on peak characteristics [106, 107].

Young et al. [50] developed a nanofluidic device that used RPS to characterize the size and shape of EV particles isolated using UC, counting at a rate of 1000 EVs per min. This group designed the device with 3 pores in series to determine the particle volume, diameter, electrophoretic mobility, and zeta potential. Nano filters were included in the device to exclude cell debris and aggregates from entering nanopore region. The analyzed EVs harvested from human breast cancer cells showed narrow EVs size distribution of 67, 82 and 115 nm in diameter.

Another challenge with RPS is its limited particle concentration limit-of-detection due to the small sample volume it interrogates. For example, a typical concentration limit-of-detection is on the order of 10^8 particles mL^{-1} for many commercial RPS machines. To address this challenge, Soper and co-workers developed an extended nano-Coulter Counter [108], which was comprised of 5-separate pores in parallel that resulted in an increase in the sampling volume. In addition, the pores were placed within a microfluidic chip, and with the proper fluidic network, the sampling efficiency of the pores was also improved. As a result, the team reported a concentration limit-of-detection of ~1,000 particles mL^{-1}.

Extracellular Vesicles' Molecular Cargo. The molecular content of EVs is diverse and varies depending on factors such as originating cell type, EV biogenesis, and the physiological condition of the originating cell [109, 110]. When secreted by a cell, EVs have surface markers that allow them either to target or be targeted by recipient cells. Upon attachment to a cell, EVs are taken into the cell by endocytosis to deliver their content into the cell, which then can modify the physiological state of the recipient cell [111].

EV RNA plays an integral role in this cell-to-cell communication and sufficient evidence has shown that EV RNA can impact cellular functionality and modulate gene expression [112]. For example, EV miRNAs were found to function as mediators of cardiac remodeling processes and phenotypes [113], contribute to motor neuron degeneration [114], and promote cancer progression [109]. The ability of EVs to transfer RNA content between cells makes them a potential biomarker for diseases or a vehicle for therapeutic purposes [111]. Therefore, understanding EV RNA composition and function is crucial for developing novel diagnostic and therapeutic strategies.

There have been studies demonstrating that EVs share a similar makeup to the cells from which they originate [118, 119]. One such study showed that in cell culture, exposure to environmental stressors, such as hypoxia and oxidative stress, caused the EV RNA profile to be altered in a manner consistent with changes seen in the parental cell's RNA [120]. Although it has been established that EV RNA reflects the differentiation and pathophysiological state of the originating cell, the total RNA

(TRNA) profiles of cells and EVs are notably different in terms of both quantity and RNA type present [64, 112]. EV RNA represents a selected portion of parent cell TRNA, including mRNA and noncoding RNAs (ncRNA). EV RNA shows strong bias toward small noncoding RNAs, which can be fragmented or truncated unlike the parental cell TRNA, as well as miRNAs. While mammalian cells' TRNA ranges in size from 20 to 12,000 nt and is typically dominated by intact ribosomal RNA (rRNA) subunits 18S and 28S (80–90% of TRNA) [121], EV RNA size can range up to 6,000 nt (Fig. 3a, b, c), with most RNA fragments shorter than 20–500 nt (Fig. 3a and c) [28]. Even if the 18S and 28S rRNA signatures are present, they appear truncated (Fig. 3c). This is manifested in the small size of the EVs, especially exosomes that can typically accommodate ~10,000 nt of nucleic acid content.

TRNA quality in mammalian cells is assessed by comparing the intensities of 28S and 18S rRNA. Typically, a 28S/18S ratio of 2:1 indicates high quality RNA in cells [122], but this is difficult to assess in EVs since the cargo consists mostly of fragmented RNA and lacks full length rRNA signatures. Therefore, EV RNA quality cannot be assessed the same way as cell TRNA. More recently, Agilent introduced the RNA Integrity Number (RIN) algorithm in tandem with their Bioanalyzer 2100 microelectrophoresis system to better assess RNA quality [123]. Briefly, a value between 1 and 10 is assigned based on the key 28S and 18S rRNA features as well as fragment size distribution throughout the rest of the trace. A value of 8 or above is generally considered high quality RNA. Unfortunately, this metric also cannot be used to evaluate EV RNA. There is not yet a consensus on the best alternative quality assessment for EV RNA as the nature of their RNA cargo is highly fragmented due to size limitations of the EVs compared to intact biological cells. Furthermore, it has not been ascertained as to whether EV RNA transcripts are fully intact, fragmented, truncated, or degraded [64]. But, it has been proposed that probing an experimentally relevant set of genes through RT-qPCR could act as a quality control approach [64]. In addition, it is important to consider the truncated nature of EV RNA when designing primers or probes for the detection of these molecules. For example, in many cases, poly dT primers are used for the reverse transcription of mRNA into cDNA with PCR primers designed to interrogate the coding segment of the mRNA and this can fail for EV mRNA due to truncation. As a result, in many cases it is advisable to use universal hexamer primers for reverse transcription followed by the need to produce short PCR products across the exon being interrogated.

The TRNA in each EV subpopulation has distinct size distributions, which could be due to differential RNA loading into EVs by the parental cell [112]. One such study performed differential centrifugation to separate ABs, MVs, and EXOs from 3 different cell lines and EV RNA from these cell lines showed different TRNA profiles. In large EVs (ABs and MVs), the 18S and 28S rRNA were predominant, while smaller RNAs without prominent rRNA peaks were enriched in small EVs (Fig. 3c) [40]. It has been shown that TRNA from MVs is the most similar to the cellular transcriptome [28].

Composition and Function of RNA Types in EVs. EVs encapsulates various RNA molecules such as mRNA, rRNA, miRNA, and several ncRNA classes such as long

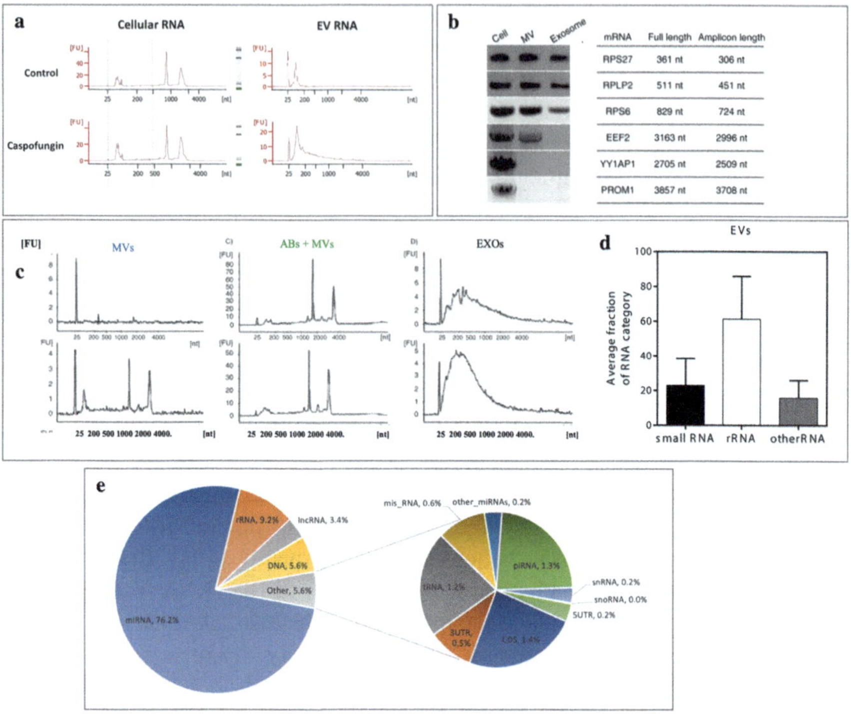

Fig. 3 TRNA content in EVs. **a** RNA profiles secured from cells and EVs from *Candida auris* in normal growth conditions or after treatment with an antifungal (caspofungin). Adapted with permission from Munhoz da Rocha et al. [115]. Licensed under CC BY 4.0. **b** End-point RT-PCR analysis of selected mRNAs transcripts showing the presence of short mRNA in both cells and EVs, but a lack of mRNAs >1,000 nt in exosomes but present in MVs. Adapted with permission from Wei et al. [28]. Licensed under CC BY 4.0. **c** Electropherogram showing the size distribution and profile of RNA of different EV subpopulations across 3 different cell lines showing a varying RNA profile by subpopulation. Adapted with permission from Crescitelli et al. [40]. **d** RNA sequencing data showing that EVs are not only enriched with small RNA but highly enriched with fragmented rRNA. Adapted with permission from Sork et al. [116]. Licensed under CC BY 4.0. **e** RNA sequencing results showing the RNA species distribution in plasma-derived exosomes. Adapted with permission from Huang et al. [117]. Licensed under CC BY 2.0

noncoding RNA (lncRNA), small nucleolar RNAs (snoRNA), piwi-interacting RNA (piRNA), vault RNA (vRNA), yRNA, and transfer RNA (tRNA) [28, 117]. Below we will give a short description of these classes of RNAs found in EVs specifically.

Messenger RNA (mRNA). mRNAs are protein-coding and are synthesized in the nucleus from the genomic DNA (transcription), which can undergo splicing and modification, and then exported to the cytosol where they function as templates for protein translation [124]. mRNA in EVs can serve as a template for novel proteins in recipient cells and several studies have shown this to occur [3, 124]. One such study showed that glioblastoma tumor cells release EVs that can modify recipient endothelial cells by translating mRNA in mediating EVs [12]. The authors demonstrated this by expressing *Gaussia luciferase* (Gluc) mRNA in glioblastoma cells that encoded a luciferase protein, which could emit fluorescence. EVs collected from glioblastoma cell cultures were introduced to recipient human brain microvascular endothelial cells (HBMVECs) and Gluc activity was monitored. Gluc activity was observed in the HBMVECs and was seen to increase continuously over a day indicating protein translation. Several other studies have reported similar protein translating capabilities of EV mRNA, corroborating that EV mRNA can be translated to functional proteins when delivered to a recipient cell [112].

The composition of EV mRNA has been reported to be mostly fragmented, while still having some full-length mRNA transcripts [28]. A study investigated EVs released by patient-derived glioma stem-like cell (GSC) cultures isolated by sequential filtration and observed that mRNA transcripts in exosomes were mostly enriched with truncated or fragmented mRNA, while those in MVs are seen to have full length transcripts (Fig. 3b). This showed that MVs are similar to the cell transcriptome while exosomes are different, which is not too surprising given differences in their biogenesis [28]. Another study investigated EVs from human umbilical vein endothelial cells (HUVEC) isolated by UC and analyzed RNA using Long Transcript EV Abundance (LoTEVA) method to identify long RNA transcript abundance in EVs. The authors found that EVs have a high diversity of long RNA transcripts present, with a significant proportion of EV-associated transcripts showing end-to-end read coverage, indicating that some may be full-length [125].

MicroRNA (miRNA). miRNAs are small noncoding RNAs [126], typically about 18–22 nt long, and are the most extensively studied RNA type in EVs due to their high abundance [127]. miRNA functions as a gene expression modulator by mRNA silencing and translational activation [8]. Certain miRNAs have been reported to be indicative of the health state of an organism and miRNA abundance profiles have been seen to show disease-specific expression patterns. Several studies have also shown EV miRNAs to be involved in a variety of pathological conditions such as cancer, neural disease, and cardiac disease (Table 1), supporting the case for using EVs' miRNA as a clinical biomarker [112]. For example, miR-193b, which is a miRNA secreted in the central nervous system, was seen to repress mRNA transcripts coding for translation of amyloid precursor protein involved in formation of amyloid plaque in Alzheimer's disease [128].

Several reports on miRNA content in EVs have been reported [117, 129, 130]. One such study analyzed EV RNA released by glioma stem cells and found that miRNA made up < 10% of non-rRNA (for reference, 64–94% of EV RNA were fragmented rRNA) [130]. Another study investigated RNA content of EVs from HEK293T, RD4, C2C12, Neuro2a and C17 cells and found the majority of RNA type across the samples were rRNA with miRNA accounting for 2–39% of EV TRNA (Fig. 3d). While these studies reported low miRNA abundance, another report that investigated RNA content in healthy plasma-derived EVs showed miRNA to be the most abundant RNA type in EVs representing 76.2% of all RNA types in these samples [117]. They

Table 1 Potential miRNA candidate biomarkers for different diseases

Disease	Sample source	miRNA biomarker candidate	References
Breast cancer	Plasma	miR-21, miR-155, and let-7a	[126]
Colorectal cancer	Serum	let-7a, miR-1229, miR-1246, miR-150, miR-21, miR-223, and miR-23a	[132]
Alzheimer's disease	Cerebrospinal fluid and serum	miR-193b	[133]
Parkinson's disease	Serum	hsa-miR-374a-5p, hsa-miR-374b-5p, hsa-miR-199a-3p, hsa-miR-28-5p, hsa-miR-22-5p and hsa-miR-151a-5p	[134]
Heart failure	Serum	miR-92b-5p	[135]
Asthma	Bronchoalveolar lavage fluid	miR-200 family and let-7 family	[136]
Rheumatoid arthritis	Serum	miR-155	[137]

also found a significant amount of other RNA species, including ribosomal RNA (9.16%), long non-coding RNA (3.36%), piwi-interacting RNA (1.31%), transfer RNA (1.24%), small nuclear RNA (0.18%), and small nucleolar RNA (0.01%) as shown in (Fig. 3e) [117]. A study carried out a stoichiometric analysis on miRNA content in EVs across different sample types (i.e., plasma, seminal fluid, conditioned media from dendritic cells, mast cells, and ovarian cancer cells) and reported miRNA was a minority of the TRNA content from isolated EVs and the stoichiometry implied that there was <1 miRNA per EV even for the most abundant miRNAs found in EVs [131]. Abundance of miRNA according to different reports varies dramatically and this could be due to the sample type, biases introduced by the EV and RNA isolation method, and/or the method for RNA analysis.

Small noncoding RNA. Other small noncoding RNAs with regulatory functions have been reported to be present in EVs [116, 117, 129, 138], which are briefly discussed here. **snoRNA** are 60–300 nt long RNA present in the nucleoli that guide chemical modifications of other RNAs, particularly rRNAs [139]. EV snoRNA has also been reported as biomarkers in pathological conditions such as cancer [139] and Alzheimer's disease [140]. **piRNA** is another small noncoding RNA, 26–31 nt in length, involved in gene expression regulation through gene silencing and epigenetic regulation [141]. Other than regulating gene expression, piRNA has also been reported to serve as a biomarker in pathological conditions such as cancer [142], Alzheimer's [120, 143], and cardiovascular disease [15]. **vRNA** is 80–150 nt long family of RNA associated with vault ribonucleoprotein protein complexes in the cytoplasm that transport molecules like mRNA from the nucleus to the cytoplasm [138]. **Y RNA** is a family of 83–110 nt long noncoding RNA that is involved in initiating DNA replication and regulating degradation of misfolded RNAs [144]. Changes in Y RNA levels in plasma have been shown to correlate with diseases such as systemic lupus erythematosus [127], Sjögren's syndrome [145], and cancer

[146, 147] making it a potential biomarker for these conditions. **tRNA** is 70–90 nt long regulatory noncoding RNA involved in protein translation [148] and reported to be an abundant small noncoding RNA type in EVs [148–150]. tRNA has also been reported to be associated with cancer progression [150].

Long noncoding RNA (lncRNA). lncRNA are RNAs > 200 nt in length and possessing diverse functions ranging from negative regulation of miRNA to degradation of mRNA and regulation of genes [124]. Several reports have confirmed the presence of lncRNAs in EVs [117, 130, 151–153]. lncRNAs were identified in EVs derived from human cervical and breast carcinomas cell lines using RT-qPCR [152]. In another study, it was found that cross talk existed between cardiomyocytes and fibroblasts and was mediated by lncRNA originating from EVs released by hypoxic cardiomyocytes during cardiac ischemia [154].

RNA Solid-phase Extraction (SPE) Methods. Accuracy and reliability of downstream analysis of EV RNA are dependent on the SPE of a sufficient amount of high-quality RNA, which is carried out following isolation of the relevant EVs for the downstream analysis and lysis of the vesicular membrane releasing the EV internal carogo. EV RNA SPE generally is carried out using a variety of commercial kits. All steps in preparation for EV RNA molecular analysis can introduce bias in terms of the size of RNA extracted. Hence, careful selection of the RNA SPE method is important.

RNA can be extracted via phase separation, spin columns, or magnetic beads [155]. Though phase separation was the earliest method used, it has largely been replaced by solid-phase methods, which use less hazardous chemicals and simpler protocols [155]. Among solid-phase methods, spin columns are commonly used with many options available including those specifically for EV RNA isolation. Common commercial kits include (mi)RNeasy and miRCURY Micro/RNA (Qiagen), mirVana (Invitrogen), and SeraMir (System Biosciences). Many of these column-type SPE methods are based on the use of silica-beads and the chaotropic salt, guanidinium chloride. Elution of the extracted material can be carried out using an aqueous buffer.

Eldh et al. [156] demonstrated how seven different RNA SPE methods can affect RNA yield and transcriptome profiling analysis. We emphasize the importance of establishing a method of standardization and quality control for a given sample type because molecular profiling requires the highest quality RNA possible to provide accurate results. The following sections will focus on the applicability of commonly utilized laboratory techniques and methods for EV RNA molecular analysis. We will provide an overview of novel approaches developed to address specific limitations in EV-related applications.

Tools for the Analysis of EV-TRNA. A major contributor to the difficulty associated with studying EV RNA is the low mass of TRNA present in EVs as compared to cells, difficulty in quantification, and RNA quality assessment [9]. For example, the reported mass of RNA isolated from 2×10^9 glioblastoma EVs using an SPE column was ~8.9 ng TRNA [28]. Therefore, stoichiometry implies that one EV contained only ~4.45 attograms of TRNA per EV [130], which is in sharp contrast to cells that can contain 10–30 pg TRNA per cell [157]. Many analytical techniques (e.g., sequencing, microarrays, or quantitative reverse transcription polymerase chain reaction (RT-qPCR)) may not have the sensitivity required to detect and quantify the low mass of RNA in EVs [9]. This challenge is compounded by the differential packaging of RNA into different sub-populations of EVs and in the case of mRNA, the presence of truncated transcripts.

2 Amplification-Based Analysis of EVs RNA

Reverse Transcription Quantitative PCR (RT-qPCR). RT-qPCR is a common approach used for quantifying the presence of certain gene fragments either through direct DNA amplification (qPCR) or employing a coupled reverse transcription step for quantifying gene transcripts (RT-qPCR). The typical workflow for RT-qPCR is illustrated in Fig. 4a, i. For RNA quantification, the attributes of RT-qPCR include its favorable limit-of-detection, many laboratories possessing the necessary machines, well-established methodology, and the relatively simple workflow (RT and PCR can be done in a single reaction tube. The technique involves quantification of cDNA transcripts at each amplification cycle based on fluorescence intensity readings, with the C_t value correlating to the amount of DNA present in the sample, but does require the inclusion of a standard calibration curve. As noted above, this method is well characterized and has many available resources and established protocols [158]. It is important to note, however, that sensitivity is highly dependent on the protocol used [159] and is not recommended for absolute quantitation in EV RNA studies [10]. The mass limits of EV cargo generally fall at the lower end of the dynamic range for most qPCR systems, and these measurements, like for all low abundance transcripts, are highly susceptible to variation between samples and non-vesicular extracellular RNA and DNA contamination [160].

When performing reverse transcription, primers must be included to allow for cDNA synthesis. These are generally poly-deoxythymine (poly-dT) oligonucleotides, random-sequence primers, or gene-specific primers. Considering that mRNA makes up only a small percentage of EV TRNA [130], it is not advisable to use poly-dT primers for whole transcriptome analysis. For qPCR primer design, two considerations are necessary: (i) Studies have shown that EV mRNAs are specifically enriched in 3' ends, meaning primers should be directed toward the sequences close to the poly-A tail [130, 161]; and (ii) as discussed, EVs generally lack full-length transcripts meaning that binding sites for analysis of typical cell RNAs may not be present and the design of EV-specific primers is necessary and essential.

However, with careful experimental design RT-qPCR can be a powerful tool for assessing EV RNA expression [12, 162–164]. RT-qPCR was used to investigate the utility of EV miRNA for determining neurological damage after cardiac arrest demonstrating a proof-of-concept analysis of several neuronal GluR2 + EV miRNAs for predicting neurological outcomes with area under the curve (AUC) values >0.75 (Fig. 4a, ii) [165]. Several studies have described protocols for the identification of reference genes specific to EVs [166–168]. Normalization to these reference genes obviates handling errors and unknown RNA quality, which we noted previously in this chapter as not assessable for EVs.

Droplet Digital PCR (ddPCR). A more recent alternative to qPCR is ddPCR, which has been shown to exhibit higher sensitivity for low-abundance targets detected at C_t values ≥29 [171] (Fig. 4b, i). Introduced in 2011 [172], the key concept involves splitting PCR reaction volume into ~20,000 separate aqueous droplets suspended in an immiscible oil, therefore creating ~20,000 individual reaction vessels. If working within the ddPCR dynamic range and due to the high degree of partitioning, each droplet should theoretically contain either zero or one cDNA/DNA template, enabling digital analysis. During processing, PCR mix/template/fluorescent dye in droplets undergoes amplification and are counted using a flow cytometer-like reader, where droplets containing amplified DNA fluoresce and are counted as positive, while non-fluorescing droplets with no amplified template are counted as negative. Based on the fraction of positive to negative droplets, as well as the PCR reaction volume of 20 μL, a copy concentration of cDNA/DNA can be calculated. If working above the ddPCR dynamic range, it is possible for more than one cDNA/DNA template to be loaded into a single droplet. Therefore, Poisson statistics must be applied to avoid such potential errors [173].

According to Hindson et al. [172], ddPCR can detect one target cDNA from picograms of mRNA input. This is an ideal situation for EV mRNA studies, where transcript copy numbers can be <100 copies/ng of TRNA [169, 174, 175]. It is worth noting that although the sensitivity of ddPCR has been found to be superior to qPCR [169, 171, 174], it is still equally susceptible to contamination and care must be taken to avoid non-vesicular RNA signals [176].

RNA sequencing. The development of next-generation sequencing (NGS) has been pivotal for characterizing EV RNA. RNA sequencing is the only analysis approach capable of quantifying both expected and previously unidentified transcripts. Hill et al. [177] and Mateescu et al. [178] summarized typical approaches for EV RNA library preparation and the potential for bias inherent to each. Generally, these workflows follow a procedure involving the addition of adapter sequences, cDNA synthesis, and PCR amplification to form double-stranded DNA as outlined in Fig. 4c (i). RNA library preparations generally involve mRNA enrichment or rRNA depletion steps. The low levels of rRNA identified in EVs [12, 40, 69], however, many times allow workflows to be performed without rRNA depletion. Because EVs can contain a variety of RNA types as delineated earlier [179], utilizing a broad whole-transcriptome approach to library preparation is of great value [180]. While it is recognized that RNA cargo consists of mostly short sequences, we have discussed the presence of medium and longer sequences [181] which, if not accounted for, will lead to biased results toward short fragments [178].

Sequencing approaches utilize sequencing-by-synthesis and nanopore methods [182, 183]. Once sequencing is completed, the sequences must be aligned to a reference genome [184]. Success of alignment to a reference directly indicates quality and abundance of reads, and this is heavily dependent on the choice of alignment

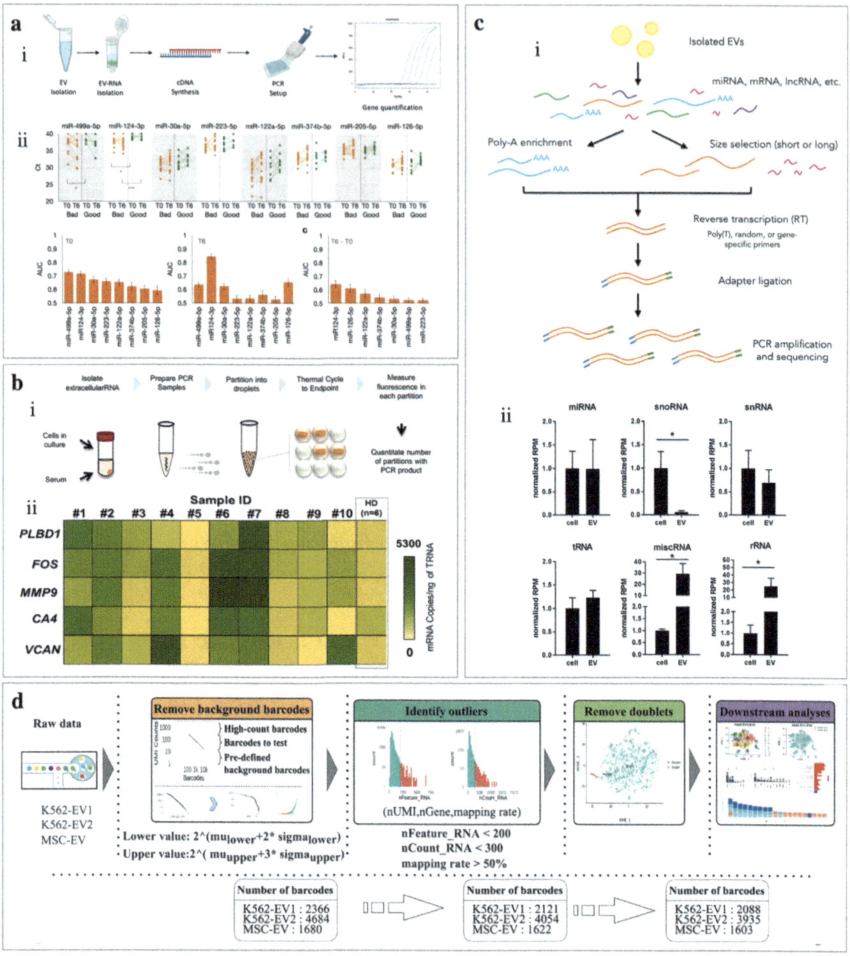

Fig. 4 Amplification-based EV RNA analysis techniques. **a** (i) RT-qPCR workflow showing the key steps involved with typical amplification curve showing gene quantification. Created with Biorender.com. (ii) RT-qPCR results for eight affinity-enriched neuron EV miRNAs to assess brain injury following cardiac arrest at two time points: T0 (directly after resuscitation) and T6 (6 h after resuscitation) comparing patients with bad ("Bad") and good ("Good") outcomes neurological outcome. Reproduced with permission from Shen et al. [165]. **b** (i) Droplet digital PCR (ddPCR) workflow. Adapted with permission from Takahashi et al. [169]. Licensed by CC BY 3.0. (ii) mRNA expression levels of affinity-isolated CD8 + EVs with a 5-gene panel for ischemic stroke diagnosis from 10 clinical samples. Adapted with permission from Wijerathne et al. [19]. Licensed under CC BY 4.0. **c** (i) General RNA sequencing workflow. (ii) Types of RNA present in primary dendritic cells and their EVs isolated by UC and analyzed by RNA sequencing. Adapted with permission from Driedonks et al. [127]. Licensed under CC BY 4.0. **d** Workflow of data analysis for single-EV RNA sequencing. Adapted with permission from Luo et al. [170]

tool [185]. One issue to be considered at this step is the occurrence of multi-mapping sequences, in which a sequence can be aligned to more than one location on the transcriptome. The prevalence of multi-mapping sequences increases with shorter reads (<200 nt) [182], which EV studies are generally limited to. Depending on the alignment process, these multi-mapping sequences are generally discarded leading to less reliable transcript quantification [183].

Another necessary step in sequencing analysis is quantifying transcript abundance, which involves converting successful alignments to reads. Once transcript abundance has been quantified, filtering can be applied to remove adapter sequences, redundancies, and low abundance transcripts. If low abundance transcripts are of interest, increasing sequencing depth will allow more sensitive detection. To account for sequencing depth and workflow bias, data normalization is necessary [186]. Many different normalization approaches have been proposed, but the best approach will depend on the specifics of the samples being analyzed and what analysis is being done—often for differential gene expression in EV studies. An extra measure for normalization and quality control increasingly recommended is inclusion of RNA spike-in controls [186].

The application of RNA sequencing to EV research has led to much of the understanding we have today of their functions. For example, Nolte-'t Hoen et al. [138] were the first to propose that EVs are differentially loaded with many noncoding RNAs being enriched in contrast to cellular levels. These findings have led researchers to investigate how RNA loading into EVs is mediated, which is being done via NGS to identify sequence motifs [187] and molecular modifications [128, 188]. This information has already been applied to understanding new disease pathways and how to utilize EVs in clinical applications [73, 189–192]. For example, Driedonks et al. [127] used RNA sequencing to investigate the types of RNAs present in EVs compared to their primary dendrite parent cells (Fig. 4c, ii) and demonstrated how their expression changes under simulated immune stimulation or suppression.

Recently, Luo et al. [170] introduced the first single-vesicle transcriptome analysis, exemplifying the powerful nature of this technique for EV analysis (Fig. 4d). The researchers analyzed EVs derived from lymphoblast and stem cell cultures and found an average of ~50 RNA molecules per EV, much higher than previous reports suggesting ≤ 1 [39, 131, 193]. Further, they speculated that the EV RNA cargo from both cell types are biologically relevant to tumor metabolism (lymphoblasts) and tissue repair (stem cells). The breadth of information obtained in this type of study is what makes RNA sequencing such a highly sought-after tool for EV research.

3 Non-amplification-Based Analysis of EV RNA

Microarray. Microarrays have been long-established as a highly specific and reliable method for RNA expression analysis [194]. Though the popularity of microarrays has decreased in recent years with the development of RNA NGS, it is still being applied for studies where many known transcripts need to be investigated across various samples in a cost-effective and high-throughput manner [195–197]. Notably, microarray studies were some of the first performed for EV RNA characterization [3, 12, 198]. Thus, it is important to understand how this technique has been used previously and the data that it has produced.

Microarrays involve detection of fluorescence from cDNA that has been reverse transcribed using dye-labeled nucleotides. Hundreds to thousands of oligonucleotide probes complementary to a gene of interest are bound to individual points on a microarray surface, and upon hybridization by target cDNA, will fluoresce and allow high-throughput investigation of gene expression (Fig. 5a, i). Though many types of studies are performed using microarrays, this approach has been most widely used for gene expression analysis, including in EVs (Fig. 5a, ii–iii) [199]. A main drawback in the use of microarrays for EV RNA analysis is that at low transcript abundance, fluorescence signal is below the limit-of-detection. Another limitation is the input requirements. Generally, micrograms of RNA are required for microarray analysis [200], which can be above the available mass of the EV-sourced RNA.

NanoString nCounter. Platforms designed to analyze cell or tissue gene expression are now being tested with EV RNA. An example of this is the NanoString technology, namely the nCounter, which allows for highly multiplexed fluorescent quantification of target mRNA transcripts. Originally designed for use with formalin-fixed, paraffin-embedded (FFPE) tissue, the approach involves incubating TRNA 100 ng with capture and reporter oligonucleotides that serve to immobilize the mRNA on a microscope slide and subsequently fluoresce in a specific "barcode" pattern unique to each target, which can be digitally counted (Fig. 5b, i–v) [201]. Though some reports have found EVs to contain insufficient yields of RNA for nCounter analysis without some type of pre-amplification [202, 203], several papers have shown that this step is not necessarily required [17, 204–208].

This technology was FDA-approved for use with the Prosigna PAM50 assay, which is a gene panel for identifying breast cancer subtypes based on gene expression profiling [209–211]. Usually, high RNA mass input requirements are posed to correct for the poor quality of FFPE RNA, but the assay can be performed with lower mass input of high quality TRNA. Hu et al. [17] demonstrated the ability to use the nCounter to perform breast cancer molecular subtyping using EV RNA based on identification of transcripts with medium to high abundance in the Prosigna test. Results demonstrated the ability of EV RNA to be used for gene expression despite their limited abundance (Fig. 5b, vi). Similarly, Bub et al. [204] utilized the nCounter to study mRNA transcripts upregulated in stroke, which they could then further map to specific cell types of origin in the brain. They demonstrated the ability to load EV lysates directly into the nCounter without the need for prior RNA isolation, preventing possible bias based on the method of isolation.

Molecular Beacons (MBs). Research involving MBs to investigate EV cargo composition in situ without disturbing the EVs' lipid bilayer can allow for single EV analysis and prevent potential bias stemming from isolation or amplification methods. MBs are designed with a fluorophore (FL) and quencher (Q) molecule, each attached to the end of a hairpin oligonucleotide capable of probing for specific nucleic acid

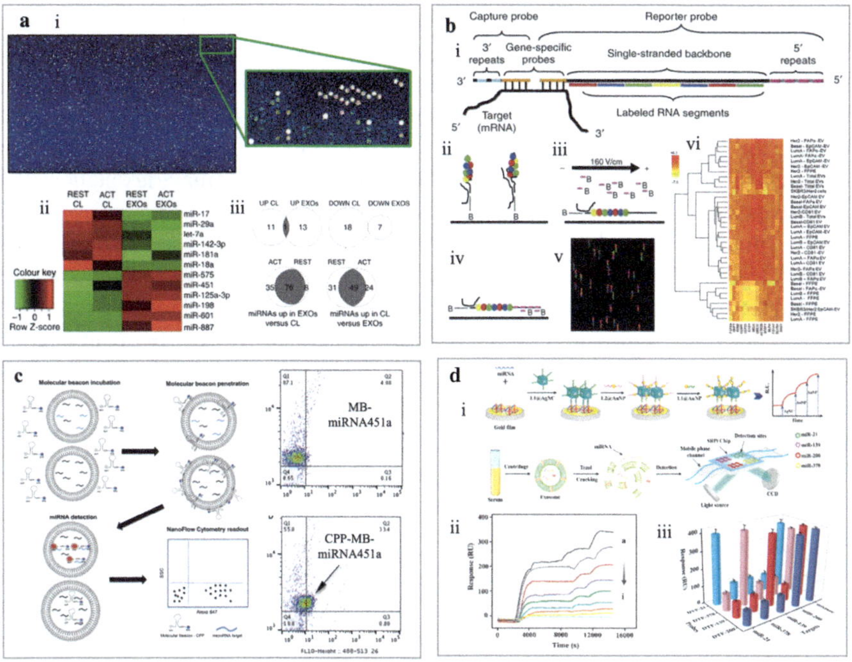

Fig. 5 Non-amplification-based methods for EV RNA analysis. **a** (i) Fluorescence image of a microarray for EV miRNA analysis from lung cancer patient plasma. Adapted with permission from Agilent Technologies [213]. (ii) Heatmap of microarray analysis of activation induced changes in cell ("CL") and exosomal (EXOs) miRNAs from human primary T lymphocytes, before ("REST") and after ("ACT") phorbol myristate acetate (PMA) and Ionomycin activated conditions. (iii) Upper panels: Microarray results show differential expression of miRNAs in cells and EVs for both upregulated and downregulated genes. Lower panels: Microarray results showed consistent loading of specific miRNAs into EVs regardless of activation state. ii–iii adapted with permission from Villarroya-Beltri et al. [199]. Licensed under CC BY NC ND 3.0. **b** Workflow of the Nanostring nCounter assay. (i) RNA sequences are incubated with capture and reporter probes in solution, (i) then transferred to a streptavidin-coated slide onto which target RNAs hybridizes to their respective biotinylated capture probe and will be bound while off-target RNAs are washed away. (iii–v) Voltage applied to the surface linearizes the complexes against the slide surface, where biotinylated oligonucleotides complementary to the 5' repeats of the reporter probe will immobilize the complex for imaging. Adapted with permission from Geiss et al. [201]. (vi) Heatmap showing expression levels for the 14 genes comprising the Exo-PAM50 panel for EVs and FFPE tissue demonstrating clear differences between cells and EVs. Adapted with permission from Hu et al. [17]. Copyright 2024 American Chemical Society. **c** Mechanism of in situ labeling of miRNAs using a membrane-penetrating peptide (MPP)-bound molecular beacons (left) and nano-flow cytometry results (right) showing that addition of the cell-penetrating peptide enabled detection of EVs carrying miRNA-451a. Adapted with permission from Oliveira et al. [102]. Licensed under CC BY NC ND 4.0. **d** Surface plasmon resonance imaging (SPRi) biosensor. (i) Workflow of SPRi biosensor surface assembly. (ii) SPR Sensogram showing dynamic range of the biosensor (2 fM to 20 nM) for EV miRNA-21. (iii) Selectivity and multiplexing capability of the biosensor. Adapted with permission from Wu et al. [214]

sequences. When in proximity, owing to the hairpin, Q quenches fluorescence. Upon hybridization to a target sequence, the Q and FL are physically separated, which triggers fluorescence [212]. De Oliviera et al. [102] applied this technology to EVs by conjugating an MB probe to a membrane-penetrating peptide, which carried the MB into the EVs without the need for chemical permeabilization (Fig. 5c). Using this approach, the authors detected and quantified miRNA-451a for malaria diagnosis via fluorescence of the MB using nano-flow cytometry.

In a similar manner, Lee et al. [215] demonstrated multiplexed in situ fluorescence detection of three cancer-associated miRNAs by way of chemical membrane permeabilization to introduce the MBs. He et al. [193] utilized total internal reflection fluorescence (TIRF) measurements to perform in situ single-vesicle quantification of EV miRNAs. Deoxyribozymes (DNAzymes) and MBs were incubated with EVs for binding to miRNAs, and the fluorescence of individual EVs was measured as they entered the evanescent field (100–300 nm in thickness). This approach enabled enumeration of EVs and quantification of miR-21 copies per EV (reporting 2–6 copies per EV), demonstrating impressive true single-vesicle analysis.

Plasmonics. Plasmonics involves collecting optical measurements of particles in close proximity to metal dielectric surfaces to take advantage of enhanced detection arising from the generation of plasmons. This can be used to measure a range of biomarkers [216, 217] and has been applied to EV physical characterization [218–220]. Studies on evaluation of EV RNA have also been performed; Wu et al. [214] developed a multiplexed surface plasmon resonance imaging (SPRi) biosensor for detection of EV-associated miRNAs for diagnosis of non-small cell lung cancer (NSCLC) (Fig. 5d). Briefly, a DNA tetrahedral framework (DTF) was immobilized onto a gold surface which gold-silver nanoparticles functionalized with ssDNA complementary to one of four miRNA targets. This design allowed for a sensitivity of 1.68 fM and clear discrimination of healthy vs. NSCLC patients. Another group developed a surface-enhanced Raman scattering (SERS)-based sensor for quantification of EV miRNAs, which demonstrated sensitivity high enough to detect expression level changes capable of subtyping breast cancer cell lines [221].

Integrated Microfluidic Systems. While the application of conventional methods for EV RNA analysis has been successful, newer technologies are evolving from our growing understanding of analytical challenges associated with the analysis of EVs and its RNA content. In many cases, it is becoming apparent that common benchtop approaches do not always possess the capabilities to detect or assess EV RNA in a useful and unbiased manner. Several groups have proposed integrated microfluidic systems for EV analysis [222] with the goal of reducing assay times, processing steps, and costs.

Shao et al. [223] introduced the immuno-magnetic exosome RNA (iMER) platform for analysis of mRNA derived from EVs on a single chip (Fig. 6a, i). The authors broke down the workflow into three main steps: (i) EV enrichment, wherein anti-CD63 and anti-EGFR antibody-coated magnetic beads affinity isolate EVs from plasma or cell culture medium and were subsequently lysed; (ii) RNA isolation, which relied on the condensation of RNA onto glass beads under precipitating conditions

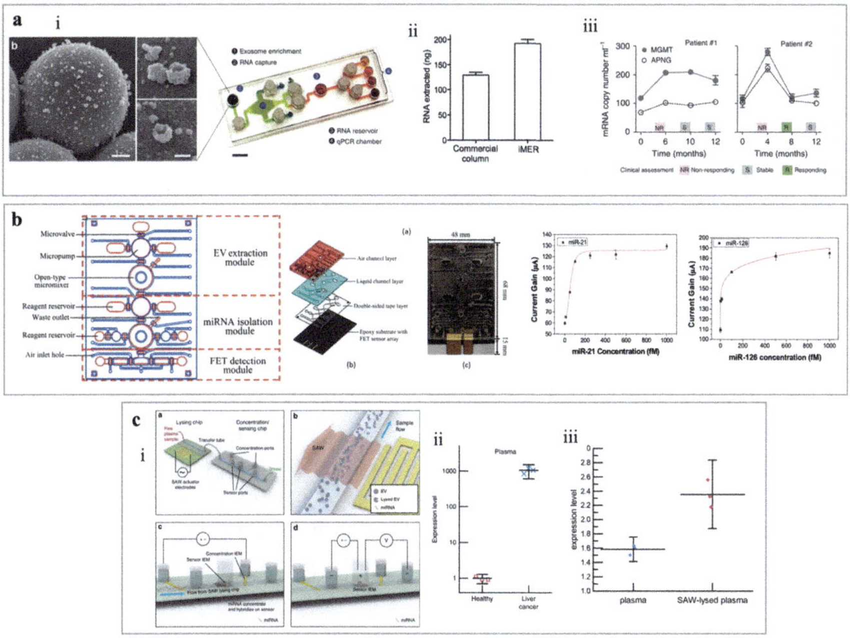

Fig. 6 Integrated microfluidic systems for EV RNA analysis. **a** Immuno-magnetic exosome RNA (iMER) microfluidic. (i) Image of the iMER chip with a close-up of EVs captured on the surface of the magnetic beads. (ii) Comparison of RNA recovery between a commercial kit (Qiagen RNeasy Micro) and the iMER platform. (iii) mRNA expression levels of two genes (MGMT and APNG) measured via the iMER compared to qualitative clinical assessments of treatment response in glioblastoma patients. Adapted with permission from Shao et al. [223]. Licensed under CC BY 4.0. **b** (i) Schematic of an integrated microfluidic equipped with AlGaN/GaN high-electron-mobility-transistor (HEMT) field-effect transistors (FET) for high-sensitivity EV-miRNA detection following on-chip EV and miRNA isolation. (ii) C_t values for EV-derived miR-21 isolated on-chip and analyzed by RT-qPCR. Negative control concentration was not provided. (iii) Stock miR-21 concentrations measured on the FET sensor using the integrated system. Adapted from Cheng et al. [73] with permission from the Royal Society of Chemistry. **c** (i) Microfluidic platform utilizing a piezoelectric to generate surface acoustic waves (SAW) for lysing EVs and subsequent collection and quantification of miRNA targets via an on-line ion-exchange membrane (IEM). (ii) On-chip analysis of miR-21 from healthy mouse EVs (left) and mouse liver cancer EVs (right). (iii) Comparison of plasma EVs lysed using TRIzol (left) and SAW (right). Adapted with permission from Ramshani et al. [224]. Licensed under CC BY 4.0

(70% ethanol and likely salt from the EV lysis buffer); (iii) analysis by on-chip RT-qPCR following elution in water from the glass beads. EV isolation was efficient at >93% for anti-EGFR and anti-CD63 monoclonal antibody enrichment, and RNA extraction on-chip was found to give 5~0% higher recoveries than commercial kits (Fig. 6a, ii). When used to evaluate drug response in glioblastoma patients, the assay could successfully detect changes in relevant mRNA transcript abundance and correlate these changes to treatment response (Fig. 6a, iii).

An integrated microfluidic capable of sensing EV microRNAs using aluminum-gallium nitride/gallium nitride (AlGaN/GaN) field effect transistors (FET) down to 15.8 fM was reported recently [73] (Fig. 6b, i–ii). The assay began with isolation of EVs using anti-CD63 antibody-coated magnetic beads on-chip, lysis, and transfer to a separate chamber in which ssDNA probe-coated magnetic beads isolated miR-21 and miR-126. After hybridization and removal of non-target molecules, miRNAs were released from the beads by denaturation and transferred to the FET to bind to complementary probes on the FET surface for detection. Once the detector was activated, the amount of miRNA hybridized changed the current magnitude of the sensor. This device demonstrated sensitivity, which was ~1 fM for purified miRNAs on the FET sensor (Fig. 6b, iii) and multiplexed detection of targets, while also reducing assay times compared to benchtop approaches.

Another study utilized surface acoustic waves (SAW) to lyse EVs on a microfluidic and subsequently decipher their miRNA content via an on-line electrokinetic sensor functionalized with a miRNA-21 complementary oligonucleotide (Fig. 6c, i–ii) [224]. A major advantage of this assay was the removal of lysing buffer and TRNA extraction reagents, which resulted in lower biases due to binding effects. The authors showed that SAW lysis has higher efficiency than TRIzol reagent resulting in 50% higher miR-21 recovery (Fig. 6c, ii).

4 Conclusions and Future Perspectives

Though EV and EV RNA analyses present unique challenges, the field is expanding with interests in both basic discovery and clinical applications (i.e., translational) with each area growing tremendously at the time of this writing. Each branch of the research endeavor related to EVs has divergent requirements, both from the isolation needs and with the analysis of the EVs' molecular cargo in particular RNA analysis due to the distinct differences between biological cells and the EVs they secrete. For example, in terms of basic discovery, non-specific isolation methods of EVs are required to allow for the interrogation of an unbiased representation of the many EV subpopulations present in an organism. Thus, methods such as UC may be preferred for EV isolation in spite of its low throughput and complicated workflow. Along the same lines, for EV RNA analysis, RNA sequencing may be preferred for basic discovery as opposed to gene-specific methods to determine the relative contribution of any RNA type to address the problem at hand. However, when considering clinical applications, different requirements are necessary including the ability to provide high throughput capabilities with exquisite analytical and clinical figures-of-merit. Thus, for the clinical isolation of EVs specific and well-defined EV subpopulations may be required to be interrogated and as such, affinity isolation methods become the method of choice due to the fact that both diseased and non-diseased EVs are found in clinical samples. In addition, certain genes relevant to the disease being tested are required and as such, a PCR-based method with gene-specific primers may be preferred rather than RNA sequencing. Indeed, in this chapter, we highlighted the need for this line of process development for EVs both from the discovery and translational perspectives.

With the evolution of many new isolation and molecular analysis technologies and their associated improved figures-of-merit, the lack of standardization within the EV-related work sets challenges with benchmarking new technologies against a "gold-standard". While the ISEV has attempted to create such standards in terms of the classification of EV types and confirmation of what is or is not an EV (for example, confirmation of an EV identity through the use or TEM, Western blotting, NTA, etc.), new discoveries sometimes refute established guidelines. For example, NTA has been considered the suggested EV enumeration method. But, recently it has been noted that NTA can generate erroneous results due to its inability to distinguish between the vast number of nanoparticles found in clinical samples from EVs, such as lipoproteins. In addition, many established biochemical and biophysical techniques have yet to be optimized for EV research, with current protocols varying widely across different laboratories. As approaches for isolating and characterizing EVs' molecular cargo evolve, the groundwork will be laid for exciting developments in clinical practice. For example, EVs are one type of liquid biopsy marker and have been found to be continuously released even during the early pathogenesis of many diseases. In conjunction with the molecular biomarkers packaged within disease-associated EVs, they may become an integral part of the early detection portfolio for many different diseases.

Acknowledgements The authors are grateful to the National Institute of Health (NIH) for financial support via NIBIB-funded Center of BioModular Multiscale Systems for Precision Medicine, CBMM (P41-EB020594), NCI: R44 CA224848 and R33 CA235597, and COBRE Kansas Institute for Precision Medicine, KIPM (P20 GM130423).

References

1. Doyle, L. M., & Wang, M. Z. (2019). Overview of extracellular vesicles, their origin, composition, purpose, and methods for exosome isolation and analysis. *Cells, 8*.
2. Wolf, P. (1967). The nature and significance of platelet products in human plasma. *British Journal of Haematology, 13*, 269–288.
3. Valadi, H., Ekstrom, K., Bossios, A., Sjostrand, M., Lee, J. J., & Lotvall, J. O. (2007). Exosome-mediated transfer of mRNAs and microRNAs is a novel mechanism of genetic exchange between cells. *Nature Cell Biology, 9*, 654–659.
4. Couch, Y., Buzas, E. I., Di Vizio, D., Gho, Y. S., Harrison, P., Hill, A. F., Lotvall, J., Raposo, G., Stahl, P. D., & Thery, C. et al. (2021). A brief history of nearly EV-erything—The rise and rise of extracellular vesicles. *Journal of Extracellular Vesicles, 10*, e12144.
5. Hadizadeh, N., Bagheri, D., Shamsara, M., Hamblin, M. R., Farmany, A., Xu, M., Liang, Z., Razi, F., & Hashemi, E. (2022). Extracellular vesicles biogenesis, isolation, manipulation and genetic engineering for potential in vitro and in vivo therapeutics: An overview. *Front Bioeng Biotechnol, 10*, 1019821.
6. Thery, C., Zitvogel, L., & Amigorena, S. (2002). Exosomes: Composition, biogenesis and function. *Nature Reviews Immunology, 2*, 569–579.

7. György, B., Szabó, T. G., Pásztói, M., Pál, Z., Misják, P., Aradi, B., László, V., Pállinger, E., Pap, E., Kittel, A., et al. (2011). Membrane vesicles, current state-of-the-art: Emerging role of extracellular vesicles. *Cellular and Molecular Life Sciences, 68*, 2667–2688.
8. O'Brien, J., Hayder, H., Zayed, Y., & Peng, C. (2018). Overview of microRNA biogenesis, mechanisms of actions, and circulation. *Frontiers in Endocrinology (Lausanne), 9*, 402.
9. Thery, C., Witwer, K. W., Aikawa, E., Alcaraz, M. J., Anderson, J. D., Andriantsitohaina, R., Antoniou, A., Arab, T., Archer, F., Atkin-Smith, G. K., et al. (2018). Minimal information for studies of extracellular vesicles 2018 (MISEV2018): A position statement of the International Society for Extracellular Vesicles and update of the MISEV2014 guidelines. *Journal of Extracellular Vesicles, 7*, 1535750.
10. Welsh, J. A., Goberdhan, D. C. I., O'Driscoll, L., Buzas, E. I., Blenkiron, C., Bussolati, B., Cai, H., Di Vizio, D., Driedonks, T. A. P., Erdbrugger, U., et al. (2024). Minimal information for studies of extracellular vesicles (MISEV2023): From basic to advanced approaches. *Journal of Extracellular Vesicles, 13*, Article e12404.
11. Cheng, L., Sharples, R. A., Scicluna, B. J., & Hill, A. F. (2014). Exosomes provide a protective and enriched source of miRNA for biomarker profiling compared to intracellular and cell-free blood. *Journal of Extracellular Vesicles, 3.*
12. Skog, J., Wurdinger, T., van Rijn, S., Meijer, D. H., Gainche, L., Sena-Esteves, M., Curry, W. T., Jr., Carter, B. S., Krichevsky, A. M., & Breakefield, X. O. (2008). Glioblastoma microvesicles transport RNA and proteins that promote tumour growth and provide diagnostic biomarkers. *Nature Cell Biology, 10*, 1470–1476.
13. Bhatnagar, S., Shinagawa, K., Castellino, F. J., & Schorey, J. S. (2007). Exosomes released from macrophages infected with intracellular pathogens stimulate a proinflammatory response in vitro and in vivo. *Blood, 110*, 3234–3244.
14. Robbins, P. D., & Morelli, A. E. (2014). Regulation of immune responses by extracellular vesicles. *Nature Reviews Immunology, 14*, 195–208.
15. Costa-Silva, B., Aiello, N. M., Ocean, A. J., Singh, S., Zhang, H., Thakur, B. K., Becker, A., Hoshino, A., Mark, M. T., Molina, H., et al. (2015). Pancreatic cancer exosomes initiate pre-metastatic niche formation in the liver. *Nature Cell Biology, 17*, 816–826.
16. Cheng, L., Vella, L. J., Barnham, K. J., McLean, C., Masters, C. L., & Hill, A. F. (2020). Small RNA fingerprinting of Alzheimer's disease frontal cortex extracellular vesicles and their comparison with peripheral extracellular vesicles. *Journal of Extracellular Vesicles, 9*, 1766822.
17. Hu, M., Brown, V., Jackson, J. M., Wijerathne, H., Pathak, H., Koestler, D. C., Nissen, E., Hupert, M. L., Muller, R., Godwin, A. K., et al. (2023). Assessing breast cancer molecular subtypes using extracellular vesicles' mRNA. *Analytical Chemistry, 95*, 7665–7675.
18. Zhao, Z., Wijerathne, H., Godwin, A. K., & Soper, S. A. (2021). Isolation and analysis methods of extracellular vesicles (EVs). *Extracellular Vesicles and Circulating Nucleic Acids, 2*, 80–103.
19. Wijerathne, H., Witek, M. A., Jackson, J. M., Brown, V., Hupert, M. L., Herrera, K., Kramer, C., Davidow, A. E., Li, Y., Baird, A. E., et al. (2020). Affinity enrichment of extracellular vesicles from plasma reveals mRNA changes associated with acute ischemic stroke. *Communications Biology, 3*, 613.
20. De Sousa, K. P., Rossi, I., Abdullahi, M., Ramirez, M. I., Stratton, D., & Inal, J. M. (2023). Isolation and characterization of extracellular vesicles and future directions in diagnosis and therapy. *Wiley Interdisciplinary Reviews: Nanomedicine and Nanobiotechnology, 15*, Article e1835.
21. Livshits, M. A., Khomyakova, E., Evtushenko, E. G., Lazarev, V. N., Kulemin, N. A., Semina, S. E., Generozov, E. V., & Govorun, V. M. (2015). Isolation of exosomes by differential centrifugation: Theoretical analysis of a commonly used protocol. *Science and Reports, 5*, 17319.
22. Akbar, A., Malekian, F., Baghban, N., Kodam, S. P., & Ullah, M. (2022). Methodologies to isolate and purify clinical grade extracellular vesicles for medical applications. *Cells, 11.*

23. Linares, R., Tan, S., Gounou, C., Arraud, N., & Brisson, A. R. (2015). High-speed centrifugation induces aggregation of extracellular vesicles. *Journal of Extracellular Vesicles, 4*, 29509.
24. Yu, L. L., Zhu, J., Liu, J. X., Jiang, F., Ni, W. K., Qu, L. S., Ni, R. Z., Lu, C. H., & Xiao, M. B. (2018). A comparison of traditional and novel methods for the separation of exosomes from human samples. *BioMed Research International, 2018*, 3634563.
25. Li, P., Kaslan, M., Lee, S. H., Yao, J., & Gao, Z. (2017). Progress in exosome isolation techniques. *Theranostics, 7*, 789–804.
26. Zhang, M., Jin, K., Gao, L., Zhang, Z., Li, F., Zhou, F., & Zhang, L. (2018). Methods and technologies for exosome isolation and characterization. *Small Methods, 2*, 1800021.
27. Cheruvanky, A., Zhou, H., Pisitkun, T., Kopp, J. B., Knepper, M. A., Yuen, P. S., & Star, R. A. (2007). Rapid isolation of urinary exosomal biomarkers using a nanomembrane ultrafiltration concentrator. *American Journal of Physiology. Renal Physiology, 292*, F1657-1661.
28. Wei, Z., Batagov, A. O., Schinelli, S., Wang, J., Wang, Y., El Fatimy, R., Rabinovsky, R., Balaj, L., Chen, C. C., Hochberg, F., et al. (2017). Coding and noncoding landscape of extracellular RNA released by human glioma stem cells. *Nature Communications, 8*, 1145.
29. Boing, A. N., van der Pol, E., Grootemaat, A. E., Coumans, F. A., Sturk, A., & Nieuwland, R. (2014). Single-step isolation of extracellular vesicles by size-exclusion chromatography. *Journal of Extracellular Vesicles, 3.*
30. Pang, B., Zhu, Y., Ni, J., Ruan, J., Thompson, J., Malouf, D., Bucci, J., Graham, P., & Li, Y. (2020). Quality assessment and comparison of plasma-derived extracellular vesicles separated by three commercial kits for prostate cancer diagnosis. *International Journal of Nanomedicine, 15*, 10241–10256.
31. Malys, M. S. S., Aigner, C., Schulz, S. M. M., Schachner, H., Rees, A. J. J., & Kain, R. (2021). Isolation of small extracellular vesicles from human sera. *International Journal of Molecular Sciences, 22.*
32. Zeringer, E., Barta, T., Li, M., & Vlassov, A. V. (2015). Strategies for isolation of exosomes. *Cold Spring Harbor Protocols, 2015*, 319–323.
33. Lee, H., He, X., Ni, K., Carnino, J. M., & Jin, Y. (2021). Low concentration of polyethylene glycol facilitates separation of extracellular vesicles from bronchoalveolar lavage fluid. *American Journal of Physiology. Lung Cellular and Molecular Physiology, 320*, L522–L529.
34. Ludwig, A. K., De Miroschedji, K., Doeppner, T. R., Borger, V., Ruesing, J., Rebmann, V., Durst, S., Jansen, S., Bremer, M., Behrmann, E., et al. (2018). Precipitation with polyethylene glycol followed by washing and pelleting by ultracentrifugation enriches extracellular vesicles from tissue culture supernatants in small and large scales. *Journal of Extracellular Vesicles, 7*, 1528109.
35. Rider, M. A., Hurwitz, S. N., & Meckes, D. G., Jr. (2016). ExtraPEG: A polyethylene glycol-based method for enrichment of extracellular vesicles. *Science and Reports, 6*, 23978.
36. Weng, Y., Sui, Z., Shan, Y., Hu, Y., Chen, Y., Zhang, L., & Zhang, Y. (2016). Effective isolation of exosomes with polyethylene glycol from cell culture supernatant for in-depth proteome profiling. *The Analyst, 141*, 4640–4646.
37. Samsonov, R., Shtam, T., Burdakov, V., Glotov, A., Tsyrlina, E., Berstein, L., Nosov, A., Evtushenko, V., Filatov, M., & Malek, A. (2016). Lectin-induced agglutination method of urinary exosomes isolation followed by mi-RNA analysis: Application for prostate cancer diagnostic. *Prostate, 76*, 68–79.
38. Zarovni, N., Corrado, A., Guazzi, P., Zocco, D., Lari, E., Radano, G., Muhhina, J., Fondelli, C., Gavrilova, J., & Chiesi, A. (2015). Integrated isolation and quantitative analysis of exosome shuttled proteins and nucleic acids using immunocapture approaches. *Methods, 87*, 46–58.
39. Kang, Y. T., Purcell, E., Palacios-Rolston, C., Lo, T. W., Ramnath, N., Jolly, S., & Nagrath, S. (2019). Isolation and profiling of circulating tumor-associated exosomes using extracellular vesicular lipid-protein binding affinity based microfluidic device. *Small (Weinheim an der Bergstrasse, Germany), 15*, Article e1903600.
40. Crescitelli, R., Lasser, C., Szabo, T. G., Kittel, A., Eldh, M., Dianzani, I., Buzas, E. I., & Lotvall, J. (2013). Distinct RNA profiles in subpopulations of extracellular vesicles: Apoptotic bodies, microvesicles and exosomes. *Journal of Extracellular Vesicles, 2.*

41. Kowal, J., Arras, G., Colombo, M., Jouve, M., Morath, J. P., Primdal-Bengtson, B., Dingli, F., Loew, D., Tkach, M., & Thery, C. (2016). Proteomic comparison defines novel markers to characterize heterogeneous populations of extracellular vesicle subtypes. *Proceedings of the National Academy of Sciences U S A, 113*, E968-977.
42. Mizenko, R. R., Brostoff, T., Rojalin, T., Koster, H. J., Swindell, H. S., Leiserowitz, G. S., Wang, A., & Carney, R. P. (2021). Tetraspanins are unevenly distributed across single extracellular vesicles and bias sensitivity to multiplexed cancer biomarkers. *Journal of Nanobiotechnology, 19*, 250.
43. Jankovicova, J., Secova, P., Michalkova, K., & Antalikova, J. (2020). Tetraspanins, more than markers of extracellular vesicles in reproduction. *International Journal of Molecular Sciences, 21*.
44. Termini, C. M., & Gillette, J. M. (2017). Tetraspanins function as regulators of cellular signaling. *Frontiers in Cell and Developmental Biology, 5*, 34.
45. Kaur, S., Elkahloun, A. G., Arakelyan, A., Young, L., Myers, T. G., Otaizo-Carrasquero, F., Wu, W., Margolis, L., & Roberts, D. D. (2018). CD63, MHC class 1, and CD47 identify subsets of extracellular vesicles containing distinct populations of noncoding RNAs. *Science and Reports, 8*, 2577.
46. Mathieu, M., Nevo, N., Jouve, M., Valenzuela, J. I., Maurin, M., Verweij, F. J., Palmulli, R., Lankar, D., Dingli, F., Loew, D., et al. (2021). Specificities of exosome versus small ectosome secretion revealed by live intracellular tracking of CD63 and CD9. *Nature Communications, 12*, 4389.
47. Mondal, A., Ashiq, K. A., Phulpagar, P., Singh, D. K., & Shiras, A. (2019). Effective visualization and easy tracking of extracellular vesicles in glioma cells. *Biological Procedures Online, 21*, 4.
48. Srinivasan, S., Duval, M. X., Kaimal, V., Cuff, C., & Clarke, S. H. (2019). Assessment of methods for serum extracellular vesicle small RNA sequencing to support biomarker development. *Journal of Extracellular Vesicles, 8*, 1684425.
49. Yang, Z., Atiyas, Y., Shen, H., Siedlik, M. J., Wu, J., Beard, K., Fonar, G., Dolle, J. P., Smith, D. H., Eberwine, J. H., et al. (2022). Ultrasensitive single extracellular vesicle detection using high throughput droplet digital enzyme-linked immunosorbent assay. *Nano Letters, 22*, 4315–4324.
50. Young, T. W., Kappler, M. P., Hockaden, N. M., Carpenter, R. L., & Jacobson, S. C. (2023). Characterization of extracellular vesicles by resistive-pulse sensing on in-plane multipore nanofluidic devices. *Analytical Chemistry, 95*, 16710–16716.
51. Woo, H. K., Sunkara, V., Park, J., Kim, T. H., Han, J. R., Kim, C. J., Choi, H. I., Kim, Y. K., & Cho, Y. K. (2017). Exodisc for rapid, size-selective, and efficient isolation and analysis of nanoscale extracellular vesicles from biological samples. *ACS Nano, 11*, 1360–1370.
52. Lee, K., Shao, H., Weissleder, R., & Lee, H. (2015). Acoustic purification of extracellular microvesicles. *ACS Nano, 9*, 2321–2327.
53. Reátegui, E., van der Vos, K. E., Lai, C. P., Zeinali, M., Atai, N. A., Aldikacti, B., Floyd, F. P., Khankhel, H. A., Thapar, V., & Hochberg, F. H., et al. (2018). Engineered nanointerfaces for microfluidic isolation and molecular profiling of tumor-specific extracellular vesicles. *Nature Communications, 9*, 175.
54. Kanwar, S. S., Dunlay, C. J., Simeone, D. M., & Nagrath, S. (2014). Microfluidic device (ExoChip) for on-chip isolation, quantification and characterization of circulating exosomes. *Lab on a Chip, 14*, 1891–1900.
55. Zhang, H., & Lyden, D. (2019). Asymmetric-flow field-flow fractionation technology for exomere and small extracellular vesicle separation and characterization. *Nature Protocols, 14*, 1027–1053.
56. Lin, A. A., Shen, H., Spychalski, G., Carpenter, E. L., & Issadore, D. (2023). Modeling and optimization of parallelized immunomagnetic nanopore sorting for surface marker specific isolation of extracellular vesicles from complex media. *Science and Reports, 13*, 13292.
57. Ku, A., Lim, H. C., Evander, M., Lilja, H., Laurell, T., Scheding, S., & Ceder, Y. (2018). Acoustic enrichment of extracellular vesicles from biological fluids. *Analytical Chemistry, 90*, 8011–8019.

58. Lane, R. E., Korbie, D., Anderson, W., Vaidyanathan, R., & Trau, M. (2015). Analysis of exosome purification methods using a model liposome system and tunable-resistive pulse sensing. *Science and Reports, 5*, 7639.
59. Antounians, L., Tzanetakis, A., Pellerito, O., Catania, V. D., Sulistyo, A., Montalva, L., McVey, M. J., & Zani, A. (2019). The regenerative potential of amniotic fluid stem cell extracellular vesicles: Lessons learned by comparing different isolation techniques. *Science and Reports, 9*, 1837.
60. Kalra, H., Adda, C. G., Liem, M., Ang, C.-S., Mechler, A., Simpson, R. J., Hulett, M. D., & Mathivanan, S. (2013). Comparative proteomics evaluation of plasma exosome isolation techniques and assessment of the stability of exosomes in normal human blood plasma. *Proteomics, 13*, 3354–3364.
61. Tauro, B. J., Greening, D. W., Mathias, R. A., Ji, H., Mathivanan, S., Scott, A. M., & Simpson, R. J. (2012). Comparison of ultracentrifugation, density gradient separation, and immunoaffinity capture methods for isolating human colon cancer cell line LIM1863-derived exosomes. *Methods, 56*, 293–304.
62. Lobb, R. J., Becker, M., Wen, S. W., Wong, C. S., Wiegmans, A. P., Leimgruber, A., & Moller, A. (2015). Optimized exosome isolation protocol for cell culture supernatant and human plasma. *Journal of Extracellular Vesicles, 4*, 27031.
63. Van Deun, J., Mestdagh, P., Sormunen, R., Cocquyt, V., Vermaelen, K., Vandesompele, J., Bracke, M., De Wever, O., & Hendrix, A. (2014). The impact of disparate isolation methods for extracellular vesicles on downstream RNA profiling. *Journal of Extracell Vesicles, 3*.
64. Mateescu, B., Kowal, E. J., van Balkom, B. W., Bartel, S., Bhattacharyya, S. N., Buzas, E. I., Buck, A. H., de Candia, P., Chow, F. W., Das, S., et al. (2017). Obstacles and opportunities in the functional analysis of extracellular vesicle RNA—an ISEV position paper. *Journal of Extracelular Vesicles, 6*, 1286095.
65. Bagci, C., Sever-Bahcekapili, M., Belder, N., Bennett, A. P. S., Erdener, S. E., & Dalkara, T. (2022). Overview of extracellular vesicle characterization techniques and introduction to combined reflectance and fluorescence confocal microscopy to distinguish extracellular vesicle subpopulations. *Neurophotonics, 9*, Article 021903.
66. Crescitelli, R., Lasser, C., & Lotvall, J. (2021). Isolation and characterization of extracellular vesicle subpopulations from tissues. *Nature Protocols, 16*, 1548–1580.
67. Ramirez, M. I., Amorim, M. G., Gadelha, C., Milic, I., Welsh, J. A., Freitas, V. M., Nawaz, M., Akbar, N., Couch, Y., Makin, L., et al. (2018). Technical challenges of working with extracellular vesicles. *Nanoscale, 10*, 881–906.
68. Caradec, J., Kharmate, G., Hosseini-Beheshti, E., Adomat, H., Gleave, M., & Guns, E. (2014). Reproducibility and efficiency of serum-derived exosome extraction methods. *Clinical Biochemistry, 47*, 1286–1292.
69. Wu, Y., Deng, W., & Klinke II, D. J. (2015). Exosomes: improved methods to characterize their morphology, RNA content, and surface protein biomarkers. *Analyst, 140*, 6631–6642.
70. Nguyen, D. B., Ly, T. B., Wesseling, M. C., Hittinger, M., Torge, A., Devitt, A., Perrie, Y., & Bernhardt, I. (2016). Characterization of microvesicles released from human red blood cells. *Cellular Physiology and Biochemistry, 38*, 1085–1099.
71. Sokolova, V., Ludwig, A. K., Hornung, S., Rotan, O., Horn, P. A., Epple, M., & Giebel, B. (2011). Characterisation of exosomes derived from human cells by nanoparticle tracking analysis and scanning electron microscopy. *Colloids and Surfaces. B, Biointerfaces, 87*, 146–150.
72. Kondratov, K. A., Petrova, T. A., Mikhailovskii, V. Y., Ivanova, A. N., Kostareva, A. A., & Fedorov, A. V. (2017). Extracellular vesicles from blood plasma studied by low voltage scanning electron microscopy. *Tsitologiia, 59*, 169–177.
73. Cheng, H. L., Fu, C. Y., Kuo, W. C., Chen, Y. W., Chen, Y. S., Lee, Y. M., Li, K. H., Chen, C., Ma, H. P., Huang, P. C., et al. (2018). Detecting miRNA biomarkers from extracellular vesicles for cardiovascular disease with a microfluidic system. *Lab on a Chip, 18*, 2917–2925.
74. Royo, F., Schlangen, K., Palomo, L., Gonzalez, E., Conde-Vancells, J., Berisa, A., Aransay, A. M., & Falcon-Perez, J. M. (2013). Transcriptome of extracellular vesicles released by hepatocytes. *PLoS ONE, 8*, Article e68693.

75. Yuana, Y., Koning, R. I., Kuil, M. E., Rensen, P. C., Koster, A. J., Bertina, R. M., & Osanto, S. (2013). Cryo-electron microscopy of extracellular vesicles in fresh plasma. *Journal of Extracellular Vesicles, 2.*
76. Anderson, W., Lane, R., Korbie, D., & Trau, M. (2015). Observations of tunable resistive pulse sensing for exosome analysis: Improving system sensitivity and stability. *Langmuir, 31*, 6577–6587.
77. Chernyshev, V. S., Rachamadugu, R., Tseng, Y. H., Belnap, D. M., Jia, Y., Branch, K. J., Butterfield, A. E., Pease, L. F., 3rd., Bernard, P. S., & Skliar, M. (2015). Size and shape characterization of hydrated and desiccated exosomes. *Analytical and Bioanalytical Chemistry, 407*, 3285–3301.
78. Thery, C., Amigorena, S., Raposo, G., & Clayton, A. (2006). Isolation and characterization of exosomes from cell culture supernatants and biological fluids. *Current Protocols in Cell Biology, 30*, 3–22.
79. Choi, H., & Mun, J. Y. (2017). Structural analysis of exosomes using different types of electron microscopy. *Applied Microscopy, 47*, 171–175.
80. Raposo, G., & Stoorvogel, W. (2013). Extracellular vesicles: Exosomes, microvesicles, and friends. *Journal of Cell Biology, 200*, 373–383.
81. Carnell-Morris, P., Tannetta, D., Siupa, A., Hole, P., & Dragovic, R. (2017). Analysis of extracellular vesicles using fluorescence nanoparticle tracking analysis. *Methods in Molecular Biology, 1660*, 153–173.
82. Weber, A., Wehmeyer, J. C., Schmidt, V., Lichtenberg, A., & Akhyari, P. (2019). Rapid fluorescence-based characterization of single extracellular vesicles in human blood with nanoparticle-tracking analysis. *Journal of Visualized Experiments.*
83. Yu, D., Mei, Y., Wang, L., Zhao, Y., Fan, X., Liang, D., Li, L., Zhu, J., Bi, S., Wang, X., et al. (2023). Nano-seq analysis reveals different functional tendency between exosomes and microvesicles derived from hUMSC. *Stem Cell Research & Therapy, 14*, 272.
84. Noble, J. M., Roberts, L. M., Vidavsky, N., Chiou, A. E., Fischbach, C., Paszek, M. J., Estroff, L. A., & Kourkoutis, L. F. (2020). Direct comparison of optical and electron microscopy methods for structural characterization of extracellular vesicles. *Journal of Structural Biology, 210*, Article 107474.
85. Kwon, Y., & Park, J. (2022). Methods to analyze extracellular vesicles at single particle level. *Micro and Nano Systems Letters, 10*, 14.
86. van der Pol, E., Coumans, F. A., Grootemaat, A. E., Gardiner, C., Sargent, I. L., Harrison, P., Sturk, A., van Leeuwen, T. G., & Nieuwland, R. (2014). Particle size distribution of exosomes and microvesicles determined by transmission electron microscopy, flow cytometry, nanoparticle tracking analysis, and resistive pulse sensing. *Journal of Thrombosis and Haemostasis, 12*, 1182–1192.
87. Tian, Y., Ma, L., Gong, M., Su, G., Zhu, S., Zhang, W., Wang, S., Li, Z., Chen, C., Li, L., et al. (2018). Protein profiling and sizing of extracellular vesicles from colorectal cancer patients via flow cytometry. *ACS Nano, 12*, 671–680.
88. Maas, S. L. N., Broekman, M. L. D., & de Vrij, J. (2017). In A. F. Hill (Ed.), *Exosomes and microvesicles: Methods and protocols* (pp. 21–33). Springer New York.
89. Fraikin, J.-L., Teesalu, T., McKenney, C. M., Ruoslahti, E., & Cleland, A. N. (2011). A high-throughput label-free nanoparticle analyser. *Nature Nanotechnology, 6*, 308–313.
90. Cimorelli, M., Nieuwland, R., Varga, Z., & van der Pol, E. (2021). Standardized procedure to measure the size distribution of extracellular vesicles together with other particles in biofluids with microfluidic resistive pulse sensing. *PLoS ONE, 16*, Article e0249603.
91. Filipe, V., Hawe, A., & Jiskoot, W. (2010). Critical evaluation of nanoparticle tracking analysis (NTA) by NanoSight for the measurement of nanoparticles and protein aggregates. *Pharmaceutical Research, 27*, 796–810.
92. Bachurski, D., Schuldner, M., Nguyen, P. H., Malz, A., Reiners, K. S., Grenzi, P. C., Babatz, F., Schauss, A. C., Hansen, H. P., Hallek, M., et al. (2019). Extracellular vesicle measurements with nanoparticle tracking analysis—an accuracy and repeatability comparison between NanoSight NS300 and ZetaView. *Journal of Extracellular Vesicles, 8*, 1596016.

93. Akers, J. C., Ramakrishnan, V., Nolan, J. P., Duggan, E., Fu, C. C., Hochberg, F. H., Chen, C. C., & Carter, B. S. (2016). Comparative analysis of technologies for quantifying extracellular vesicles (EVs) in clinical cerebrospinal fluids (CSF). *PLoS ONE, 11*, Article e0149866.
94. Griffiths, D., Carnell-Morris, P., & Wright, M. (2020). In E. Ferrari & M. Soloviev (Eds.), *Nanoparticles in biology and medicine: Methods and protocols* (pp. 289–303). Springer US.
95. Kim, A., Ng, W. B., Bernt, W., & Cho, N. J. (2019). Validation of size estimation of nanoparticle tracking analysis on polydisperse macromolecule assembly. *Science and Reports, 9*, 2639.
96. McKinnon, K. M. (2018). Flow cytometry: An overview. *Current Protocols Immunology, 120*, 1–5.
97. Witwer, K. W., Buzas, E. I., Bemis, L. T., Bora, A., Lasser, C., Lotvall, J., Nolte-'t Hoen, E. N., Piper, M. G., Sivaraman, S., & Skog, J. et al. (2013) Standardization of sample collection, isolation and analysis methods in extracellular vesicle research. *Journal of Extracellular Vesicles, 2*.
98. Zhao, Z., Yang, Y., Zeng, Y., & He, M. (2016). A microfluidic ExoSearch chip for multiplexed exosome detection towards blood-based ovarian cancer diagnosis. *Lab on a Chip, 16*, 489–496.
99. Morales-Kastresana, A., & Jones, J. C. (2017). Flow cytometric analysis of extracellular vesicles. *Methods in Molecular Biology, 1545*, 215–225.
100. Gorgens, A., Bremer, M., Ferrer-Tur, R., Murke, F., Tertel, T., Horn, P. A., Thalmann, S., Welsh, J. A., Probst, C., Guerin, C., et al. (2019). Optimisation of imaging flow cytometry for the analysis of single extracellular vesicles by using fluorescence-tagged vesicles as biological reference material. *Journal of Extracellular Vesicles, 8*, 1587567.
101. Choi, D., Montermini, L., Jeong, H., Sharma, S., Meehan, B., & Rak, J. (2019). Mapping subpopulations of cancer cell-derived extracellular vesicles and particles by nano-flow cytometry. *ACS Nano, 13*, 10499–10511.
102. Oliveira, G. P. D., Zigon, E., Rogers, G., Davodian, D., Lu, S., Jovanovic-Talisman, T., Jones, J., Tigges, J., Tyagi, S., & Ghiran, I. C. (2020). Detection of extracellular vesicle RNA using molecular beacons. *iScience, 23*, 100782.
103. Liu, H., Tian, Y., Xue, C., Niu, Q., Chen, C., & Yan, X. (2022). Analysis of extracellular vesicle DNA at the single-vesicle level by nano-flow cytometry. *Journal of Extracellular Vesicles, 11*, Article e12206.
104. van der Vlist, E. J., Nolte-'t Hoen, E. N., Stoorvogel, W., Arkesteijn, G. J., & Wauben, M. H. (2012). Fluorescent labeling of nano-sized vesicles released by cells and subsequent quantitative and qualitative analysis by high-resolution flow cytometry. *Nature Protocols, 7*, 1311–1326.
105. Nelson, B. C., Maragh, S., Ghiran, I. C., Jones, J. C., DeRose, P. C., Elsheikh, E., Vreeland, W. N., & Wang, L. (2020). Measurement and standardization challenges for extracellular vesicle therapeutic delivery vectors. *Nanomedicine (London, England), 15*, 2149–2170.
106. Thery, C., Witwer, K. W., Aikawa, E., Alcaraz, M. J., Anderson, J. D., Andriantsitohaina, R., Antoniou, A., Arab, T., Archer, F., Atkin-Smith, G. K., et al. (2018). Minimal information for studies of extracellular vesicles 2018 (MISEV2018): A position statement of the international society for extracellular vesicles and update of the MISEV2014 guidelines. *Journal of Extracellular Vesicles, 7*, 1535750.
107. Sodar, B. W., Kittel, A., Paloczi, K., Vukman, K. V., Osteikoetxea, X., Szabo-Taylor, K., Nemeth, A., Sperlagh, B., Baranyai, T., Giricz, Z., et al. (2016). Low-density lipoprotein mimics blood plasma-derived exosomes and microvesicles during isolation and detection. *Science and Reports, 6*, 24316.
108. Vaidyanathan, S., Wijerathne, H., Gamage, S. S. T., Shiri, F., Zhao, Z., Choi, J., Park, S., Witek, M. A., McKinney, C., Verber, M., et al. (2023). High sensitivity extended nano-coulter counter for detection of viral particles and extracellular vesicles. *Analtycal Chemistry, 96*, 9892–9900.
109. Abels, E. R., & Breakefield, X. O. (2016). Introduction to extracellular vesicles: Biogenesis, RNA cargo selection, content, release, and uptake. *Cellular and Molecular Neurobiology, 36*, 301–312.

110. Mir, B., & Goettsch, C. (2020). Extracellular vesicles as delivery vehicles of specific cellular cargo. *Cells, 9*.
111. Tkach, M., & Thery, C. (2016). Communication by extracellular vesicles: Where we are and where we need to go. *Cell, 164*, 1226–1232.
112. O'Brien, K., Breyne, K., Ughetto, S., Laurent, L. C., & Breakefield, X. O. (2020). RNA delivery by extracellular vesicles in mammalian cells and its applications. *Nature Reviews Molecular Cell Biology, 21*, 585–606.
113. Danielson, K. M., Shah, R., Yeri, A., Liu, X., Camacho Garcia, F., Silverman, M., Tanriverdi, K., Das, A., Xiao, C., Jerosch-Herold, M., et al. (2018). Plasma circulating extracellular RNAs in left ventricular remodeling post-myocardial infarction. *eBioMedicine, 32*, 172–181.
114. Varcianna, A., Myszczynska, M. A., Castelli, L. M., O'Neill, B., Kim, Y., Talbot, J., Nyberg, S., Nyamali, I., Heath, P. R., Stopford, M. J., et al. (2019). Micro-RNAs secreted through astrocyte-derived extracellular vesicles cause neuronal network degeneration in C9orf72 ALS. *eBioMedicine, 40*, 626–635.
115. Rocha, I. F. M. D., Martins, S. T., Amatuzzi, R. F., Zamith-Miranda, D., Nosanchuk, J. D., Rodrigues, M. L., & Alves, L. R. (2021). Cellular and extracellular vesicle RNA analysis in the global threat fungus *Candida auris. Microbiology Spectrum, 9*, e01538–e01521.
116. Sork, H., Corso, G., Krjutskov, K., Johansson, H. J., Nordin, J. Z., Wiklander, O. P. B., Lee, Y. X. F., Westholm, J. O., Lehtiö, J., Wood, M. J. A., et al. (2018). Heterogeneity and interplay of the extracellular vesicle small RNA transcriptome and proteome. *Scientific Reports, 8*, 10813.
117. Huang, X., Yuan, T., Tschannen, M., Sun, Z., Jacob, H., Du, M., Liang, M., Dittmar, R. L., Liu, Y., Liang, M., et al. (2013). Characterization of human plasma-derived exosomal RNAs by deep sequencing. *BMC Genomics, 14*, 319.
118. Kumar, M. A., Baba, S. K., Sadida, H. Q., Marzooqi, S. A., Jerobin, J., Altemani, F. H., Algehainy, N., Alanazi, M. A., Abou-Samra, A.-B., Kumar, R., et al. (2024). Extracellular vesicles as tools and targets in therapy for diseases. *Signal Transduction and Targeted Therapy, 9*, 27.
119. Collino, F., Deregibus, M. C., Bruno, S., Sterpone, L., Aghemo, G., Viltono, L., Tetta, C., & Camussi, G. (2010). Microvesicles derived from adult human bone marrow and tissue specific mesenchymal stem cells shuttle selected pattern of miRNAs. *PLoS ONE, 5*, Article e11803.
120. de Jong, O. G., Verhaar, M. C., Chen, Y., Vader, P., Gremmels, H., Posthuma, G., Schiffelers, R. M., Gucek, M., & van Balkom, B. W. (2012). Cellular stress conditions are reflected in the protein and RNA content of endothelial cell-derived exosomes. *Journal of Extracellular Vesicles, 1*.
121. Deng, Z. L., Munch, P. C., Mreches, R., & McHardy, A. C. (2022). Rapid and accurate identification of ribosomal RNA sequences via deep learning. *Nucleic Acids Research, 50*, Article e60.
122. Skrypina, N. A., Timofeeva, A. V., Khaspekov, G. L., Savochkina, L. P., & Beabealashvilli, R. S. (2003). Total RNA suitable for molecular biology analysis. *Journal of Biotechnology, 105*, 1–9.
123. Schroeder, A., Mueller, O., Stocker, S., Salowsky, R., Leiber, M., Gassmann, M., Lightfoot, S., Menzel, W., Granzow, M., & Ragg, T. (2006). The RIN: An RNA integrity number for assigning integrity values to RNA measurements. *BMC Molecular Biology, 7*, 3.
124. Kim, K. M., Abdelmohsen, K., Mustapic, M., Kapogiannis, D., & Gorospe, M. (2017). RNA in extracellular vesicles. *Wiley Interdisciplinary Reviews: RNA, 8*.
125. O'Grady, T., Njock, M. S., Lion, M., Bruyr, J., Mariavelle, E., Galvan, B., Boeckx, A., Struman, I., & Dequiedt, F. (2022). Sorting and packaging of RNA into extracellular vesicles shape intracellular transcript levels. *BMC Biology, 20*, 72.
126. Khalighfard, S., Alizadeh, A. M., Irani, S., & Omranipour, R. (2018). Plasma miR-21, miR-155, miR-10b, and Let-7a as the potential biomarkers for the monitoring of breast cancer patients. *Scientific Reports, 8*, 17981.
127. Driedonks, T. A. P., van der Grein, S. G., Ariyurek, Y., Buermans, H. P. J., Jekel, H., Chow, F. W. N., Wauben, M. H. M., Buck, A. H., 't Hoen, P. A. C., & Nolte-'t Hoen, E. N. M. (2018). Immune stimuli shape the small non-coding transcriptome of extracellular vesicles released by dendritic cells. *Cellular and Molecular Life Sciences*, 75, 3857–3875.

128. Liu, X., Zheng, Q., Vrettos, N., Maragkakis, M., Alexiou, P., Gregory, B. D., & Mourelatos, Z. (2014). A MicroRNA precursor surveillance system in quality control of MicroRNA synthesis. *Molecular Cell, 55*, 868–879.
129. O'Grady, T., Njock, M.-S., Lion, M., Bruyr, J., Mariavelle, E., Galvan, B., Boeckx, A., Struman, I., & Dequiedt, F. (2022). Sorting and packaging of RNA into extracellular vesicles shape intracellular transcript levels. *BMC Biology, 20*, 72.
130. Wei, Z., Batagov, A. O., Schinelli, S., Wang, J., Wang, Y., El Fatimy, R., Rabinovsky, R., Balaj, L., Chen, C. C., Hochberg, F., et al. (2017). Coding and noncoding landscape of extracellular RNA released by human glioma stem cells. *Nature Communications, 8*, 1145.
131. Chevillet, J. R., Kang, Q., Ruf, I. K., Briggs, H. A., Vojtech, L. N., Hughes, S. M., Cheng, H. H., Arroyo, J. D., Meredith, E. K., Gallichotte, E. N., et al. (2014). Quantitative and stoichiometric analysis of the microRNA content of exosomes. *Proceedings of the National Academy of Sciences U S A, 111*, 14888–14893.
132. Ogata-Kawata, H., Izumiya, M., Kurioka, D., Honma, Y., Yamada, Y., Furuta, K., Gunji, T., Ohta, H., Okamoto, H., Sonoda, H., et al. (2014). Circulating exosomal microRNAs as biomarkers of colon cancer. *PLoS ONE, 9*, Article e92921.
133. Liu, C. G., Song, J., Zhang, Y. Q., & Wang, P. C. (2014). MicroRNA-193b is a regulator of amyloid precursor protein in the blood and cerebrospinal fluid derived exosomal microRNA-193b is a biomarker of Alzheimer's disease. *Molecular Medicine Reports, 10*, 2395–2400.
134. He, S., Huang, L., Shao, C., Nie, T., Xia, L., Cui, B., Lu, F., Zhu, L., Chen, B., & Yang, Q. (2021). Several miRNAs derived from serum extracellular vesicles are potential biomarkers for early diagnosis and progression of Parkinson's disease. *Translational Neurodegeneration, 10*, 25.
135. Wu, T., Chen, Y., Du, Y., Tao, J., Li, W., Zhou, Z., & Yang, Z. (2018). Circulating exosomal miR-92b-5p is a promising diagnostic biomarker of heart failure with reduced ejection fraction patients hospitalized for acute heart failure. *Journal of Thoracic Disease, 10*, 6211–6220.
136. Levänen, B., Bhakta, N. R., Torregrosa Paredes, P., Barbeau, R., Hiltbrunner, S., Pollack, J. L., Sköld, C. M., Svartengren, M., Grunewald, J., Gabrielsson, S., et al. (2013). Altered microRNA profiles in bronchoalveolar lavage fluid exosomes in asthmatic patients. *The Journal of Allergy and Clinical Immunology, 131*, 894–903.
137. Kmiołek, T., & Paradowska-Gorycka, A. (2022). miRNAs as biomarkers and possible therapeutic strategies in rheumatoid arthritis. *Cells, 11*.
138. Nolte-'t Hoen, E. N., Buermans, H. P., Waasdorp, M., Stoorvogel, W., Wauben, M. H., & t Hoen, P. A. (2012). Deep sequencing of RNA from immune cell-derived vesicles uncovers the selective incorporation of small non-coding RNA biotypes with potential regulatory functions. *Nucleic Acids Research, 40*, 9272–9285.
139. Huang, Z.-H., Du, Y.-P., Wen, J.-T., Lu, B.-F., & Zhao, Y. (2022). SnoRNAs: Functions and mechanisms in biological processes, and roles in tumor pathophysiology. *Cell Death Discovery, 8*, 259.
140. Fitz, N. F., Wang, J., Kamboh, M. I., Koldamova, R., & Lefterov, I. (2021). Small nucleolar RNAs in plasma extracellular vesicles and their discriminatory power as diagnostic biomarkers of Alzheimer's disease. *Neurobiology of Diseases, 159*, Article 105481.
141. Goh, T. X., Tan, S. L., Roebuck, M. M., Teo, S. H., & Kamarul, T. (2022). A systematic review of extracellular vesicle-derived piwi-interacting RNA in human body fluid and its role in disease progression. *Tissue Engineering. Part C, Methods, 28*, 511–528.
142. Liu, Y., Dou, M., Song, X., Dong, Y., Liu, S., Liu, H., Tao, J., Li, W., Yin, X., & Xu, W. (2019). The emerging role of the piRNA/piwi complex in cancer. *Molecular Cancer, 18*, 123.
143. Jain, G., Stuendl, A., Rao, P., Berulava, T., Pena Centeno, T., Kaurani, L., Burkhardt, S., Delalle, I., Kornhuber, J., Hüll, M., et al. (2019). A combined miRNA–piRNA signature to detect Alzheimer's disease. *Translational Psychiatry, 9*, 250.
144. Kowalski, M. P., & Krude, T. (2015). Functional roles of non-coding Y RNAs. *International Journal of Biochemistry & Cell Biology, 66*, 20–29.
145. Driedonks, T. A. P., & Nolte-'t Hoen, E. N. M. (2018). Circulating Y-RNAs in extracellular vesicles and ribonucleoprotein complexes; Implications for the immune system. *Frontiers in Immunology, 9*, 3164.

146. Dhahbi, J. M., Spindler, S. R., Atamna, H., Boffelli, D., & Martin, D. I. (2014). Deep sequencing of serum small RNAs identifies patterns of 5′ tRNA half and YRNA fragment expression associated with breast cancer. *Biomark Cancer, 6*, 37–47.
147. Lovisa, F., Di Battista, P., Gaffo, E., Damanti, C. C., Garbin, A., Gallingani, I., Carraro, E., Pillon, M., Biffi, A., Bortoluzzi, S., et al. (2020). RNY4 in circulating exosomes of patients with pediatric anaplastic large cell lymphoma: An active player? *Frontiers in Oncology, 10*, 238.
148. Weng, Q., Wang, Y., Xie, Y., Yu, X., Zhang, S., Ge, J., Li, Z., Ye, G., & Guo, J. (2022). Extracellular vesicles-associated tRNA-derived fragments (tRFs): Biogenesis, biological functions, and their role as potential biomarkers in human diseases. *Journal of Molecular Medicine (Berlin, Germany), 100*, 679–695.
149. Shurtleff, M. J., Yao, J., Qin, Y., Nottingham, R. M., Temoche-Diaz, M. M., Schekman, R., & Lambowitz, A. M. (2017). Broad role for YBX1 in defining the small noncoding RNA composition of exosomes. *Proceedings of the National Academy of Sciences U S A, 114*, E8987-e8995.
150. Goodarzi, H., Liu, X., Nguyen, H. C., Zhang, S., Fish, L., & Tavazoie, S. F. (2015). Endogenous tRNA-derived fragments suppress breast cancer progression via YBX1 displacement. *Cell, 161*, 790–802.
151. Ahadi, A., Brennan, S., Kennedy, P. J., Hutvagner, G., & Tran, N. (2016). Long non-coding RNAs harboring miRNA seed regions are enriched in prostate cancer exosomes. *Scientific Reports, 6*, 24922.
152. Gezer, U., Ozgur, E., Cetinkaya, M., Isin, M., & Dalay, N. (2014). Long non-coding RNAs with low expression levels in cells are enriched in secreted exosomes. *Cell Biology International, 38*, 1076–1079.
153. Jenjaroenpun, P., Kremenska, Y., Nair, V. M., Kremenskoy, M., Joseph, B., & Kurochkin, I. V. (2013). Characterization of RNA in exosomes secreted by human breast cancer cell lines using next-generation sequencing. *PeerJ, 1*, Article e201.
154. Kenneweg, F., Bang, C., Xiao, K., Boulanger, C. M., Loyer, X., Mazlan, S., Schroen, B., Hermans-Beijnsberger, S., Foinquinos, A., Hirt, M. N., et al. (2019). Long noncoding RNA-enriched vesicles secreted by hypoxic cardiomyocytes drive cardiac fibrosis. *Molecular Therapy Nucleic Acids, 18*, 363–374.
155. Tan, S. C., & Yiap, B. C. (2009). DNA, RNA, and protein extraction: The past and the present. *Journal of Biomedicine & Biotechnology, 2009*, Article 574398.
156. Eldh, M., Lotvall, J., Malmhall, C., & Ekstrom, K. (2012). Importance of RNA isolation methods for analysis of exosomal RNA: Evaluation of different methods. *Molecular Immunology, 50*, 278–286.
157. Palazzo, A. F., & Lee, E. S. (2015). Non-coding RNA: What is functional and what is junk? *Frontiers in Genetics, 6*.
158. Arya, M., Shergill, I. S., Williamson, M., Gommersall, L., Arya, N., & Patel, H. R. H. (2005). Basic principles of real-time quantitative PCR. *Expert Review of Molecular Diagnostics, 5*, 209–219.
159. Bustin, S., & Huggett, J. (2017). QPCR primer design revisited. *Biomol Detect Quantif, 14*, 19–28.
160. Levesque-Sergerie, J. P., Duquette, M., Thibault, C., Delbecchi, L., & Bissonnette, N. (2007). Detection limits of several commercial reverse transcriptase enzymes: Impact on the low- and high-abundance transcript levels assessed by quantitative RT-PCR. *BMC Molecular Biology, 8*, 93.
161. Batagov, A. O., & Kurochkin, I. V. (2013). Exosomes secreted by human cells transport largely mRNA fragments that are enriched in the 3′-untranslated regions. *Biology Direct, 8*, 12.
162. Sueta, A., Yamamoto, Y., Tomiguchi, M., Takeshita, T., Yamamoto-Ibusuki, M., & Iwase, H. (2017). Differential expression of exosomal miRNAs between breast cancer patients with and without recurrence. *Oncotarget, 8*, 69934–69944.
163. Kumar, S. R., Kimchi, E. T., Manjunath, Y., Gajagowni, S., Stuckel, A. J., & Kaifi, J. T. (2020). RNA cargos in extracellular vesicles derived from blood serum in pancreas associated conditions. *Science and Reports, 10*, 2800.

164. Bellingham, S. A., Coleman, B. M., & Hill, A. F. (2012). Small RNA deep sequencing reveals a distinct miRNA signature released in exosomes from prion-infected neuronal cells. *Nucleic Acids Research, 40*, 10937–10949.
165. Shen, H., Zaitseva, D., Yang, Z., Forsythe, L., Joergensen, S., Zone, A. I., Shehu, J., Maghraoui, S., Ghorbani, A., Davila, A., et al. (2023). Brain-derived extracellular vesicles as serologic markers of brain injury following cardiac arrest: A pilot feasibility study. *Resuscitation, 191*, Article 109937.
166. Pinheiro, C., Guilbert, N., Lippens, L., Roux, Q., Boiy, R., Fischer, S., Van Dorpe, S., De Craene, B., Berx, G., Boterberg, T., et al. (2024). Identification and validation of extracellular vesicle reference genes for the normalization of RT-qPCR data. *Journal of Extracellular Vesicles, 13*, Article e12421.
167. Crossland, R. E., Norden, J., Bibby, L. A., Davis, J., & Dickinson, A. M. (2016). Evaluation of optimal extracellular vesicle small RNA isolation and qRT-PCR normalisation for serum and urine. *Journal of Immunological Methods, 429*, 39–49.
168. Gouin, K., Peck, K., Antes, T., Johnson, J. L., Li, C., Vaturi, S. D., Middleton, R., de Couto, G., Walravens, A. S., Rodriguez-Borlado, L., et al. (2017). A comprehensive method for identification of suitable reference genes in extracellular vesicles. *Journal of Extracellular Vesicles, 6*, 1347019.
169. Takahashi, K., Yan, I. K., Kim, C., Kim, J., & Patel, T. (2014). Analysis of extracellular RNA by digital PCR. *Frontiers in Oncology, 4*, 129.
170. Luo, T., Chen, S. Y., Qiu, Z. X., Miao, Y. R., Ding, Y., Pan, X. Y., Li, Y., Lei, Q., & Guo, A. Y. (2022). Transcriptomic features in a single extracellular vesicle via single-cell RNA sequencing. *Small Methods, 6*, Article e2200881.
171. Taylor, S. C., Laperriere, G., & Germain, H. (2017). Droplet digital PCR versus qPCR for gene expression analysis with low abundant targets: From variable nonsense to publication quality data. *Science and Reports, 7*, 2409.
172. Hindson, B. J., Ness, K. D., Masquelier, D. A., Belgrader, P., Heredia, N. J., Makarewicz, A. J., Bright, I. J., Lucero, M. Y., Hiddessen, A. L., Legler, T. C., et al. (2011). High-throughput droplet digital PCR system for absolute quantitation of DNA copy number. *Analytical Chemistry, 83*, 8604–8610.
173. Vogelstein, B., & Kinzler, K. W. (1999). Digital PCR. *Proceedings of the National Academy of Sciences U S A, 96*, 9236–9241.
174. Wang, C., Ding, Q., Plant, P., Basheer, M., Yang, C., Tawedrous, E., Krizova, A., Boulos, C., Farag, M., Cheng, Y., et al. (2019). Droplet digital PCR improves urinary exosomal miRNA detection compared to real-time PCR. *Clinical Biochemistry, 67*, 54–59.
175. Barberis, E., Vanella, V. V., Falasca, M., Caneapero, V., Cappellano, G., Raineri, D., Ghirimoldi, M., De Giorgis, V., Puricelli, C., Vaschetto, R., et al. (2021). Circulating exosomes are strongly involved in SARS-CoV-2 infection. *Frontiers in Molecular Biosciences, 8*, Article 632290.
176. de Voogt, W. S., Tanenbaum, M. E., & Vader, P. (2021). Illuminating RNA trafficking and functional delivery by extracellular vesicles. *Advanced Drug Delivery Reviews, 174*, 250–264.
177. Hill, A. F., Pegtel, D. M., Lambertz, U., Leonardi, T., O'Driscoll, L., Pluchino, S., Ter-Ovanesyan, D., & Nolte-'t Hoen, E. N. (2013). ISEV position paper: Extracellular vesicle RNA analysis and bioinformatics. *Journal of Extracellular Vesicles, 2*.
178. Mateescu, B., Kowal, E. J., van Balkom, B. W., Bartel, S., Bhattacharyya, S. N., Buzas, E. I., Buck, A. H., de Candia, P., Chow, F. W., Das, S., et al. (2017). Obstacles and opportunities in the functional analysis of extracellular vesicle RNA—An ISEV position paper. *Journal of Extracellular Vesicles, 6*, 1286095.
179. van Balkom, B. W., Eisele, A. S., Pegtel, D. M., Bervoets, S., & Verhaar, M. C. (2015). Quantitative and qualitative analysis of small RNAs in human endothelial cells and exosomes provides insights into localized RNA processing, degradation and sorting. *Journal of Extracellular Vesicles, 4*, 26760.
180. Raabe, C. A., Tang, T. H., Brosius, J., & Rozhdestvensky, T. S. (2014). Biases in small RNA deep sequencing data. *Nucleic Acids Research, 42*, 1414–1426.

181. Zhou, R., Chen, K. K., Zhang, J., Xiao, B., Huang, Z., Ju, C., Sun, J., Zhang, F., Lv, X.-B., & Huang, G. (2018). The decade of exosomal long RNA species: An emerging cancer antagonist. *Molecular Cancer, 17*, 75.
182. Deshpande, D., Chhugani, K., Chang, Y., Karlsberg, A., Loeffler, C., Zhang, J., Muszyńska, A., Munteanu, V., Yang, H., & Rotman, J. et al. (2023). RNA-seq data science: From raw data to effective interpretation. *Frontiers in Genetics, 14*.
183. Stark, R., Grzelak, M., & Hadfield, J. (2019). RNA sequencing: The teenage years. *Nature Reviews Genetics, 20*, 631–656.
184. Makarova, J., Maltseva, D., & Tonevitsky, A. (2023). Challenges in characterization of transcriptomes of extracellular vesicles and non-vesicular extracellular RNA carriers. *Frontiers in Molecular Biosciences, 10*, 1327985.
185. Robert, C., & Watson, M. (2015). Errors in RNA-Seq quantification affect genes of relevance to human disease. *Genome Biology, 16*, 177.
186. Conesa, A., Madrigal, P., Tarazona, S., Gomez-Cabrero, D., Cervera, A., McPherson, A., Szczesniak, M. W., Gaffney, D. J., Elo, L. L., Zhang, X., et al. (2016). A survey of best practices for RNA-seq data analysis. *Genome Biology, 17*, 13.
187. Garcia-Martin, R., Wang, G., Brandao, B. B., Zanotto, T. M., Shah, S., Kumar Patel, S., Schilling, B., & Kahn, C. R. (2022). MicroRNA sequence codes for small extracellular vesicle release and cellular retention. *Nature, 601*, 446–451.
188. Koppers-Lalic, D., Hackenberg, M., Bijnsdorp, I. V., van Eijndhoven, M. A. J., Sadek, P., Sie, D., Zini, N., Middeldorp, J. M., Ylstra, B., de Menezes, R. X., et al. (2014). Nontemplated nucleotide additions distinguish the small RNA composition in cells from exosomes. *Cell Reports, 8*, 1649–1658.
189. Berardocco, M., Radeghieri, A., Busatto, S., Gallorini, M., Raggi, C., Gissi, C., D'Agnano, I., Bergese, P., Felsani, A., & Berardi, A. C. (2017). RNA-seq reveals distinctive RNA profiles of small extracellular vesicles from different human liver cancer cell lines. *Oncotarget, 8*, 82920–82939.
190. Vojtech, L., Woo, S., Hughes, S., Levy, C., Ballweber, L., Sauteraud, R. P., Strobl, J., Westerberg, K., Gottardo, R., Tewari, M., et al. (2014). Exosomes in human semen carry a distinctive repertoire of small non-coding RNAs with potential regulatory functions. *Nucleic Acids Research, 42*, 7290–7304.
191. Yang, Z., LaRiviere, M. J., Ko, J., Till, J. E., Christensen, T., Yee, S. S., Black, T. A., Tien, K., Lin, A., Shen, H., et al. (2020). A Multianalyte panel consisting of extracellular vesicle miRNAs and mRNAs, cfDNA, and CA19-9 shows utility for diagnosis and staging of pancreatic ductal adenocarcinoma. *Clinical Cancer Research, 26*, 3248–3258.
192. Zayakin, P., Sadovska, L., Eglītis, K., Romanchikova, N., Radoviča-Spalviņa, I., Endzeliņš, E., Liepniece-Karele, I., Eglītis, J., & Linē, A. (2023). Extracellular vesicles-a source of RNA biomarkers for the detection of breast cancer in liquid biopsies. *Cancers (Basel), 15*.
193. He, D., Wang, H., Ho, S. L., Chan, H. N., Hai, L., He, X., Wang, K., & Li, H. W. (2019). Total internal reflection-based single-vesicle in situ quantitative and stoichiometric analysis of tumor-derived exosomal microRNAs for diagnosis and treatment monitoring. *Theranostics, 9*, 4494–4507.
194. Sealfon, S. C., & Chu, T. T. (2011). In A. Khademhosseini, K.-Y. Suh, & M. Zourob (Eds.), *Biological microarrays: Methods and protocols* (pp. 3–34). Humana Press.
195. Savolainen, K., Scaravilli, M., Ilvesmäki, A., Staff, S., Tolonen, T., Mäenpää, J. U., Visakorpi, T., & Auranen, A. (2020). Expression of the miR-200 family in tumor tissue, plasma and urine of epithelial ovarian cancer patients in comparison to benign counterparts. *BMC Research Notes, 13*, 311.
196. Tian, X., Hu, Y., Liu, Y., Yang, Z., Xie, H., Zhou, L., & Zheng, S. (2021). Circular RNA microarray analyses in hepatic ischemia-reperfusion injury with ischemic preconditioning prevention. *Frontiers in Medicine (Lausanne), 8*, Article 626948.
197. Wang, L., Zeng, L., Jiang, H., Li, Z., & Liu, R. (2020). Microarray profile of long noncoding RNA and messenger RNA expression in a model of Alzheimer's disease. *Life (Basel), 10*.

198. Ekström, K., Valadi, H., Sjöstrand, M., Malmhäll, C., Bossios, A., Eldh, M., & Lötvall, J. (2012). Characterization of mRNA and microRNA in human mast cell-derived exosomes and their transfer to other mast cells and blood CD34 progenitor cells. *Journal of Extracellular Vesicles, 1*, 18389.
199. Villarroya-Beltri, C., Gutierrez-Vazquez, C., Sanchez-Cabo, F., Perez-Hernandez, D., Vazquez, J., Martin-Cofreces, N., Martinez-Herrera, D. J., Pascual-Montano, A., Mittelbrunn, M., & Sanchez-Madrid, F. (2013). Sumoylated hnRNPA2B1 controls the sorting of miRNAs into exosomes through binding to specific motifs. *Nature Communications, 4*, 2980.
200. Lowe, R., Shirley, N., Bleackley, M., Dolan, S., & Shafee, T. (2017). Transcriptomics technologies. *PLoS Computational Biology, 13*, Article e1005457.
201. Geiss, G. K., Bumgarner, R. E., Birditt, B., Dahl, T., Dowidar, N., Dunaway, D. L., Fell, H. P., Ferree, S., George, R. D., Grogan, T., et al. (2008). Direct multiplexed measurement of gene expression with color-coded probe pairs. *Nature Biotechnology, 26*, 317–325.
202. Bracht, J. W. P., Gimenez-Capitan, A., Huang, C. Y., Potie, N., Pedraz-Valdunciel, C., Warren, S., Rosell, R., & Molina-Vila, M. A. (2021). Analysis of extracellular vesicle mRNA derived from plasma using the nCounter platform. *Science and Reports, 11*, 3712.
203. Dong, L., Huang, C. Y., Johnson, E. J., Yang, L., Zieren, R. C., Horie, K., Kim, C. J., Warren, S., Amend, S. R., Xue, W., et al. (2021). High-throughput simultaneous mRNA profiling using ncounter technology demonstrates that extracellular vesicles contain different mRNA transcripts than their parental prostate cancer cells. *Analytical Chemistry, 93*, 3717–3725.
204. Bub, A., Brenna, S., Alawi, M., Kugler, P., Gui, Y., Kretz, O., Altmeppen, H., Magnus, T., & Puig, B. (2022). Multiplexed mRNA analysis of brain-derived extracellular vesicles upon experimental stroke in mice reveals increased mRNA content with potential relevance to inflammation and recovery processes. *Cellular and Molecular Life Sciences, 79*, 329.
205. Lo, T. W., Figueroa-Romero, C., Hur, J., Pacut, C., Stoll, E., Spring, C., Lewis, R., Nair, A., Goutman, S. A., Sakowski, S. A., et al. (2021). Extracellular vesicles in serum and central nervous system tissues contain microRNA signatures in sporadic amyotrophic lateral sclerosis. *Frontiers in Molecular Neuroscience, 14*, Article 739016.
206. Crossland, R. E., Albiero, A., Sanjurjo-Rodriguez, C., Reis, M., Resteu, A., Anderson, A. E., Dickinson, A. M., Pratt, A. G., Birch, M., McCaskie, A. W., et al. (2023). MicroRNA profiling of low concentration extracellular vesicle RNA utilizing NanoString nCounter technology. *Journal of Extracellular Biology, 2*, Article e72.
207. Vicentini, C., Calore, F., Nigita, G., Fadda, P., Simbolo, M., Sperandio, N., Luchini, C., Lawlor, R. T., Croce, C. M., Corbo, V., et al. (2020). Exosomal miRNA signatures of pancreatic lesions. *BMC Gastroenterology, 20*, 137.
208. Pantano, F., Zalfa, F., Iuliani, M., Simonetti, S., Manca, P., Napolitano, A., Tiberi, S., Russano, M., Citarella, F., Foderaro, S. et al. (2022). Large-scale profiling of extracellular vesicles identified miR-625–5p as a novel biomarker of immunotherapy response in advanced non-small-cell lung cancer patients. *Cancers (Basel), 14.*
209. Nielsen, T., Wallden, B., Schaper, C., Ferree, S., Liu, S., Gao, D., Barry, G., Dowidar, N., Maysuria, M., & Storhoff, J. (2014). Analytical validation of the PAM50-based prosigna breast cancer prognostic gene signature assay and nCounter analysis system using formalin-fixed paraffin-embedded breast tumor specimens. *BMC Cancer, 14*, 177.
210. Parker, J. S., Mullins, M., Cheang, M. C. U., Leung, S., Voduc, D., Vickery, T., Davies, S., Fauron, C., He, X., Hu, Z., et al. (2009). Supervised risk predictor of breast cancer based on intrinsic subtypes. *Journal of Clinical Oncology, 27*, 1160–1167.
211. Wallden, B., Storhoff, J., Nielsen, T., Dowidar, N., Schaper, C., Ferree, S., Liu, S., Leung, S., Geiss, G., Snider, J., et al. (2015). Development and verification of the PAM50-based prosigna breast cancer gene signature assay. *BMC Medical Genomics, 8*, 54.
212. Tyagi, S., & Kramer, F. R. (1996). Molecular beacons: Probes that fluoresce upon hybridization. *Nature Biotechnology, 14*, 303–308.
213. Coco, S., Vanni, I., Carenzo, A., Savasta, F. L., & Kisakesen, I. (2019). Exosome miRNA profiling with the agilent miRNA microarray platform. *Agilent Technologies Application Note*, Publication number 5994-0777EN.

214. Wu, W., Yu, X., Wu, J., Wu, T., Fan, Y., Chen, W., Zhao, M., Wu, H., Li, X., & Ding, S. (2021). Surface plasmon resonance imaging-based biosensor for multiplex and ultrasensitive detection of NSCLC-associated exosomal miRNAs using DNA programmed heterostructure of Au-on-Ag. *Biosensors & Bioelectronics, 175*, Article 112835.
215. Lee, J. H., Kim, J. A., Jeong, S., & Rhee, W. J. (2016). Simultaneous and multiplexed detection of exosome microRNAs using molecular beacons. *Biosensors & Bioelectronics, 86*, 202–210.
216. Chin, L. K., Son, T., Hong, J.-S., Liu, A.-Q., Skog, J., Castro, C. M., Weissleder, R., Lee, H., & Im, H. (2020). Plasmonic sensors for extracellular vesicle analysis: From scientific development to translational research. *ACS Nano, 14*, 14528–14548.
217. Soler, M., Huertas, C. S., & Lechuga, L. M. (2019). Label-free plasmonic biosensors for point-of-care diagnostics: A review. *Expert Review of Molecular Diagnostics, 19*, 71–81.
218. Yang, Y., Zhai, C., Zeng, Q., Khan, A. L., & Yu, H. (2020). Multifunctional detection of extracellular vesicles with surface plasmon resonance microscopy. *Analytical Chemistry, 92*, 4884–4890.
219. Im, H., Yang, K., Lee, H., & Castro, C. M. (2017). Characterization of extracellular vesicles by surface plasmon resonance. *Methods in Molecular Biology, 1660*, 133–141.
220. Hosseinkhani, B., van den Akker, N., D'Haen, J., Gagliardi, M., Struys, T., Lambrichts, I., Waltenberger, J., Nelissen, I., Hooyberghs, J., Molin, D. G. M., et al. (2017). Direct detection of nano-scale extracellular vesicles derived from inflammation-triggered endothelial cells using surface plasmon resonance. *Nanomedicine, 13*, 1663–1671.
221. Lee, J. U., Kim, W. H., Lee, H. S., Park, K. H., & Sim, S. J. (2019). Quantitative and specific detection of exosomal miRNAs for accurate diagnosis of breast cancer using a surface-enhanced raman scattering sensor based on plasmonic head-flocked gold nanopillars. *Small (Weinheim an der Bergstrasse, Germany), 15*, Article e1804968.
222. Khashayar, P., Al-Madhagi, S., Azimzadeh, M., Scognamiglio, V., & Arduini, F. (2022). New frontiers in microfluidics devices for miRNA analysis. *TrAC Trends in Analytical Chemistry, 156*, Article 116706.
223. Shao, H., Chung, J., Lee, K., Balaj, L., Min, C., Carter, B. S., Hochberg, F. H., Breakefield, X. O., Lee, H., & Weissleder, R. (2015). Chip-based analysis of exosomal mRNA mediating drug resistance in glioblastoma. *Nature Communications, 6*, 6999.
224. Ramshani, Z., Zhang, C., Richards, K., Chen, L., Xu, G., Stiles, B. L., Hill, R., Senapati, S., Go, D. B., & Chang, H. C. (2019). Extracellular vesicle microRNA quantification from plasma using an integrated microfluidic device. *Communications Biology, 2*, 189.

High-Throughput Analysis of Single Extracellular Vesicles

Seung-Ryoung Jung, James Kim, Lucia Vojtech, and Daniel T. Chiu

Abstract For the past several decades, conventional flow cytometry has been continuously improved for analyzing single cells and providing quantifying information such as surface protein copy numbers. However, using this tool, it is challenging to detect small extracellular vesicles (EVs) such as exosomes, which are ~30–150 nm in diameter, ~100-fold smaller in diameter, and one-million-fold smaller in mass than cells. Despite improvements in instrument sensitivity, flow cytometers continue to struggle to detect small EVs due to the limits of detection of sheath flow systems. Nevertheless, the number of publications related to EVs continues to increase exponentially because of the potential utility of EVs in drug delivery and disease diagnostics/liquid biopsy applications. New, high-sensitivity tools for phenotyping small EVs are urgently needed. Here we provide an overview of our efforts to overcome the obstacles to identifying and phenotyping small EVs.

Keywords Extracellular vesicles · Single-molecule flow cytometry · Exosome sizing · Counting protein number · Bottom-up approach

1 Introduction

Traditional flow cytometers are designed to detect cells that are typically ~10 μm in diameter. Flow cytometers measure scattered light and fluorescence from cells passing one by one through channels which are ~100–200 μm in inner diameter, and 10–40 μm in flow-focused diameter. Cells in a flow channel are hydrodynamically

S.-R. Jung · J. Kim · D. T. Chiu (✉)
Department of Chemistry, University of Washington, Seattle, WA, USA
e-mail: chiu@uw.edu

D. T. Chiu
Department of Bioengineering, University of Washington, Seattle, WA, USA

L. Vojtech
Department of Obstetrics and Gynecology, University of Washington, Seattle, WA, USA
e-mail: luciav@uw.edu

R. K. Marcus and D. Lubman (eds.), *Exosomes and Extracellular Vesicles*,
https://doi.org/10.1007/978-3-031-89270-7_8

focused by sheath flow. Focused laser beams excite fluorophores attached to the cell surface via antibodies bound to membrane proteins, and the emitted fluorescence and scattered signals are collected [11]. These fluorescence and scatter signals are converted to protein copy numbers and cell size by using calibration curves, with external calibration using beads.

Extracellular vesicles (EVs) are lipid bilayer membrane vesicles ranging in size from nanometers to a few micrometers. EVs are categorized into three main subtypes based on their biogenesis: (1) exosomes (30–150 nm in diameter), which originate from multivesicular bodies (MVBs); (2) microvesicles (100–1000 nm in diameter), which are generated by budding off the plasma membrane; and (3) apoptotic bodies (500–5000 nm in diameter), which are produced by blebbing of the cell membrane. However, the biogenesis of an EV sample is often unknown, so EVs are also categorized based on size or method of preparation. In this chapter, we focus on detecting small EVs (<200 nm) using a line-confocal microfluidic detection system with single-molecule sensitivity and compare results with conventional flow cytometry methods. However, there have been multiple challenges to achieving the goal of detecting and phenotyping every EV in a sample. Here we summarize what we have done to overcome these challenges.

Conventional Flow Cytometry (FCM). The commercial flow cytometers CytoFLEX LX by Beckman Coulter, CellStream by Cytek, BD FACSymphony A1 Cell Analyzer, and Aurora by Cytek were originally designed for single cell analysis but have been optimized for detection of single particles in the range of approximately 80–200 nm. Recently, higher sensitivity systems—CytoFLEX nano and NanoFCM [12]—have been developed, which can detect smaller particles (e.g., ~40 nm polystyrene beads). It is challenging to detect small vesicles using conventional flow cytometry methods due to the limited amount of signal scattered by particles relative to background noise, i.e., the limited signal-to-noise ratio (SNR). A signal scattered by a single cell (~10 μm) is very large, well above background noise. However, a 100-nm particle has a 10^{12}-fold smaller scatter intensity than a 10 μm cell, because the scattering signal is dependent on the six power of particle size [12]. Thus, the scattered signal by a small EV is often close to the background noise. Various approaches have been employed to improve SNR by increasing the scatter signal and reducing background noise. Slower flow rates allow collecting more photons to increase signals. Lower particle concentrations reduce the photon mixing effect to reduce the background. However, small EVs remain difficult or impossible to detect. In addition, scattering-based sizing is problematic due to contamination of EV samples with lipoproteins and other non-membranous particles. Thus, fluorescence-based approaches for EV detection were developed. For example, a dual-color scheme was used to distinguish real EVs from antibodies or membrane dye aggregates. For counting EV cargo such as proteins and nucleic acids, bright fluorophores such as PE and APC are used to enhance SNR [11]. Nevertheless, detecting every dye molecule—including low dye emissions (since fluorescence emission is stochastic)—cannot be achieved by conventional FCM due to the high background inherent in these large flow-focused systems (10–40 μm).

Single-Molecule Flow Cytometry (SMFC). Rather than adapting flow cytometry systems originally designed for the study of cells to much smaller entities (EVs)—a "top-down" approach—we have adapted methods originally designed for the study of single molecules to the analysis of EVs. We employed this "bottom-up" approach because small EVs can have even a single copy of an important protein or nucleic acid, and even the highest sensitivity commercial flow cytometers cannot detect every dye molecule in a sample. In our proof-of-concept paper [6], we describe a flow platform for quantifying the number of biomolecules on individual fluorescent nanoparticles. To increase SNR, we designed a line-confocal detection geometry (Fig. 8.1a). 10 pM of biotinylated Alexa Fluor 647 dye was introduced into the microfluidic channel via an inlet (sample channel). We observed single spikes generated by the dye (Fig. 8.1b, lower trace). We then scaled up to nanometer-sized particles that mimic biological particles. We quantified the number of streptavidins on individual semiconducting polymer dots (Pdots) (Fig. 8.2). Streptavidin was labeled with biotin-Alexa Fluor 647.

Statistical Method for Quantifying Protein Copy Numbers on Extracellular Vesicles. The number of biomolecules present on each nanoparticle was determined by deconvolving the fluorescence intensity distribution of single-nanoparticle-biomolecule complexes with the distribution of single biomolecules (Fig. 8.3).

This deconvolution method is described in [8]. Briefly, we (1) normalize the entire intensity distribution with the total number of events, (2) obtain a basis set to simulate

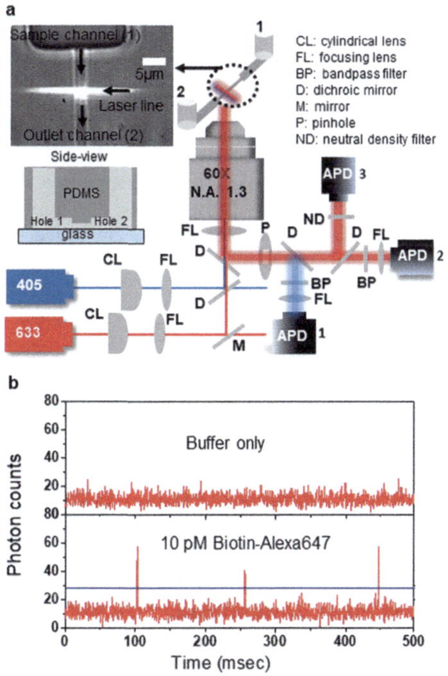

Fig. 8.1 Single-molecule flow platform. **a** Schematic figure for single-molecule flow cytometry (SMFC). The dimension of detection channel was 2–4 μm which is ~10-fold smaller size than that in conventional flow cytometers. **b** Real time measurement of single spikes of biotinylated Alexa Fluor 647 dye using the SMFC [6]. The blue line is threshold to distinguish noise and real signals in lower panel

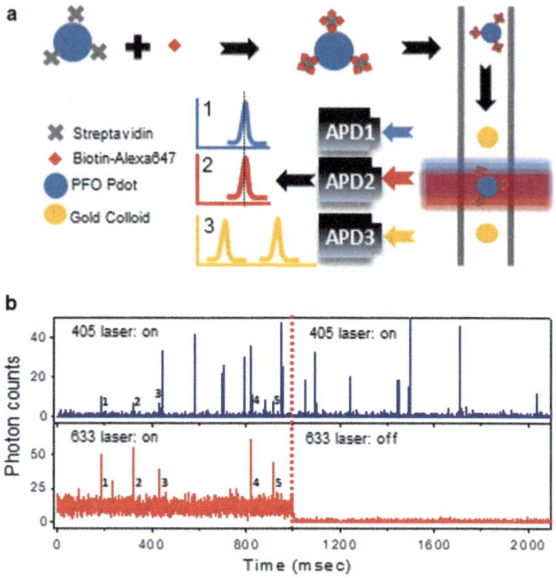

Fig. 8.2 Mimicking of biological particles using streptavidin coated PFO polymer dots (Pdots). **a** Scheme of experiments with Pdots. Streptavidin coated PFO Pdots can be labeled with biotinylated Alexa Fluor 647 dye. **b** Dual-color experiments using 405 nm laser for excitation of PFO Pdots and 633 nm laser for Alexa Fluor 647 dye. No crosstalk between blue channel and red channel when the 633 nm laser was turned off to check whether it has bleed-through [6]

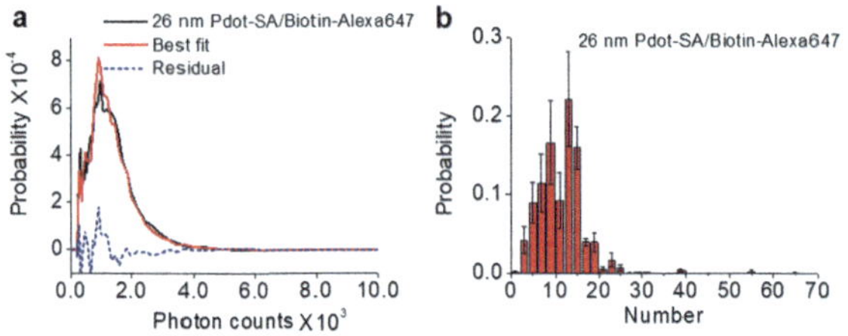

Fig. 8.3 Deconvolution to obtain the distribution of the number of streptavidins on single Pdots. **a** The Alexa Fluor 647 intensity distribution of individual Pdots was normalized with the total number of events. To obtain the best fit iteratively, a basis set was generated by multiplying the singly labeled distribution by an integer. We added a weighting factor to each basis set to simulate the distribution using the best fit. **b** The weighting factor represents the probability of each streptavidin number on the x-axis. The data are averages of multiple experiments [6]

the normalized distribution by multiplying the singly labeled distribution by integers; and (3) obtain a weighting factor for each basis, which is the probability of each copy number. This method was originally developed for imaging data and was adapted for flow data. In using this method for single-EV analysis, we encountered two hurdles to obtaining quantitative information on EV size: (1) selecting a lipophilic dye and labeling conditions that allow detection of all membranous particles; and (2) measuring EV size from the intensity of the lipophilic dye.

Selection of a Membrane Dye for Detecting and Sizing EVs. We demonstrated sizing of EVs using our fluorescence-based flow analyzer with single-molecule sensitivity [1].

Using a dye that selectively partitions into the membrane of EVs, we showed that the fluorescence intensity of a vesicle is proportional to its surface area. For calibration, we used either transmission electron microscopy (TEM) or stochastic optical reconstruction microscopy (STORM) imaging as the reference. In the paper above, we discuss constraints in sample preparation that are inherent to sizing nanoscale vesicles with a fluorescent membrane dye and propose several guidelines to improve data consistency. After optimizing staining conditions, we were able to measure the size of vesicles that are as small as ~30 nm.

In a comparison of the membrane dyes di-8-ANEPPS, Dioc14(3), FM1-43, and Membright-488, we found that di-8-ANEPPS best labeled the entire population of EVs (Fig. 8.4). This finding was consistent with a study that compared the number of events detected by scattered and fluorescence signals [2]. One issue was self-aggregation of di-8-ANEPPS and other lipophilic dyes due to their low solubility and tendency to form micelles in solution. One advantage of di-8-ANEPPS is that the aggregates of this dye are smaller than those of other dyes (~30 nm for di-8-ANEPPS versus 70–80 nm for MEMbright dyes [1, 3]), allowing the use of a 35-nm-cutoff size exclusion column (SEC, IZON) to remove dye aggregates. MEMbright has larger aggregates because it has a longer hydrophobic tail. In addition to filtering out membrane dye aggregates, it is also possible to distinguish dye aggregates from EVs based on the absence of surface proteins on the dye aggregates when labeling a sample with both membrane dye and antibody-dye conjugates.

Quantifying Protein Copy Numbers on EVs Using SMFC. We found that over 80% of di-8-ANEPPS-labeled human colorectal cancer (DiFi) cell-derived exosomes were

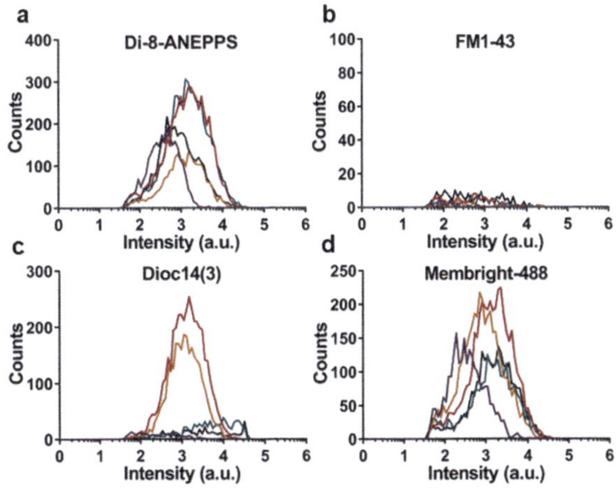

Fig. 8.4 EV staining with different lipophilic dyes to choose the dye that best labels all vesicles in the sample. The concentrations of dyes were 5 nM (violet), 20 nM (orange), 50 nM (red), 100 nM (blue), and 250 nM (black). The x-axis is logarithmic scale, with counts (number of detected events) plotted against fluorescence intensity. Di-8-ANEPPS was the best dye for labeling EVs. Of note, if we use >250 nM di-8-ANEPPS, it starts self-quenching in the EV membrane [1]

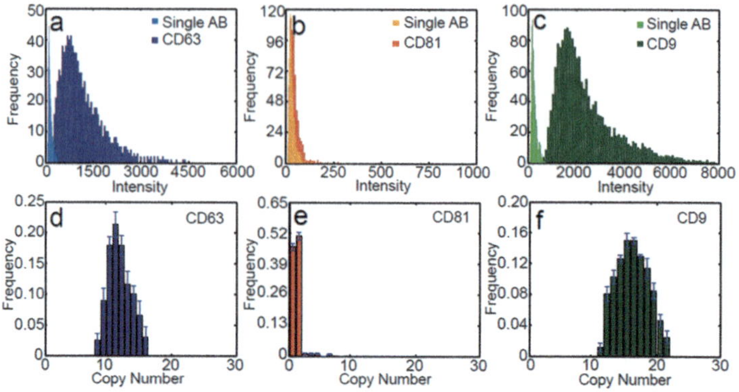

Fig. 8.5 The copy number distributions of three tetraspanins on single seminal EVs. The CD63 (**a**, **d**), CD81 (**b**, **e**), and CD9 (**c**, **f**) intensity distributions from single EVs were deconvolved with the intensity distribution of single antibodies [5]

labeled when using a mixture of PE-tagged anti-CD9, CD63, and CD81 tetraspanin antibodies after filtering with a 35-nm-cutoff size exclusion column (IZON). This labeling efficiency varies for other sources of exosomes and for different types of EV preparations due to the differential expression of tetraspanins on different types of EVs, due to contamination of samples with other particles. We labeled seminal exosomes with the membrane dye di-8-ANEPPS and with Brilliant Violet 510 (BV510)-anti-CD9, phycoerythrin (PE)-anti-CD63, and Alexa Fluor 647-anti-CD81 [5]. We observed three distributions of antibody-labeled EVs, along with distributions of single antibody intensity (Fig. 8.5a–c, top row) which were used for deconvolution (Fig. 8.5d–f, bottom row). The CD81 expression on seminal EVs was very low.

Sizing of Extracellular Vesicles Using SMFC and 3D Single-Molecule Imaging. Single-molecule imaging is another method for quantifying the number of proteins on extracellular vesicles [9]. However, data collection for single-molecule imaging is much slower than for flow cytometry, resulting in much slower throughput, and the degree of multiplexing is limited. Nevertheless, superresolution imaging methods provide spatial information such as interprotein distances on EVs and direct sizing of vesicles (Fig. 8.6). We performed superresolution mapping of EVs labeled with novel transistor-like Pdots [5], which exhibit spontaneous blinking, with <5 nm localization error (the best performing Alexa Fluor 647 achieves a localization error ~10 nm [4]) and a broad range of optical-adjustable duty cycles. Based on the copy numbers from flow analysis, the switch-on frequency of the Pdots was finely adjusted to improve imaging throughput. At the same time, we obtained EV size information from the positions of antibodies.

High-Throughput Calibration of SMFC Fluorescence Intensity Using 2D Single-Molecule Imaging with Error Correction. Sizing particles smaller than

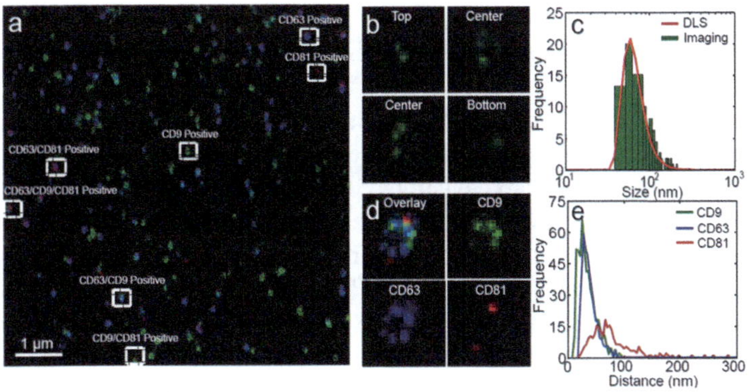

Fig. 8.6 Multicolor superresolution imaging of EVs. **a** Three-color superresolution imaging of CD63, CD81, and CD9 on seminal EVs. **b** Lateral slices of a 3D superresolution image of an EV. **c** EV size distribution from image analysis (green bars) and from DLS (red line). **d** Magnified image of an EV in different imaging channels. **e** Interprotein distance distributions [5]

~50 nm can be inaccurate due to single-molecule localization error caused by distortion of the point spread function by moving particles or dyes and the limited photon number emitted by a single dye. We demonstrated a method to correct localization errors when sizing vesicles and other spherical nanoparticles [7]. We used mean approximation theory to derive a simple equation using full width at half-maximum (FWHM) for correcting particle sizes measured by two-dimensional single-molecule localization microscopy (SMLM), validated the method by sizing streptavidin-coated polystyrene nanobeads with the SMLM technique direct STORM (dSTORM) with and without error correction, using transmission electron microscopy (TEM) for comparison, then applied the method to sizing small seminal EVs [7].

Nanobead sizes measured by dSTORM became increasingly less accurate (larger than TEM values) for beads smaller than 50 nm. The error-correction method reduced the size difference versus TEM from 15% without error correction to 7% with error correction for 40 nm beads, from 44 to 9% for 30 nm beads, and from 66 to 15% for 20 nm beads. For this study, we used MEMbright700 dye to label the lipid membranes of EVs. Because this dye has a cyanine structure, it blinks quite well, like Alexa Fluor 647. Correction of localization error was achieved as follows. Localization of the position of fluorophores was done by fitting the emitted photon distribution by a Gaussian function to obtain a peak point to assign as the center of fluorescence. During this step, the localization precision can be affected by the quality of fitting, depending on the smoothness of the photon profile emitted from single fluorophores. Thus, localization precision is a parameter that describes how much the localization is off from the real positions of dyes. Mathematically, the localization precision is the standard deviation of Gaussian fitting for single fluorophores. When we apply this to many blinking events, we can reconstruct images to obtain the distribution of dyes in 2D. During this step, we can also superimpose position information of dyes to obtain another Gaussian distribution. Thus, we can use the two standard deviations from:

(1) the distribution of individual dyes, and (2) the photon distribution of each dye. The first term is related to the size of particles, and the second term to the localization precision.

According to Cremer's decomposition theorem, the square of the standard deviation of measured size is equal to the square of the standard deviation of actual size plus the square of the standard deviation of localization precision [7].

As shown in Fig. 8.7, we simulated the effect of localization precision for sizing nanometer-sized vesicles using the SuReSim algorithm, which was originally developed for analyzing 40-nm synaptic vesicles [10]. The results show that the measured size increases nonlinearly depending on the square root of the localization precision of the dyes. From this simulation, we realized that at least 10 blinking events are required for accurate estimates of size. If the localization precision is close to the actual size, the measured size can be significantly distorted. On the other hand, if the localization precision is much smaller than the actual size, then we can ignore this effect on measured size. For example, if the localization precision of MEMbright700 is 10 nm and the actual size 100 nm, then the effect of localization error induced by uncertainty of localization will be marginal. We found that the localization error should be considered when the particle size is <50 nm.

After correction of localization errors between MEMbright700 and CD63-AF488 antibodies (Fig. 8.8), we compared EV size between two different labeling conditions. Interestingly, we found that the measured sizes of antibody-labeled EVs were ~12 nm larger than the measured sizes of EVs using the lipophilic dye. The likely main sources of this difference were (1) the antibody size; (2) the uncertainty of the dye labeling position on a single antibody and the binding position of the antibody to

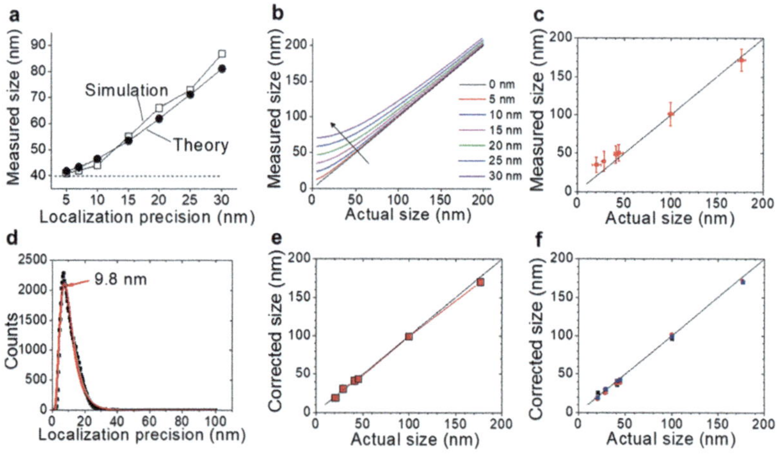

Fig. 8.7 Validation of the error-correction method. **a** Simulation (SuReSim) and theory. **b**, **c** Relationship between measured size and actual size in simulations (**b**) and in experiments (**c**). **d** An example of localization precision of Alexa Fluor 647. **e**, **f** Corrected size versus actual size from single experiment (**e**) and three independent experiments (**f**). All data are from [7]

the target proteins on the EV membrane; and (3) the binding angle of antibodies, which can generate another factor that increases the measured size during STORM imaging. All three effects can be additive when we perform dSTORM imaging. With the assumption that any antibody binding angle is possible and with no weighting factor for a certain angle, we found that if the linker size is 10 nm, then the measured size of EVs can be 10 nm larger than the actual size in simulations (Fig. 8.9).

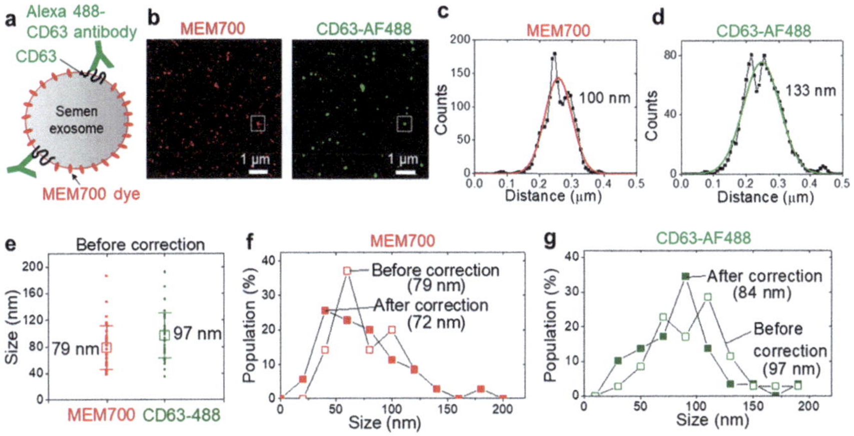

Fig. 8.8 Dual-color labeled seminal EVs using MEMbright700 (MEM700) and CD63-Alexa Fluor 488 (CD63-AF488). **a**, **b** Simplified scheme and images. **c**, **d** Profile of 2D superimposed images marked in b. **e–g** Actual size before and after error correction for MEM700 and CD63-AF488 [7]

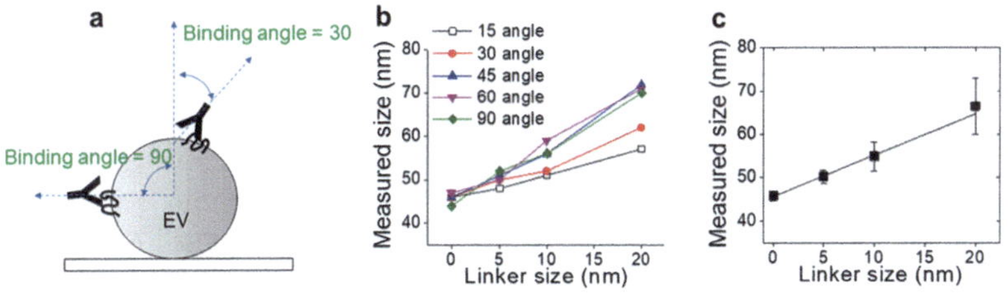

Fig. 8.9 Measured EV size was proportional to linker size. SuReSim was used for this simulation. **a**, **b** Binding angle-dependent measured size versus linker sizes. **c**, Measured size average in 5 different angles versus linker sizes [7]

2 Conclusion

With the above capabilities, it is now possible to phenotype single EVs in terms of the presence or absence of multiple biomarkers, with quantification of protein copy numbers and EV size, at high throughput and with unprecedented single-molecule sensitivity for the analysis of all EVs present in the sample.

References

1. Andronico, L. A., Jiang, Y., Jung, S.-R., Fujimoto, B. S., Vojtech, L., & Chiu, D. T. (2021). Sizing extracellular vesicles using membrane dyes and a single molecule-sensitive flow analyzer. *Analytical Chemistry, 93*(14), 5897–5905. https://doi.org/10.1021/acs.analchem.1c00253
2. Chen, C., Cai, N., Niu, Q., Tian, Y., Hu, Y., & Yan, X. (2023). Quantitative assessment of lipophilic membrane dye-based labelling of extracellular vesicles by nano-flow cytometry. *Journal of Extracellular Vesicles, 12*(8), Article e12351. https://doi.org/10.1002/jev2.12351
3. Collot, M., Ashokkumar, P., Anton, H., Boutant, E., Faklaris, O., Galli, T., Mely, Y., Danglot, L., & Klymchenko, A. S. (2019). MemBright: A family of fluorescent membrane probes for advanced cellular imaging and neuroscience. *Cell Chemical Biology, 26*(4), 600-614.e7. https://doi.org/10.1016/j.chembiol.2019.01.009
4. Dempsey, G. T., Vaughan, J. C., Chen, K. H., Bates, M., & Zhuang, X. (2011). Evaluation of fluorophores for optimal performance in localization-based super-resolution imaging. *Nature Methods, 8*(12), 1027–1036.
5. Jiang, Y., Andronico, L. A., Jung, S.-R., Chen, H., Fujimoto, B., Vojtech, L., & Chiu, D. T. (2021). High-throughput counting and superresolution mapping of tetraspanins on exosomes using single-molecule sensitive flow techniques and transistors-like semiconducting polymer dots. *Angewandte Chemie (International ed. in English), 60*(24), 13470–13475. https://doi.org/10.1002/anie.202103282
6. Jung, S.-R., Han, R., Sun, W., Jiang, Y., Fujimoto, B. R., Yu, J., Kuo, C.-T., Rong, Y., Zhou, X.-H., & Chiu, D. T. (2018). Single-molecule flow platform for the quantification of biomolecules attached to single nanoparticles. *Analytical Chemistry, 90*(10), 6089–6095. https://doi.org/10.1021/acs.analchem.8b00024
7. Jung, S.-R., Kim, J., Vojtech, L., Vaughan, J. C., & Chiu, D. T. (2023). Error-correction method for high-throughput sizing of nanoscale vesicles with single-molecule localization microscopy. *The Journal of Physical Chemistry B, 127*(12), 2701–2707. https://doi.org/10.1021/acs.jpcb.2c09053
8. Mutch, S. A., Fujimoto, B. S., Kuyper, C. L., Kuo, J. S., Bajjalieh, S. M., & Chiu, D. T. (2007). Deconvolving single-molecule intensity distributions for quantitative microscopy measurements. *Biophysical Journal, 92*(8), 2926–2943. https://doi.org/10.1529/biophysj.106.101428
9. Saftics, A., Abuelreich, S., Romano, E., Ghaeli, I., Jiang, N., Spanos, M., Lennon, K. M., Singh, G., Das, S., Keuren-Jensen, K. V., & Jovanovic-Tailsman, T. (2023). Single extracellular VEsicle nanoscopy. *Journal of Extracellular Vesicles, 12*(7), Article e12346. https://doi.org/10.1002/jev2.12346
10. Venkataramani, V., Herrmannsdorfer, F., Heilemann, M., & Kuner, T. (2016). SuReSim: Simulating localization microscopy experiments from ground truth models. *Nature Methods, 13*(4), 319–321. https://doi.org/10.1038/nmeth.3775
11. Welsh et al. (2023). A compendium of single extracellular vesicle flow cytometry. *Journal of Extracellular Vesicles, 12*(2), e12299. https://doi.org/10.1002/jev2.12299

12. Zhu, S., Ma, L., Wang, S., Chen, C., Zhang, W., Yang, L., Hang, W., Nolan, J. P., Wu, L., & Yan, X. (2014). Light-scattering detection below the level of single fluorescent molecules for high-resolution characterization of functional nanoparticles. *ACS Nano, 8*(10), 10998–11006. https://doi.org/10.1021/nn505162u

Leveraging Single Molecule Localization Microscopy for Enhanced Characterization of Exosomes

Shaked Ashkenazi, Colbie R. Chinowsky, James H. Felce, and Laura Woythe

Abstract The importance of exosomes and other extracellular vesicles (EVs) in cell–cell communication cannot be overstated. Thus, many efforts are being invested in intercepting exosomes in order to analyze their content and gain information both on the cells that produce them and on the cells that receive them. The holy grail of EV research is to associate specific EV biomarkers with the tissues that produced them, as this will open endless opportunities for diagnostics. One of the challenges in this type of research is to identify different subpopulations within a biological sample, which can be addressed using a comprehensive analysis at a single exosome level. Single-molecule localization microscopy (SMLM) is a term used to describe several super-resolution microscopy methods, all of which are based on the common principle of identifying the precise localization of individual fluorophores that are usually associated with a target of interest. SMLM is based on the ability of certain fluorophores to stochastically switch from an OFF, dark state, to an ON, bright state, so that at any given moment only some of the fluorophores are ON while others are OFF. Temporally resolving nearby fluorophores allows scientists to obtain a resolution of $\geq$20 nm, well within the range of single exosomes. By analyzing thousands of individual exosomes, you gain strong statistical confidence and the ability to detect even rare subpopulations. In this chapter, we discuss the unique features of SMLM in exosome research and we provide recent examples where these cutting-edge methods were applied in diagnostics and therapeutics.

Keywords SMLM · dSTORM · Super-resolution · Nanoimager · Microscopy · Exosomes · Extracellular vesicles · Imaging · DBSCAN · Single molecule ·

S. Ashkenazi (✉) · C. R. Chinowsky · J. H. Felce · L. Woythe
Oxford Nanoimaging Limited (ONI), Oxford OX2 8TA, UK
e-mail: sashkenazi@oni.bio

C. R. Chinowsky
e-mail: cchinowsky@oni.bio

J. H. Felce
e-mail: jfelce@oni.bio

L. Woythe
e-mail: lwoythe@oni.bio

R. K. Marcus and D. Lubman (eds.), *Exosomes and Extracellular Vesicles*,
https://doi.org/10.1007/978-3-031-89270-7_9

Single molecule localization microscopy · Immunofluorescence · Fluorescence microscopy · Tetraspanins · Pan-EV

1 Introduction

Extracellular vesicles (EVs), and exosomes among them, play an important role in cell–cell communication and are sometimes referred to as "instant bio-messages" thanks to the fact that they contain active biomolecules that can directly affect recipient cells [1–4]. Some of the key open questions regarding EVs revolve around the two ends of this communication—the cell that produces the vesicles and the cells that receive them. Specifically, the control and the selectivity of the vesicle content, as well as the specificity of its targets and how this specificity is achieved. One of the key steps in the journey to understand the regulation of exosome biogenesis and their effects on other cells is to correctly identify the nature of their cargo [5–7]. Many methods are available to characterize the proteins, nucleic acids, and lipids that compose exosomes, however, most of them rely on bulk measurements. While obtaining information on the composition of the entire vesicle population is extremely valuable, additional information can be learned from studying individual exosomes, such as the identification of rare subpopulations within a sample, due to the heterogeneous nature of exosomes.

Whilst properly controlled flow cytometry can provide high-throughput information on both target positivity and sizing, it lacks the possibility to visualize and assess individual exosomes. Additionally, its sizing capabilities often work under the assumption that exosomes are perfectly spherical [8–10]. Microscopy is an excellent tool for looking at subcellular structures, and exosomes are no exception, although this approach has a few significant challenges: imaging exosomes inside cells is not at all trivial, which is why most imaging techniques are based on isolated vesicles. Moreover, the small size of exosomes poses a challenge for standard light microscopes: They are limited in their ability to resolve nearby objects by the constraints of the diffraction limit of light, whereby in most light microscopes is approximately 200–300 nm. Therefore, they cannot resolve individual exosomes, which are usually from 30 nm and up to 200 nm, averaging at about 100 nm [5, 7]. Consequently, identifying features of single exosomes, including their cargo, is impractical. To put it simply, two exosomes of 100 nm that are 100 nm apart, each carrying its own cargo, will appear as one blurry bright spot, and it will not be possible to examine each one individually using this method (Fig. 9.1a). This will be particularly problematic when studying rare subpopulations.

The two options currently available to overcome the challenges posed by the diffraction limit are electron microscopy and super-resolution microscopy. Both provide a resolution that enables visualizing single exosomes, although the sample preparation is substantially different, especially around fixing the sample and labeling the targets. One feature of electron microscopy is that it is done in vacuum, which requires an extensive sample preparation process that includes the removal of water

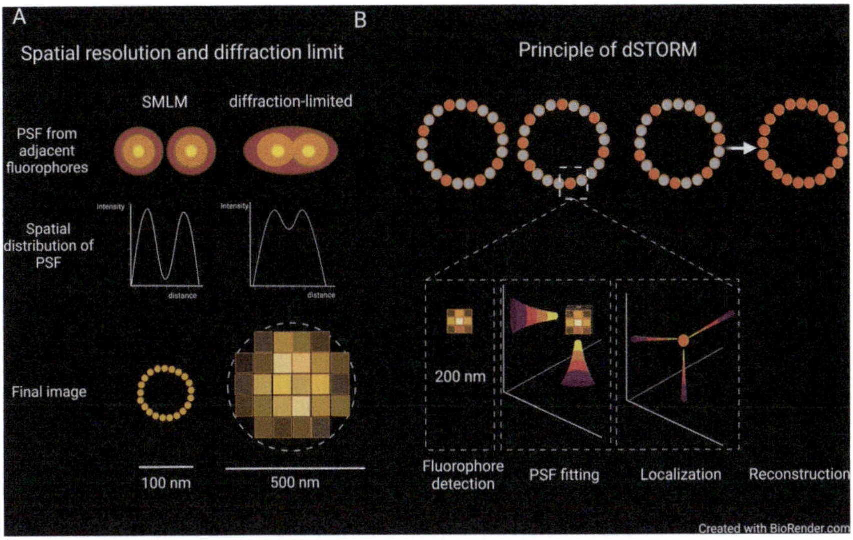

Fig. 9.1 Breaking the resolution barriers. **a** Spatial resolution is defined by the ability to separate between two distinct nearby objects. Conventional light microscopes are diffraction-limited, where the resolution is usually limited to 200–300 nm. If inspecting objects that are smaller than the diffraction limit, a conventional microscope will not be able to resolve the structural details but instead will show a larger blurry object. **b** The principles of dSTORM: a structural object with a diameter of 100 nm is labeled with dSTORM-compatible fluorophores, which stochastically blink between an ON state and an OFF state. The sensitive camera records the ON states of the fluorophores. A point-spread function (PSF) is fitted to determine the precise spatial localization. Once each blinking fluorophore is localized, the structure of the object can be reconstructed

from the sample. When sizing exosomes in electron microscopy images, it is important to consider the drying out and the shrinking that occurs during the preparation. Compared with super-resolution microscopy, transmission electron microscopy (TEM) and scanning electron microscopy (SEM) are more limited in the number of specific targets they can detect in a single sample: Even though theoretically they can be used for multiple targets, under the current limitation of the methods, in practice they are compatible with detecting a single target. Cryo-electron microscopy (Cryo-EM) can be used for detailed analyses of single exosomes, with some limitations, primarily due to the complexed sample preparation [8, 11–15].

In this chapter, we will focus on super-resolution, and primarily the concept of single-molecule localization microscopy (SMLM), which allows scientists to visualize single particles, with the addition of substantial quantitative data on the sample. Other methods, like stimulated emission depletion microscopy (STED), can image in super-resolution below the diffraction limit, but their technology cannot yield single-molecule data [9, 16, 17].

2 Single-Molecule Localization Microscopy (SMLM)

SMLM is an umbrella term for several distinct methods that share common principles, most importantly defining the accurate spatial localization of individual fluorophores. The most common SMLM methods are direct stochastic optical reconstruction microscopy (dSTORM), photo-activated localization microscopy (PALM) and point accumulation in nanoscale topography (PAINT). Each one uses different groups of fluorophores with different modes of action, but all capitalize on their photoswitching from dark states to bright, emissive states, a phenomenon known as "blinking". With only some fluorophores emitting light at any given moment, it is possible to record each of these events and later fit their point spread function (PSF), from which the exact localization coordinates can be deduced. While the chemical nature of the fluorophore depends on the SMLM method, the principle of individual fluorophores found in either an "ON" or an "OFF" state remains in all [16–19].

dSTORM is a common SMLM method that requires blinking fluorophores. dSTORM-compatible fluorophores have a long half-life and a chemical structure that allows them to blink [9, 17–25]. Examples of dSTORM-compatible fluorophores include Alexa Fluor647™, ATTO655, Cy3 and Cy5 [16, 17]. dSTORM allows for fluorophores of multiple colors, as long as the peaks of the emission spectrums do not overlap [16, 18, 19, 24, 26].

In a dSTORM experiment, a biological sample is first chemically fixed to a coverslip, and then stained with dSTORM-compatible fluorescent dyes, usually conjugated to antibodies or other equivalent fluorescently labeled affinity-based molecules [21–24]. When setting up dSTORM experiments, several controls should be considered, including the specificity of the antibody and the presence of the target in the sample.

Sizing can also be done with dyes that are specific to different membrane components, such as fluorogenic dimers or the commercial dye ExoBrite™. Their broad specificity to membranes, rather than to specific protein targets, makes them useful for exosome sizing, because identifying them within the sample is independent of the expression of a specific biomarker [27, 28].

If using antibodies to label targets, it is important to consider the overall size of the antigen-antibodies complex. Using a fluorescently-labeled primary antibody introduces less sizing-bias compared with a complex of primary and secondary antibodies since the latter is becoming substantially large relative to the size of the exosome itself. Thus, while electron microscopy images result in shrinking of exosomes, dSTORM imaging leads to increased size measurements.

In order to image the labeled exosomes or cells using dSTORM, a compatible imaging buffer must be applied. These buffers typically include thiols and oxygen-scavenging agents which are there to help the fluorophores blink. Once oxygen level is substantially reduced from the sample, it is ready for dSTORM imaging using a super-resolution microscope [16–20, 22, 25, 26]. Powerful lasers excite the fluorophores in the sample at different wavelengths, and the camera records the emitted fluorescence. It is essential to balance the intensity of the lasers and the duration of the excitation in order to make sure that the fluorophores can photoswitch and will not be bleached [16,

20, 22–24, 26, 29]. Since dSTORM is based on stochastic photoswitching, under the right conditions random fluorophores will blink at any given moment, which allows for temporal resolution to overcome the barriers of spatial resolution [16, 18–20, 26, 30]. The X and Y coordinates of each fluorophore are calculated based on the PSF to give its localization, usually accurate to 20 nm or less, a size that is well within the range of a single exosome [9, 10, 16, 30–32] (Fig. 9.1b). Accurately identifying the localization of the fluorophore is one of the most important features behind SMLM. Generally, a single localization stems from a single molecule, however, due to the stochastic nature of the blinking, it is possible to have multiple events from one target, effectively resulting in a number of localizations. Therefore, it is inaccurate to count the number of molecules based only on the number of localizations. However, the overall number of localizations does correlate with the abundance of labeled targets.

dSTORM is capable of detecting 1–3 targets and is limited by the availability of specific antibodies, as well as the availability of dSTORM compatible dyes of different wavelengths [16, 23, 24, 30]. Provided that these are available, it may be possible to detect 3 protein targets on the external surface of fixed exosomes, although due to the small size of exosomes, there could be a steric hindrance between the different antigen–antibody complexes. Gently permeabilizing exosomes also facilitates the detection of internal cargo, allowing scientists to specifically detect a handful of targets, irrespective of their location within the exosome, which can be immensely valuable.

In order to go from a field of view full of individual exosomes to statistically meaningful high-throughput data, additional analyses are needed. Clustering of individual localizations is the first step in the path to interpreting SMLM results. First, clustering is used to extract meaningful information about the morphology, size, composition, circularity, etc. of the sample. One of the most popular clustering algorithms for SMLM analysis is density-based spatial clustering of applications with noise (DBSCAN) [17, 33]. For each localization, the algorithm builds an outwards tree, connecting it with all other neighboring localizations in a hierarchical manner, so that the connected localizations are then grouped together to form a cluster (Fig. 9.2).

Since the essence of DBSCAN is centered around density, it is possible to optimize the clustering parameters for exosome analysis: in imaging, exosomes often appear as round, high density objects of a very similar size, making them highly compatible with this clustering approach. Additionally, simultaneous imaging of thousands of exosomes in one sample provides substantial statistical power in the downstream analyses, especially at the single exosome level: For example, reconstructing the distribution of biomarkers in the population could indicate their source tissues, including in the cases of rare subpopulations. Several recent studies used dSTORM to specifically detect CNS-derived extracellular vesicles in plasma, despite their relatively low abundance in the sample [34, 35]. General membrane-associated fluorescent dyes provide robust information about the size and the morphology of single exosomes, which can also be cross-referenced with the data on biomarker positivity to add an additional layer of analysis (Fig. 9.3). It is important to remember that relying exclusively on data from the protein targets for sizing can be misleading, because their distribution on the exosome surface may not be even, and could be distributed

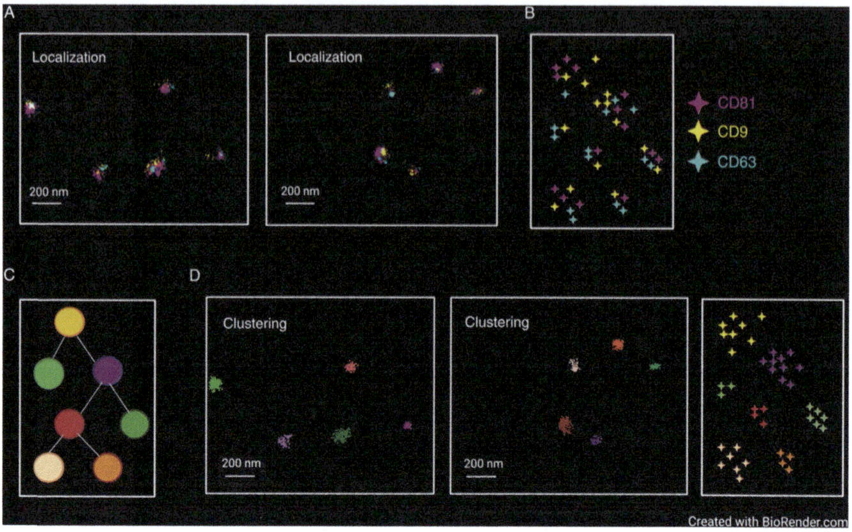

Fig. 9.2 Density-based clustering. The sensitive camera records the blinking of fluorophores at different wavelengths. Algorithms like DBSCAN cluster each localization in an outward tree, connecting it with neighboring localizations. **a** Localizations of EVs isolated from colorectal cancer, stained with CD81-647 (magenta), CD63-561 (cyan), and CD9-488 (yellow). **b** Schematic illustration of localizations using the same fluorophores as panel A. **c** Schematics of an outward tree DBSCAN clustering. **d** Corresponding DBSCAN clustering for each image from panel A, and schematics of the localization clustering corresponding to panel B

into specific rafts, thus providing inaccurate information on the morphology of the entire vesicle. Combining information from membrane-associated dyes with information from protein targets results in more accurate information on sizing and the distribution of the targets in and on the exosome.

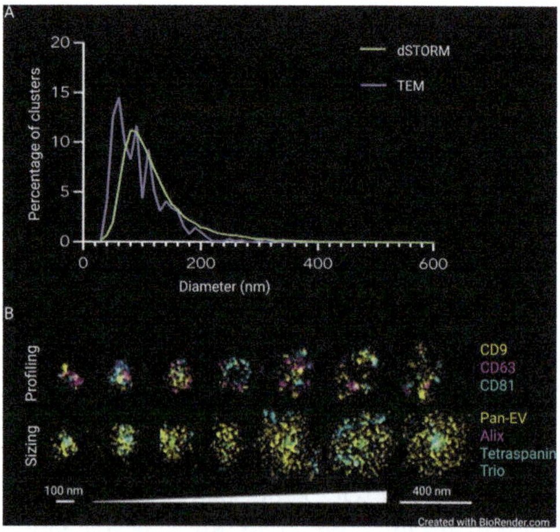

Fig. 9.3 EV sizing and characterization using dSTORM. **a** Comparing EVs sizing with dSTORM and with TEM. **b** Images of single EVs at different sizes, stained either with individual tetraspanins (upper row) or with the membrane-associated dye Pan-EV, Alix, and a mix of 3 tetraspanins (lower row). Data collected by ONI using its Application Kit™: EV Profiler 2

Alongside its strengths, it is also important to discuss some of the limitations of using SMLM to study exosomes. Perhaps most obviously, it requires a super-resolution microscope with powerful lasers, which is far less common in research labs than flow-cytometers or equipment for SDS-PAGE. In addition, SMLM is a relatively new technology for EV research, so protocols, guidelines, and standards for both sample preparation and analyses are only beginning to emerge, and therefore less widely available compared with technologies like TEM. Even though preparing a sample for SMLM is simpler than for electron microscopy, it is not trivial, and it can be quite challenging in the absence of reliably optimized reagents. Some examples for such challenges are striking the right balance of fluorescence labeling, as well as getting the precise composition of the imaging buffer to get a good blinking, which is essential for the success of the assay. SMLM generates large amounts of information, and processing it into meaningful insights requires big-data tools which are not always readily available, especially in reasonably new fields like EV research. To this end, there are only a handful of commercially available products specifically designed for SMLM in EV research, so scientists may have to invest significant efforts to set up an experiment and analyze the results.

Nevertheless, SMLM is a powerful technology for exosome research: Its main strength resides in the fact that once it is optimized, it enables researchers to inspect thousands of individual exosomes and simultaneously gain quantitative data on the population level. Its main weaknesses are that it requires highly specific fluorescent detection reagents and that the experimental setup often requires multiple controls, although it is important to note that both these weaknesses are relevant to other techniques as well.

3 Recent Applications of SMLM Imaging in EV Research

The fields of virology and infectious diseases provide a number of examples of the power of dSTORM in detecting single particles [7, 36]. Interestingly, two recent papers harnessed SMLM to tackle the ongoing Covid-19 challenges: Singh et al. developed a method to detect SARS-CoV-2 proteins inside exosomes that were extracted from patients' plasma. This approach can be used to detect persistent infection, including in tissues other than the upper respiratory tract. Thanks to its high sensitivity, SMLM can detect proteins even at low concentrations and perhaps could be adapted for further research that aims to correlate viral persistence with ongoing disease symptoms [31].

Meanwhile, Wang et al. established a method to load exosomes with ACE2, the receptor used by SARS-CoV-2 to penetrate cells, thus trapping free virions in the airways of rodents and depleting the pool of active viruses. They used an array of methods to characterize their exosomes, among them dSTORM, which here provided confirmation of expected sizing, as well as tetraspanin positivity [37].

Using exosomes in liquid biopsies to characterize or diagnose diseases is an attractive idea, especially when they involve tissues that are not easily accessible, like the nervous system or the lungs [6, 38–40]. Under the right conditions, SMLM enables the screening of thousands of individual exosomes and identifying those which carry specific biomarkers, including rare subpopulations. You et al. [34] identified ATP1A3 as a neuronal marker that can be identified in cerebrospinal fluid as well as in the plasma of Alzheimer's Disease patients, and furthermore, its elevated levels correlate with higher levels of amyloid-β. Accordingly, isolating exosomes from plasma and using ATP13A as a marker for their neuronal origin has a potential to be a fairly non-invasive approach for diagnostics, for what otherwise requires either challenging biopsies or imaging-based diagnostics, for example, MRI, which are of far lower throughput than laboratory-based plasma analysis.

In another study, Roseborough et al. identified TMEM119 and CD14 as biomarkers on the membrane of exosomes which correlate with microglia activation following stroke. While the acute early microglia activation can be beneficial for the removal of damaged cells and for tissue repair, chronic neuroinflammation can be harmful. Therefore, developing non-invasive tools to assess microglia activation can be very helpful in preventing secondary damage in the nervous system. Rosebourough et al. used super-resolution microscopy as one tool in their diagnostic arsenal to characterize their biomarker-positive exosomes: they assessed the size distribution and the concentration of their biomarker-positive exosomes by applying several methods, including single-particle tracking (SPT) on membrane-associated dyed exosomes [35].

Thanks to their natural ability to deliver cargo between cells, EVs are considered a promising approach in drug delivery and therapeutics. Malle et al [41] developed a method to fuse EVs with synthetic liposomes, hence keeping their desired cargo while eliminating the immune response against the synthetic liposomes. They used super-resolution microscopy to characterize their newly loaded EVs and confirm the presence of the cargo.

Likewise, Wiklander et al [42] recently described a new method to pack drugs inside EVs which are then loaded with an antibody of choice, a process that directs them to solid tumors. Remarkably, the researchers were able to show that their engineered EVs can indeed reach their melanoma tumor target and furthermore, they are effective in reducing tumor burden and they improve survival. To obtain that effect, they took several approaches to optimize their antibody-loading process, when super-resolution microscopy provided direct and compelling evidence for the success of antibody-loading in one fraction. This group took several other approaches to study EVs for therapeutic purposes, and they frequently used SMLM to characterize them [43, 44]. Some of their work was done in collaboration with the biotechnology industry, highlighting not only the potential that EVs have in biomedical innovation, but also the role of super-resolution microscopy in future technology development.

4 Concluding Remarks

Super-resolution microscopy is an excellent tool in the exosome research toolbox. The main strengths of super-resolution localization microscopy (SMLM) are in its ability to image thousands of individual exosomes and provide biologically meaningful information about them, both at a single exosome level and at a population level. SMLM is a sensitive, accurate quantitative method, providing resolution $\geq$20 nm and occasionally even less than that. This resolution is well within the range of single exosomes, which makes the technology very attractive for exosome research.

Like all methods that rely on affinity binding for detection, the success of SMLM depends on the availability of good, specific primary antibodies. The preparation of the samples is highly similar to the preparation needed for standard immunofluorescence microscopy assays, which makes this approach quite appealing. Nonetheless, when setting up an experiment, it is essential to validate the specificity of the antibodies, alongside other important controls, such as controlling for the success of the exosome purification, if such is required. One of the more challenging aspects of SMLM in exosome research is the image acquisition and analysis, which could require some calibrations until the right conditions are found. To this end, a small number of commercial products have been developed to make super-resolution microscopy of exosomes more accessible to non-experts.

Analyzing the content of thousands of individual exosomes is similar, in essence, to the many other biological questions addressed with big data. While there are still many challenges around sample preparation, primarily around identifying exosomes within cells, the main challenge with using SMLM is the analysis of the quantitative data obtained by the imaging process.

This requires significant statistical tools in order to gain meaningful information on the heterogeneity of the different components in the sample. One of the most important future prospects in this field is the development of intuitive, easy-to-use software tools to analyze large amounts of data generated by SMLM, which combines information on size and morphology with information on the presence of biomarkers. Similar to many other fields in biological research, the study of exosomes is also departing from classical cell biology methods toward more computational approaches. Thanks to their fairly consistent size and shape, it is possible to capitalize on artificial intelligence to automatically identify exosomes within the microscope's field of view, although that requires substantial computational skills, which are not necessarily available to the average research scientist. The availability of automatic analyses, together with efforts around intuitive user-interface software, will be a huge leap forward in terms of making SMLM more accessible.

In recent years there are more and more research articles using SMLM to study exosomes. The availability of real-world examples, alongside clear guidelines (e.g., MISEV 2023 [8]) are likely to place SMLM at the forefront of tools to study exosomes and analyze their content. With more and more available literature, planning and carrying out experiments will become more trivial. As this technology gains popularity, the biotechnology industry will likely join the research efforts, and develop specialized products, which, in turn, will help popularize the technology even further. Good software tools will support the transition of SMLM from an expert technology to a common lab technique, not too different from confocal microscopy or cell sorting.

References

1. Fry, C. S., Kirby, T. J., Kosmac, K., McCarthy, J. J., & Peterson, C. A. (2017). Myogenic progenitor cells control extracellular matrix production by fibroblasts during skeletal muscle hypertrophy. *Cell Stem Cell, 20*, 56–69.
2. Spinazzola, J. M., & Gussoni, E. (2017). Exosomal small talk carries strong messages from muscle stem cells. *Cell Stem Cell, 20*, 1–3.
3. Braicu, C., et al. (2015). Exosomes as divine messengers: Are they the Hermes of modern molecular oncology? *Cell Death and Differentiation, 22*, 34–45.
4. Kumari, M., & Anji, A. (2022). Small but mighty—exosomes, novel intercellular messengers in neurodegeneration. *Biology, 11*, 413.
5. Gurung, S., Perocheau, D., Touramanidou, L., & Baruteau, J. (2021). The exosome journey: From biogenesis to uptake and intracellular signalling. *Cell Communication and Signaling: CCS, 19*, 47.
6. Zhang, Y., Liu, Y., Liu, H., & Tang, W. H. (2019). Exosomes: Biogenesis, biologic function and clinical potential. *Cell & Bioscience, 9*, 19.
7. Kalluri, R., & LeBleu, V. S. (2020). The biology, function, and biomedical applications of exosomes. *Science,* 367, eaau6977.
8. Welsh, J. A., et al. (2024). Minimal information for studies of extracellular vesicles (MISEV2023): From basic to advanced approaches. *Journal of Extracellular Vesicles, 13*, Article e12404.
9. Wu, S., et al. (2024). The advances and applications of characterization technique for exosomes: From dynamic light scattering to super-resolution imaging technology. *Photonics, 11*, 101.
10. Panagopoulou, M. S., Wark, A. W., Birch, D. J. S., & Gregory, C. D. (2020). Phenotypic analysis of extracellular vesicles: A review on the applications of fluorescence. *Journal of Extracellular Vesicles, 9*, 1710020.
11. Emelyanov, A., et al. (2020). Cryo-electron microscopy of extracellular vesicles from cerebrospinal fluid. *PLoS ONE, 15*, Article e0227949.
12. Koifman, N., Biran, I., Aharon, A., Brenner, B., & Talmon, Y. (2017). A direct-imaging cryo-EM study of shedding extracellular vesicles from leukemic monocytes. *Journal of Structural Biology, 198*, 177–185.
13. Morandi, M. I. et al (2022). Extracellular vesicle fusion visualized by cryo-electron microscopy. *PNAS Nexus, 1*, pgac156.
14. Lai, J. J., et al. (2022). Exosome processing and characterization approaches for research and technology development. *Advancement of Science, 9*, 2103222.
15. Zhou, M., Weber, S. R., Zhao, Y., Chen, H., Sundstrom, J. M. (2020). Methods for exosome isolation and characterization. In *Exosomes* (pp. 23–38). Elsevier. https://doi.org/10.1016/B978-0-12-816053-4.00002-X
16. Lelek, M., et al. (2021). Single-molecule localization microscopy. *Nature Reviews Methods Primer, 1*, 39.
17. Khater, I. M., Nabi, I. R., & Hamarneh, G. (2020). A review of super-resolution single-molecule localization microscopy cluster analysis and quantification methods. *Patterns, 1*, Article 100038.
18. Wu, Y.-L., Tschanz, A., Krupnik, L., & Ries, J. (2020). Quantitative data analysis in single-molecule localization microscopy. *Trends in Cell Biology, 30*, 837–851.
19. Hugelier, S., Colosi, P. L., & Lakadamyali, M. (2023). Quantitative single-molecule localization microscopy. *Annual Review of Biophysics, 52*, 139–160.

20. Schermelleh, L., et al. (2019). Super-resolution microscopy demystified. *Nature Cell Biology, 21*, 72–84.
21. Huang, B., Jones, S. A., Brandenburg, B., & Zhuang, X. (2008). Whole-cell 3D STORM reveals interactions between cellular structures with nanometer-scale resolution. *Nature Methods, 5*, 1047–1052.
22. Huang, B., Wang, W., Bates, M., & Zhuang, X. (2008). Three-dimensional super-resolution imaging by stochastic optical reconstruction microscopy. *Science, 319*, 810–813.
23. Huang, B., Babcock, H., & Zhuang, X. (2010). Breaking the diffraction barrier: Super-resolution imaging of cells. *Cell, 143*, 1047–1058.
24. Bates, M., Huang, B., Dempsey, G. T., & Zhuang, X. (2007). Multicolor super-resolution imaging with photo-switchable fluorescent probes. *Science, 317*, 1749–1753.
25. Van De Linde, S., et al. (2011). Direct stochastic optical reconstruction microscopy with standard fluorescent probes. *Nature Protocols, 6*, 991–1009.
26. Jung, S.-R., Fujimoto, B. S., & Chiu, D. T. (2017). Quantitative microscopy based on single-molecule fluorescence. *Current Opinion in Chemical Biology, 39*, 64–73.
27. Collot, M., et al. (2019). MemBright: A family of fluorescent membrane probes for advanced cellular imaging and neuroscience. *Cell Chemical Biology, 26*, 600-614.e7.
28. Aparin, I. O., et al. (2022). Fluorogenic dimers as bright switchable probes for enhanced super-resolution imaging of cell membranes. *Journal of the American Chemical Society, 144*, 18043–18053.
29. Li, H., & Vaughan, J. C. (2018). Switchable fluorophores for single-molecule localization microscopy. *Chemical Reviews, 118*, 9412–9454.
30. Khater, I. M., Nabi, I. R., & Hamarneh, G. (2020). A review of super-resolution single-molecule localization microscopy cluster analysis and quantification methods. *Patterns, 1*, Article 100038.
31. Singh, R. K., et al. (2023). Detection by super-resolution microscopy of viral proteins inside bloodborne extracellular vesicles. *Extracellular Vesicles and Circulating Nucleic Acids, 4*, 557–567.
32. Chuo, S.T.-Y., Chien, J.C.-Y., & Lai, C.P.-K. (2018). Imaging extracellular vesicles: Current and emerging methods. *Journal of Biomedical Science, 25*, 91.
33. Ram, A., Jalal, S., Jalal, A. S., & Kumar, M. (2010). A density based algorithm for discovering density varied clusters in large spatial databases. *International Journal of Computers and Applications, 3*, 1–4.
34. You, Y. et al. (2023). ATP1A3 as a target for isolating neuron-specific extracellular vesicles from human brain and biofluids. *Science Advance, 9*, eadi3647.
35. Roseborough, A. D., et al. (2023). Plasma derived extracellular vesicle biomarkers of microglia activation in an experimental stroke model. *Journal of Neuroinflammation, 20*, 20.
36. McNamara, R. P., & Dittmer, D. P. (2020). Extracellular vesicles in virus infection and pathogenesis. *Current Opinion in Virology, 44*, 129–138.
37. Wang, Z., et al. (2024). Inhalation of ACE2-expressing lung exosomes provides prophylactic protection against SARS-CoV-2. *Nature Communications, 15*, 2236.
38. Mitchell, M. I., et al. (2024). Non-invasive detection of orthotopic human lung tumors by microRNA expression profiling of mouse exhaled breath condensates and exhaled extracellular vesicles. *Extracellular Vesicles and Circulating Nucleic Acids, 5*, 138–164.
39. Mitchell, M. I., et al. (2024). Exhaled breath condensate contains extracellular vesicles (EVs) that carry miRNA cargos of lung tissue origin that can be selectively purified and analyzed. *Journal of Extracellular Vesicles, 13*, Article e12440.
40. Zhang, Z., et al. (2023). Comprehensive characterization of human brain-derived extracellular vesicles using multiple isolation methods: Implications for diagnostic and therapeutic applications. *Journal of Extracellular Vesicles, 12*, 12358.
41. Malle, M. G., et al. (2024). Programmable RNA loading of extracellular vesicles with toehold-release purification. *Journal of the American Chemical Society, 146*, 12410–12422.
42. Wiklander, O. P. B., et al. (2024). Antibody-displaying extracellular vesicles for targeted cancer therapy. *Nature Biomedical Engineering*. https://doi.org/10.1038/s41551-024-01214-6

43. Zheng, W., et al. (2023). Surface display of functional moieties on extracellular vesicles using lipid anchors. *Journal of Controlled Release, 357*, 630–640.
44. Hagey, D. W. et al. (2023). The cellular response to extracellular vesicles is dependent on their cell source and dose. *Science Advance,* 9, eadh1168.

Internal Standard for Quantification of Extracellular Vesicles in Biomarker Studies

Sehong Min, Tong Zhang, Komal Abhange, Reta Birhanu Kitata, Matthew J. Gaffrey, David M. Lubman, and Tujin Shi

Abstract Extracellular vesicles (EVs) are cell-derived particles that carry proteins, RNAs, lipids, metabolites, and other molecules. EVs play vital roles in intercellular communication and are implicated in disease and cancer, positioning them as a promising resource for non-invasive diagnostics and therapeutics. However, there are technical challenges in rigor and reproducibility for analysis of EVs. To address these challenges, we have been working on the development of super-SILAC based EVs as universal internal standard (UIS) for robust reproducible quantification of EV proteins. In this chapter, we have discussed different EV isolation methods and updated our recent progress in the proteome characterization of a mixture of EVs derived from different cancer cell lines and SILAC-labeled HeLa EVs. Our data suggest that it is highly promising to use super-SILAC EVs as UIS for reliable accurate quantification of EVs across large cohorts of clinical samples. Lastly, we have provided perspectives for future directions in EV proteomics research toward clinical application.

Keywords Extracellular vesicles · Universal internal standard · SILAC · Proteomics · Biomarker

1 Introduction

Extracellular vesicles (EVs) are small blebs released by all living cells and formed from bilayer lipid membranes. There are various types of EVs including exosomes, microvesicles (MVs), and apoptotic bodies which differ based on their mechanism of formation and size [1, 2]. EVs carry various cell-derived molecules, including proteins, nucleic acids, lipids, and metabolites that reflect the cell origin [1, 3, 4].

S. Min · T. Zhang · R. B. Kitata · M. J. Gaffrey · T. Shi
Biological Sciences Division, Pacific Northwest National Laboratory, Richland, WA, USA
e-mail: tujin.shi@pnnl.gov

K. Abhange · D. M. Lubman (✉)
Department of Surgery, University of Michigan Medical Center, Ann Arbor, MI, USA
e-mail: dmlubman@umich.edu

R. K. Marcus and D. Lubman (eds.), *Exosomes and Extracellular Vesicles*,
https://doi.org/10.1007/978-3-031-89270-7_10

EVs circulate in biofluids such as blood, ascites, urine, and saliva and carry their contents to other sites within the body [1]. EVs can combine with other cells, and thus act as mediators of cell-to-cell interactions resulting in the alteration of the functions of recipient cells [4]. EVs are also secreted by disease cells and play an important role in the development and progression of cancer and various neurological diseases [5, 6]. In addition to eukaryotic cells, they are secreted by bacterial and plant cells where bacterial EVs may play important roles in health and disease [7].

Exosomes are a specific subclass of EVs typically 30–150 nm in diameter [8]. Exosomes distinguish them from other types of EVs by their formation via endosomal biogenesis, in which the invagination of the endosomal membrane forms multivesicular bodies [8]. The analysis of exosomes for their biomolecule content as biomarkers has been actively pursued for early detection of diseases and the progression and response to therapies [3, 7, 9–11]. Exosomes are also being explored as therapeutic agents due to unique properties including low immunogenicity, the ability to transport their contents to target cells, and the possibility of modifying their surface proteins or glycans to target specific disease cells [12]. In addition, there are efforts to modify their contents for targeted drug delivery and the development of exosome-based regenerative medicine [13].

The major challenge in EV proteomics lies in the isolation and purification of EVs [14, 15]. While a variety of methods exist, they often yield inconsistent results, complicating the comparison of findings across studies [16]. Furthermore, these techniques can co-isolate non-EV proteins that can potentially skew subsequent analyses. Protein extraction presents another critical hurdle. Diverse extraction methods may recover different sets of proteins, impacting the reproducibility of protein identification and quantification [17]. This variability necessitates meticulous optimization and validation of extraction protocols to ensure consistent and reliable outcomes. The subsequent analysis of EV proteins via liquid chromatography-mass spectrometry (LC–MS) is also susceptible to multiple influencing factors. The MS type, MS settings for data acquisition, and the choice of data analysis software can all significantly affect protein identification and quantification. Consequently, these factors introduce considerable variability among studies, underscoring the need for standardized protocols in MS-based EV proteomics. To address these challenges, standardization of EV proteomics methods is essential for robust reproducible quantification of EV proteins. Furthermore, validation of EV discovery results across different cohorts of samples is crucial to enhance the reliability and robustness of EV proteomics research. By adopting standardized protocols and rigorous validation practices, the field can achieve more consistent and comparable results to advance our understanding of EV biology and its potential applications.

The stable isotope labeling by amino acids in cell culture (SILAC) strategy has emerged as a powerful technique for robust reproducible MS-based proteomics since its first use to cell lines by the Mann's group in 2002 [18]. Super-SILAC uses a mixture of SILAC-labeled cells as a spike-in standard, allowing for analysis of tissues and bodily fluids [19–25]. Prior to sample preparation, the addition of SILAC-labeled cells to biological samples maximally reduces variations introduced by sample preparation and LC–MS analysis because SILAC-labeled peptides have the same physical and chemical properties as those from endogenous proteins in the biological samples and they follow the same experimental workflow [26–28]. The heavy isotope-labeled peptides ("heavy") from SILAC provide a mass shift, allowing them to be distinguishable from their corresponding endogenous peptides ("light") from samples. Super-SILAC labeled cells serve as internal standard for broad accurate quantification of proteins in different types of biological samples including cell lines, tissues, and bodily fluids [29–31].

Similarly, super-SILAC labeled cells have been used as internal standard for analysis of EVs derived from cell lines and bodily fluids by the addition of super-SILAC labeled cell lysates to EV lysates after EV isolation (i.e., the conventional super-SILAC EV method) (Fig. 1a). The conventional method has been applied to understand the biological pathways perturbed by a certain treatment [32–35], to study the EV protein synthesis rates [36], and to identify EV biomarkers [37, 38]. However, the use of SILAC-labeled cell lysates as EV internal standard does not address the rigor and reproducibility issue for EV isolation because they are introduced after EV isolation (Fig. 1a). For robust reproducible quantification of EV proteins, the use of super-SILAC labeled EVs as truly universal internal standard (UIS) is needed because it can maximally reduce all analytical variations including EV isolation, sample preparation, and LC–MS analysis (Fig. 1b).

With this in mind, we have recently undertaken a comprehensive study of the proteomic content of EVs from several cancer cell lines as standards and compared them to the protein content of EVs isolated from the serum of patients with hepatocellular carcinoma or pancreatic cancers [39]. EVs from three different cancer cell lines were mixed at an equal amount and the mixed EVs served as internal standard to detect EV cancer markers in human patient serum samples. Furthermore, we have generated and characterized SILAC-labeled EVs derived from HeLa cells. The overarching goal is to develop super-SILAC labeled EV UIS for broad reproducible quantification of EV proteins (Fig. 1b). In this chapter, we have summarized our

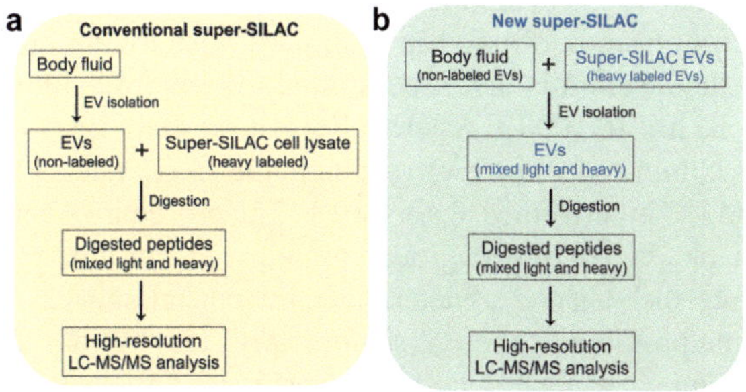

Fig. 1 Use of SILAC labeling for quantitative proteomics analysis of EVs. **a** Conventional method with super-SILAC labeled cell lysates as internal standard; **b** New method with super-SILAC labeled EVs as UIS. Note that super-SILAC labeled cell lyates are spiked into EV samples *after* isolation in the conventional method. By contrast, super-SILAC labeled EV standards are added to samples *before* EV isolation in the new method

recent progress in generation of EV internal standard and outlooked future directions in the EV research.

2 Methods

A critical challenge in the analysis of EVs is the lack of rigor and reproducibility for EV isolation. The isolation method selection can introduce biases in the types and quantities of EVs and consequently impact EV protein identification and quantitation. The selection of EV isolation method is primarily dependent on detection of target EV proteins and types of EVs of interest. Commonly used EV isolation methods are listed in Table 1 with description of their advantages and disadvantages. In addition, other EV isolation techniques have recently emerged including polymer-coated magnetic nanoparticles and acoustic nanofiltering [40, 41]. In this context, we focus on discussion of several EV isolation methods routinely used by our group.

2.1 Ultracentrifugation (UC)

Ultracentrifugation (UC) is generally considered as the "gold standard" for isolation of EVs. The EVs are isolated based on their size which is typically between 20 and 150 nm for exosomes and small EVs (sEVs). In the case of serum, the fluid is initially spun at a low speed, usually 5,000–10,000 g, to eliminate larger particles and filtered through a 0.22 μm filter [70]. The fluid is then centrifuged at 100,000 g for several hours to precipitate the EVs while the contaminants in the remaining fluid is removed. However, many proteins stick to the EVs, and lipoproteins are especially difficult to remove [71]. Chylomicrons, the balls of lipids with similar sizes as the EVs and viruses, may also contaminate the EVs.

UC has recently been performed in multiple cycles to eliminate contaminants [70]. In this method, serum is first processed at low centrifugation speed, filtered, and then spun for 2 h at 100,000 g. A gel analysis shows that a large amount of protein contaminants is eliminated while EVs remain in the precipitates. PBS is then added to the pellets and UC is performed again for 1 h. The procedure is repeated 4–5 times until the supernatant is free of background proteins. The resulting EVs are between 20 and 120 nm and disk-shaped, as determined by transmission electron microscopy (TEM). Proteome profiling of the EV samples with LC–MS has shown that many common EV markers (e.g., CD9, CD81, and CD63) can be detected with the removal of most of contaminant proteins though APOA remains difficult to eliminate entirely. To further eliminate APOA and other contaminants, alternative methods may need to be combined with UC.

Table 1 EV enrichment methods with their advantages and disadvantages

Methods	Descriptions	Advantages	Disadvantages	References
Ultracentrifugation (UC)	Isolate by sequential centrifugation steps to remove debris and larger vesicles	• Low expense • Large sample volume can be processed	• Low throughput, co-isolation of particles with similar size • Time-consuming • Potential damage to EVs due to high centrifugal forces	[42–44]
Size exclusion chromatography	Separate by size using a column filled with porous beads	• Gentle on EVs • Low risk of aggregation • Ease of use • Less expensive	• Dilution of EV isolates • Co-isolation of particles with similar size	[44–46]
Immunoaffinity-based isolation	Capture EVs using antibodies or lectins that specifically bind to EV surface markers	• High specificity • High purity	• Costly • Potential for non-specific binding • Limited by the availability of specific antibodies	[47–51]
Hydrophobic interaction chromatography	Separate by hydrophobicity using a column with hydrophobic ligands	• Good separation of EVs from soluble proteins • Scalable	• Can be complex • May require optimization for different samples	[52–54]
Ultrafiltration	Use membrane filters with defined pore sizes to separate EVs from larger particles	• Fast • Easy to perform	• Risk of filter clogging • Possible loss of EVs • Variable recovery	[55–59]
Polymeric precipitation	Precipitate EVs from solution using water-excluding polymers	• Simple • Does not require specialized equipment	• Co-precipitation of contaminants • Requires additional purification steps	[60–62]

(continued)

Table 1 (continued)

Methods	Descriptions	Advantages	Disadvantages	References
Sucrose density gradient	Separate by density through layering samples on a sucrose gradient followed by centrifugation	• High purity • Differential separation	• Labor-intensive • Time-consuming	
Asymmetric flow field-flow fractionation (AF4)	Laminar flow	• High resolution • Gentle on EVs	• May require optimization for different samples • Complex instrumentation	[63–65]
Microfluidics	Separate based on size, charge, and/or affinity	• Low sample volumes • High specificity and purity	• Risk of clogging • Costly	[66–69]

2.2 *Size Exclusion Chromatography (SEC)*

Size exclusion chromatography (SEC) separates EVs from other components based on size. SEC columns are packed by beads with pores of a defined size distribution. SEC columns that can accommodate different volumes of fluids are commercially available including those from Izon used in our work. In SEC, large particles bypass the pores and directly pass through the column, while smaller particles go into the pores and diffuse according to their size. Consequently, large particles are eluted first, and smaller particles (e.g., EVs) are eluted in the later fractions. These EV fractions can be characterized using common methods such as nanoparticle tracking analysis, TEM, MS, and Western blotting.

The SEC method is convenient, easy to use, and relatively inexpensive. However, when analyzing complex fluids, there is a notable challenge of incomplete separation, such as blood components eluting with EVs from the column. Therefore, a secondary method is often required to isolate the EVs with sufficient purity.

2.3 *The Combined UC and SEC*

An alternative to the methods described above is the combined UC and SEC for EV isolation. This has been performed by several groups where UC can be performed either before or after SEC (i.e., UC-SEC or SEC-UC) [72–75]. In many cases, the combined method can eliminate the need for multiple cycles of tedious UC and overcome the issue of protein bleeding into the fractions of EV. The advantage of UC-SEC is that UC can clean up the serum remarkably before loading the samples onto the SEC column, thus preserving the SEC column for additional runs [72]. Other work has used SEC first to isolate EV fractions so that the UC does not need to deal with the large amount of contaminants [74]. In both cases, TEM analysis has shown that the combined UC and SEC can provide a similar size distribution as UC. Common EV protein markers are observed for the combined method with similar protein content as UC.

2.4 *Immunoaffinity-Based Isolation*

Immunoaffinity-based approaches can be used to isolate EVs based on the EV surface markers. Immunoaffinity-based isolation can be performed by bead- or HPLC-based methods. Bead-based immunoaffinity isolation is more generally used due to the availability of commercial antibodies against EV surface marker proteins, which can be conjugated to magnetic beads. However, the beads may also bind serum proteins and they cannot be used for multiple times. To address this issue, we have recently developed an CD9 affinity column for EV isolation from patient serum [76]. This

CD9 immunoaffinity column (CD9-IC) method enables the isolation of EVs from microscale amounts of serum within 30 min and can be used for more than 50 runs. The immunoaffinity isolation is highly effective in eliminating unwanted background proteins and can provide relatively pure EVs with greatly reduced lipoprotein background compared to other EV isolation methods. This method can also isolate specific subpopulations of EVs for further study, which is of great value in studies of EVs originating from specific cell types closely associated with cancer or diseases. The drawback is that immunoaffinity isolation can only isolate distinct EV populations, and thus other EV populations will be lost unless a common EV surface marker is used for EV isolation.

3 Results and Discussion

3.1 In-Depth Proteome Profiling of EVs from Different Cell Lines and Patient Plasma

We have recently undertaken a comprehensive study of the proteomic content of the EVs from several cancer cell lines as standards and compared them to the protein content of EVs isolated from the serum of patients with hepatocellular carcinoma (HCC) and pancreatic cancer [39]. The detailed experimental workflow is shown in Fig. 2. For cancer cell EV analysis, sEV standards derived from HT29 (ab239690), PC3 (ab239689), and MCF7 (ab239691) were purchased from Abcam (Waltham, MA). The serum samples were initially processed using low-speed centrifugation to eliminate larger particles. The serum EVs were then isolated using either multiple cycles of UC or CD9-IC. The EVs were characterized by TEM, NanoSight light scattering, and ZetaView, revealing that most of the EVs were in the sEV or exosome regime. The sEVs from the commercial cell lines and patient serum were then lysed and digested using the S-trap method with a mixture of Lys-C and trypsin. The digested sEVs were subjected to LC–MS analysis with MS operation in the DIA mode on an Orbitrap Fusion Lumos MS (Thermo Scientific).

Several thousand proteins were identified, including many classic EV markers such as CD9, CD81, CD63, TSG101, Alix, and Syntenin-1. In addition, distinctive cancer-related proteins were detected in at least one cancer cell line, and several of them have been previously identified as potential markers for cancer detection and monitoring of cancer progression. Among these EV proteins, 44 important EV proteins were listed in Table 2, and they include common EV marker proteins (CD9, CD81, CD63, Syntenin-1 (SDCBP), TSG101, and Alix (PDCD6IP)) and cancer-related EV proteins (e.g., vimentin, CD29 (ITGB1), CD73 (NT5E), LYZ, PARVB, fibronectin, pleckstrin, and HGFAC). Besides these EV marker proteins, there are several annexins and many Ras and Rab proteins detected, which are known to be related to cancer. Further comparative proteomics showed differential serum EV proteins between normal controls and pancreatic cancer or HCC patients. Of note,

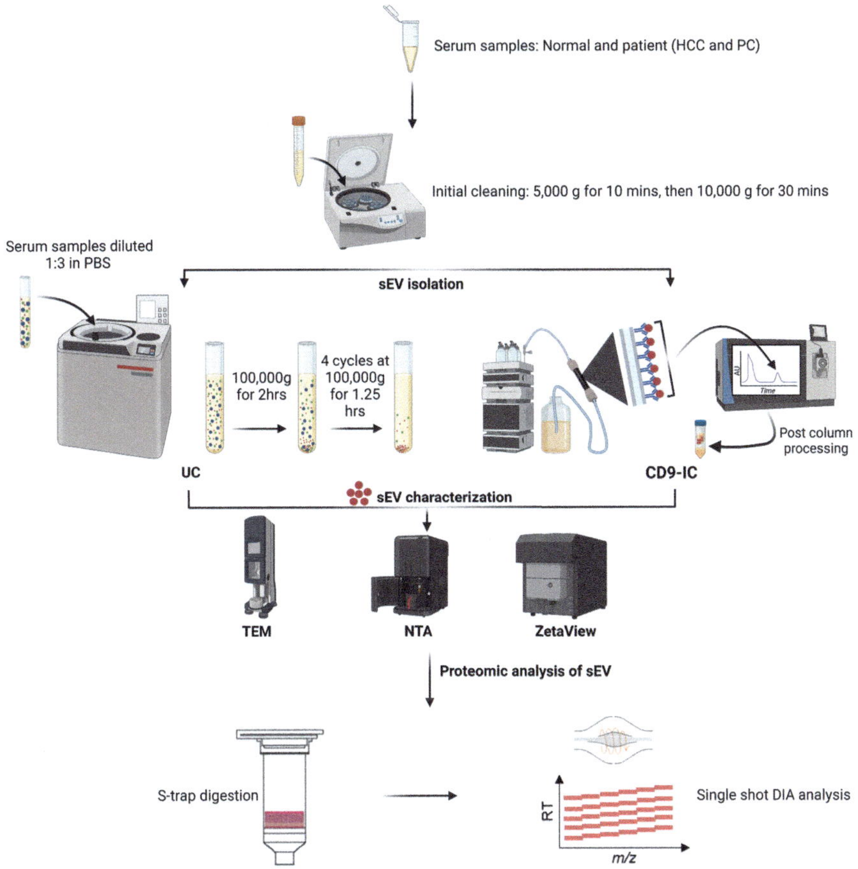

Fig. 2 Proteomic workflow for the isolation, characterization, and processing of small EVs (sEVs) followed by direct DIA-MS analysis. Reprinted with permission from (Abhange et al. [39]). Copyright 2024 American Chemical Society

500 – 1500 proteins could be identified, and most of them overlapped with the EV proteins identified from the cell line standards. Except for some EV proteins specific to pancreatic cancer and HCC, many other EV proteins can be reliably detected in all these serum EV samples.

Table 2 Comparison of standard cancer cell EVs with the mixed standard

Genes	Accessions	HT29	MCF7	PC3	MixedStd	Genes	Accessions	HT29	MCF7	PC3	MixedStd
CD9	P21926	6.3	5.7	6.2	6.3	TSG101	Q99816	5.0	4.0	4.4	4.4
CD81	P60033	6.8	5.3	6.6	6.2	SDC1	P18827	4.8	4.2	4.3	4.3
ANXA2	P07355	6.5	5,6	6.1	5.9	MAPRE2	Q15555	4.1	4.4	4.7	4.3
LGALS3BP	Q08380	5.2	4.7	5.1	5.4	ALDH1A3	P47895	4.3	3.8	4.4	4.2
SDCBP	O00560	5.9	5.0	5.4	5.4	CDH1	P12830	4.9	4.0	3.9	4.1
ITGB1	P05556	4.7	4.5	5.7	5.3	ALDH1L1	O75891	3.8	4.7	5.0	4.1
FN1	P02751	5.0	5.3	5.5	5.1	ALDH2	P05091	5.5	4.2	4.2	4.1
VIM	P08670	4.8	3.2	5.3	5.1	HGFAC	Q04756	3.6	4.1	4.3	4.0
PDCD6IP	Q8WUM4	6.4	4.8	5.3	5.1	ALDH18A1	P54886	5.2	3.7	4.7	4.0
ANXA1	P04083	5.3	4.6	5.2	5.1	ALDH7A1	P49419	4.5	4.2	4.4	4.0
ANXA4	P09525	6.1	5.0	5.1	5.1	ALDH3A2	P51648	4.5	–	4.5	4.0
ANXA11	P50995	5.5	4.9	5.5	5.0	ALDH3A1	P30838	5.1	–	4.1	3.9
ANXA7	P20073	4.8	5.0	5.3	5.0	ALDH16A1	Q8IZ83	4.5	4.3	3.9	3.9
TFRC	P02786	5.0	4.7	5.4	5.0	PARVB	Q9HBI1	3.8	4.4	4.5	3.9
ANXA5	P08758	5.0	4.9	5.1	4.9	ALDH9A1	P49189	4.4	3.9	4.4	3.9
ANXA6	P08133	4.5	4.8	5.3	4.8	ALDH3B1	P43353	4.4	–	4.5	3.8
LYZ	P61626	5.6	4.7	4.4	4.7	CDH2	P19022	–	–	4.2	3.7
PF4	P02776	–	4.8	–	4.7	PLEK	P08567	4.3	4.3	4.7	3.6
ANXA3	P12429	5.3	4.7	4.2	4.6	CDH11	P55287	–	3.9	4.0	3.1
CD63	P08962	5.0	4.6	5.4	4.5	ALDH4A1	P30038	4.0	–	–	–
ALDH1A1	P00352	5.3	4.1	4.1	4.5	CDH6	P55285	–	–	3.6	–
NT5E	P21589	5.3	3.4	4.9	4.4	ALDH8A1	Q9H2A2	–	3.3	4.3	–

3.2 A Mixture of EVs from Multiple Cell Lines for Internal Standard Controls

As shown in Table 2, some important EV protein markers may be observed in EVs from one cancer cell line but not from another cell line. For example, PF4, an important EV marker for HCC and pancreatic cancer, was not detected in the EVs from the PC3 or HT29 cell lines but in MCF7. Thus, a mixture of the EVs from multiple cancer cell lines is necessary to cover all the EV marker proteins of interest. This mixture can serve as internal standard controls for reproducible accurate quantification of EV proteins of interest in future EV studies. By spiking SILAC-labeled EVs derived from multiple cancer cell lines into the EV samples, more reliable quantification can be achieved (Fig. 1b).

As an initial demonstration, a mixture of EV standard was generated by mixing EVs from three cancer cell lines (PC3, HT29, and MCF7) at an equal amount. Similar to EV standard from individual cell lines, several thousand proteins were detected from the mixed EV standard. We then focused on a limited number of essential EV and cancer markers that have been of interest in our previous studies. The key markers for EVs including CD9, CD81, Alix, CD63, Syntenin-1, and TSG101 are detected where CD9 and CD81 are at the top of this list. A quantitative comparison between the proteins detected in this mixed standard versus EVs from each of the individual cell lines has shown that the mixed standard provides coverage of the EV protein markers of interest, whereas such coverage is often not complete for the individual samples. Furthermore, the quantitative expression of individual proteins in the mixed EV sample was comparable to that in each individual EV cancer cell line sample. The mixed EV standard was also compared to serum EVs isolated through multiple cycles of UC. The EV markers were found to be quantitatively similar for the normal and cancer samples compared with the mixed standard samples. Most of the target cancer markers selected in this work were found to be quantitatively similar to those detected in the mixed EV sample. All these results suggest that a mixture of EVs from multiple cell lines can be used as internal standard controls for proteomic analysis of EVs from cell lines and bodily fluids.

3.3 Generation and Characterization of SILAC-Based UIS

Besides its broad applications to cell lines, tissues, and bodily fluids, SILAC labeling has recently been used for quantitative proteomic analysis of EVs by mixing SILAC-labeled cell lysates with EV lysates (Fig. 1a). It can greatly reduce variations from sample digestion and MS analysis. However, the use of SILAC-labeled cell lysates as internal standard cannot address the issue in rigor and reproducibility for EV isolation because they are introduced after EV isolation. To achieve robust reproducible quantification of EV proteins, SILAC-labeled EV is needed and can be used as universal internal standard (UIS) with its addition to samples before EV isolation

(Fig. 1b). SILAC-labeled EV UIS could maximally reduce all possible variations including EV isolation.

To generate high-quality SILAC-labeled EVs, we have optimized and evaluated SILAC incorporation efficiency for EVs derived from HeLa cells (Fig. 3a). We have cultured HeLa cells in exosome-free DMEM medium supplemented with heavy isotope-labeled Arg10 and Lys8. EVs were purified using the combined SEC (qEVsingle 70 nm Gen 2 column, Izon, MA) and UC at 100,000 g. The purified EV samples underwent protein extraction and digestion using S-Trap, and the resulting peptides were analyzed by LC–MS. There are ~ 30,000 heavy isotope-labeled peptides with ~ 4,000 proteins identified from SILAC-labeled HeLa EVs. To evaluate SILAC incorporation efficiency, selected reaction monitoring (SRM)-based targeted proteomics was used for accurate quantification of both "heavy" and "light" versions of common EV marker proteins (e.g., CD9, CD63, CD81, and ANAX4) (Fig. 3b). Based on the abundance ratios of light over heavy peptides, we have calculated the SILAC incorporation efficiency for each EV protein. On average, > 95% of incorporation efficiency can be achieved for $\geq$ 10 cell doublings. Further optimization of cell culture medium and the number of cell doublings is needed to achieve > 98% of incorporation efficiency. The characterized SILAC-labeled EVs can be used as UIS by the addition of them to cell culture media or bodily fluids for reproducible quantification of EV proteins (Fig. 1b). The SILAC-based EV UIS method offers several potential advantages over conventional EV analysis methods in terms of measurement rigor and reproductivity, quantitation accuracy, broad EV protein quantification, being independent of EV isolation methods, and easy adoption and standardization.

4 Future Directions

Advancing the application of SILAC-labeled EV UIS to human biofluids will be pivotal in achieving robust and reproducible quantification of EV proteins with MS-based proteomics. With demonstrated high incorporation efficiency for SILAC-labeled EVs from a single HeLa cell line, the next focus is to generate super-SILAC labeled EVs derived from multiple cell lines as UIS at different EV mixing ratios, and then optimize and evaluate the UIS spike-in levels for reliable accurate quantification of EV proteins in human bodily fluids. A quantitative comparison between with and without UIS will be conducted to demonstrate the enhancement in robustness and reproducibility for proteomic analysis of EVs. Finally, the broad applications of this EV UIS based method will be assessed by testing its efficacy in diverse clinical samples, aiming to standardize the EV proteomic analysis workflow for reliable biomarker development and therapeutic monitoring.

UC or the combined UC and other isolation methods are typically used to generate high-purity exosomes for discovery proteomics, but they are time-consuming and cannot be used for proteomic analysis of large cohorts of clinical samples. Future developments will focus on optimization and evaluation of easily automatable, and high-throughput EV isolation methods (e.g., SEC, Mag-Net, or affinity isolation)

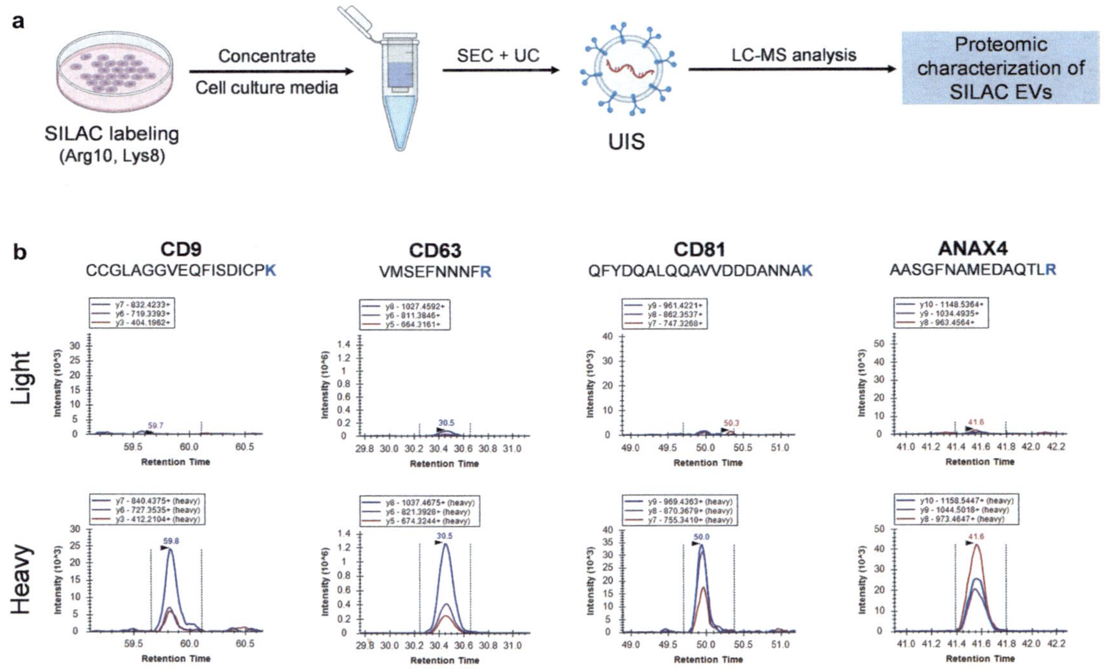

Fig. 3 Generation and characterization of SILAC-labeled EVs as universal internal standard (UIS). **a** Generation of UIS. Cells are cultured in the exosome-free medium supplemented with heavy isotope-labeled arginine (Arg10) and lysine (Lys8). Cell-derived EVs are concentrated using a molecular weight cutoff filter, followed by purification by the combined SEC and UC. The purified SILAC EVs can be used as UIS. **b** Quantification of selected EV marker proteins. Extracted ion chromatograms of surrogate peptides for common EV marker proteins. Top and bottom panels show SRM signals for unlabeled (light) and labeled (heavy) surrogate peptides, respectively. Heavy isotope incorporation efficiency was calculated as the percentage of heavy signal over the sum of heavy and light signals, where were 98.2, 93.1, 92.8, and 96.2% for CD9, CD63, CD81, and ANXA4, respectively

[77, 78]. Issues with EV impurity could be effectively addressed by high-specificity targeted proteomics (i.e., MS-based "western blots") [79–81]. The combination of rapid EV isolation methods and super-SILAC EV UIS is highly promising for broad, accurate, targeted quantification of 100s of EV protein markers in large clinical cohorts.

With improved rigor and reproducibility by the introduction of super-SILAC EV UIS, future developments will focus on improvements in detection sensitivity for proteome profiling of rare subpopulations of cell type specific EVs to improve understanding of the biology of EV biogenesis and sample throughput for rapid analysis of large cohorts of clinical samples toward translational applications. Enhancing detection sensitivity could be achieved by effective integration of ultralow-flow LC or capillary electrophoresis (CE) and a high-efficiency ion source/ion transmission interface with the most advanced MS platform. Sample throughput could be increased by using ultrafast high-resolution ion mobility-based gas-phase separation (e.g., SLIM [82]) to replace current slow liquid-phase (LC or CE) separation, and effective integration of liquid- and gas-phase separations (e.g., SLIM [82] or FAIMS [83]) for greatly reducing separation time but without trading off separation resolution. All these improvements together with super-SILAC EV UIS could lead to a powerful EV proteomics workflow in terms of robustness, sensitivity, reproducibility, and throughput to facilitate translation of EV protein markers into clinical use.

5 Conclusions

Extracellular vesicles (EVs) hold significant promise as biomarkers for non-invasive diagnostics and therapeutics due to their ability to carry and protect various biomolecules that are reflective of their cells of origin. However, the field faces challenges in rigor and reproducibility of EV isolation and standardization of EV proteomics workflow, which severely impact the translation of EV markers for clinical use. This chapter underscores the importance of developing super-SILAC EVs as UIS to address these challenges for robust reproducible quantification of EV proteins. The super-SILAC EV UIS method capitalizes on the addition of EV UIS to samples before EV isolation, and thus maximally reduce variations in sample preparation and LC–MS analysis. We showed a mixture of EVs derived from multiple cancer cell lines to be promising as internal standard controls because EV marker proteins of interest can be covered in the mixed EVs by combining various cell line-derived EVs at different ratios. High-quality SILAC-labeled EVs can be achieved by optimization of cell culture medium and the number of cell doublings. Thus, super-SILAC EVs derived from multiple cell lines can be used as UIS for reliable analysis of EVs across large cohorts of samples. Lastly, we have outlooked future directions in the EV proteomics research toward clinical applications.

Acknowledgements This work was funded by the National Cancer Institute under grant 1R01 CA258240 (to T.S. and D.M.L.). The research was also supported by NIH R01GM139858 (to T.S.).

A portion of this research was performed in the Environmental Molecular Sciences Laboratory, Pacific Northwest National Laboratory, a national scientific user facility sponsored by the United States Department of Energy under Contract DE-AC05-76RL0 1830. We also acknowledge the assistance of the Wayne State University Proteomics Core that is supported through NIH grants P30ES020957, P30CA022453, and S10OD030484. Research reported in this publication was also supported by the National Cancer Institutes of Health under Award Number P30 CA046592. The content is solely the responsibility of the authors and does not necessarily represent the official views of the National Institutes of Health.

Competing Interests The authors declare no competing interest.

References

1. Yanez-Mo, M., et al. (2015). Biological properties of extracellular vesicles and their physiological functions. *Journal of Extracellular Vesicles, 4*, 27066. https://doi.org/10.3402/jev.v4.27066
2. van Niel, G., D'Angelo, G., & Raposo, G. (2018). Shedding light on the cell biology of extracellular vesicles. *Nature Reviews Molecular Cell Biology, 19*(4), 213–228. https://doi.org/10.1038/nrm.2017.125
3. Kalluri, R., & LeBleu, V. S. (2020). The biology, function, and biomedical applications of exosomes. *Science, 367*(6478), eaau6977. https://doi.org/10.1126/science.aau6977
4. Mathieu, M., Martin-Jaular, L., Lavieu, G., & Thery, C. (2019). Specificities of secretion and uptake of exosomes and other extracellular vesicles for cell-to-cell communication. *Nature Cell Biology, 21*(1), 9–17. https://doi.org/10.1038/s41556-018-0250-9
5. Xu, R., et al. (2018). Extracellular vesicles in cancer–implications for future improvements in cancer care. *Nature Reviews Clinical Oncology, 15*(10), 617–638. https://doi.org/10.1038/s41571-018-0036-9
6. Hill, A. F. (2019). Extracellular vesicles and neurodegenerative diseases. *Journal of Neuroscience, 39*(47), 9269–9273. https://doi.org/10.1523/JNEUROSCI.0147-18.2019
7. Kumar, M. A., et al. (2024). Extracellular vesicles as tools and targets in therapy for diseases. *Signal Transduction and Targeted Therapy, 9*(1), 27. https://doi.org/10.1038/s41392-024-01735-1
8. Colombo, M., Raposo, G., & Thery, C. (2014). Biogenesis, secretion, and intercellular interactions of exosomes and other extracellular vesicles. *Annual Review of Cell and Developmental Biology, 30*, 255–289. https://doi.org/10.1146/annurev-cellbio-101512-122326
9. Thery, C., et al. (2018). Minimal information for studies of extracellular vesicles 2018 (MISEV2018): A position statement of the International society for extracellular vesicles and update of the MISEV2014 guidelines. *Journal of Extracellular Vesicles, 7*(1), 1535750. https://doi.org/10.1080/20013078.2018.1535750
10. An, M., et al. (2017). Quantitative proteomic analysis of serum exosomes from patients with locally advanced pancreatic cancer undergoing chemoradiotherapy. *Journal of Proteome Research, 16*(4), 1763–1772. https://doi.org/10.1021/acs.jproteome.7b00024
11. Czystowska-Kuzmicz, M., & Whiteside, T. L. (2021). The potential role of tumor-derived exosomes in diagnosis, prognosis, and response to therapy in cancer. *Expert Opinion on Biological Therapy, 21*(2), 241–258. https://doi.org/10.1080/14712598.2020.1813276
12. Johnsen, K. B., et al. (1846). (2014) A comprehensive overview of exosomes as drug delivery vehicles–endogenous nanocarriers for targeted cancer therapy. *Biochimica et Biophysica Acta, 1*, 75–87. https://doi.org/10.1016/j.bbcan.2014.04.005
13. Ha, D., Yang, N., & Nadithe, V. (2016). Exosomes as therapeutic drug carriers and delivery vehicles across biological membranes: Current perspectives and future challenges. *Acta Pharmaceutica Sinica B, 6*(4), 287–296. https://doi.org/10.1016/j.apsb.2016.02.001

14. Jalaludin, I., Lubman, D. M., & Kim, J. (2023). A guide to mass spectrometric analysis of extracellular vesicle proteins for biomarker discovery. *Mass Spectrometry Reviews, 42*(2), 844–872. https://doi.org/10.1002/mas.21749
15. Wang, Y. T., et al. (2020). Proteomic analysis of exosomes for discovery of protein biomarkers for prostate and bladder cancer. *Cancers (Basel), 12*(9), 2335. https://doi.org/10.3390/cancers12092335
16. Burton, J. B., Carruthers, N. J., & Stemmer, P. M. (2023). Enriching extracellular vesicles for mass spectrometry. *Mass Spectrometry Reviews, 42*(2), 779–795. https://doi.org/10.1002/mas.21738
17. Askeland, A., et al. (2020). Mass-spectrometry based proteome comparison of extracellular vesicle isolation methods: Comparison of ME-kit, size-exclusion chromatography, and high-speed centrifugation. *Biomedicines, 8*(8), 246. https://doi.org/10.3390/biomedicines8080246
18. Ong, S. E., et al. (2002). Stable isotope labeling by amino acids in cell culture, SILAC, as a simple and accurate approach to expression proteomics. *Molecular and Cellular Proteomics, 1*(5), 376–386. https://doi.org/10.1074/mcp.m200025-mcp200
19. Ishihama, Y., et al. (2005). Quantitative mouse brain proteomics using culture-derived isotope tags as internal standards. *Nature Biotechnology, 23*(5), 617–621. https://doi.org/10.1038/nbt1086
20. Geiger, T., et al. (2010). Super-SILAC mix for quantitative proteomics of human tumor tissue. *Nature Methods, 7*(5), 383–385. https://doi.org/10.1038/nmeth.1446
21. Ong, S. E. (2010). Whole proteomes as internal standards in quantitative proteomics. *Genome Medicine, 2*(7), 49. https://doi.org/10.1186/gm170
22. Geiger, T., et al. (2011). Use of stable isotope labeling by amino acids in cell culture as a spike-in standard in quantitative proteomics. *Nature Protocols, 6*(2), 147–157. https://doi.org/10.1038/nprot.2010.192
23. Shenoy, A., & Geiger, T. (2015). Super-SILAC: Current trends and future perspectives. *Expert Review of Proteomics, 12*(1), 13–19. https://doi.org/10.1586/14789450.2015.982538
24. Lindemann, C., et al. (2017). Strategies in relative and absolute quantitative mass spectrometry based proteomics. *Biological Chemistry, 398*(5–6), 687–699. https://doi.org/10.1515/hsz-2017-0104
25. Overmyer, K. A., et al. (2018). Multiplexed proteome analysis with neutron-encoded stable isotope labeling in cells and mice. *Nature Protocols, 13*(2), 293–306. https://doi.org/10.1038/nprot.2017.121
26. Chen, X., et al. (2015). Quantitative proteomics using SILAC: Principles, applications, and developments. *Proteomics, 15*(18), 3175–3192. https://doi.org/10.1002/pmic.201500108
27. Mann, M. (2006). Functional and quantitative proteomics using SILAC. *Nature Reviews Molecular Cell Biology, 7*(12), 952–958. https://doi.org/10.1038/nrm2067
28. Everley, P. A., et al. (2006). Enhanced analysis of metastatic prostate cancer using stable isotopes and high mass accuracy instrumentation. *Journal of Proteome Research, 5*(5), 1224–1231. https://doi.org/10.1021/pr0504891
29. Neubert, T. A., & Tempst, P. (2010). Super-SILAC for tumors and tissues. *Nature Methods, 7*(5), 361–362. https://doi.org/10.1038/nmeth0510-361
30. Deeb, S. J., et al. (2012). Super-SILAC allows classification of diffuse large B-cell lymphoma subtypes by their protein expression profiles. *Molecular and Cellular Proteomics, 11*(5), 77–89. https://doi.org/10.1074/mcp.M111.015362
31. Boersema, P. J., Geiger, T., Wisniewski, J. R., & Mann, M. (2013). Quantification of the N-glycosylated secretome by super-SILAC during breast cancer progression and in human blood samples. *Molecular and Cellular Proteomics, 12*(1), 158–171. https://doi.org/10.1074/mcp.M112.023614
32. Zhao, X., et al. (2014). Quantitative proteomic analysis of exosome protein content changes induced by hepatitis B virus in Huh-7 cells using SILAC labeling and LC-MS/MS. *Journal of Proteome Research, 13*(12), 5391–5402. https://doi.org/10.1021/pr5008703
33. Fricke, F., et al. (2019). SILAC-based quantification of TGFBR2-regulated protein expression in extracellular vesicles of microsatellite unstable colorectal cancers. *International Journal of Molecular Sciences, 20*(17), 4162. https://doi.org/10.3390/ijms20174162

34. Clark, D. J., Fondrie, W. E., Yang, A., & Mao, L. (2016). Triple SILAC quantitative proteomic analysis reveals differential abundance of cell signaling proteins between normal and lung cancer-derived exosomes. *Journal of Proteomics, 133*, 161–169. https://doi.org/10.1016/j.jprot.2015.12.023
35. Martin-Jaular, L., et al. (2021). Unbiased proteomic profiling of host cell extracellular vesicle composition and dynamics upon HIV-1 infection. *The EMBO Journal, 40*(8), e105492. https://doi.org/10.15252/embj.2020105492
36. Tan, C. F., et al. (2020). Exploring extracellular vesicles biogenesis in hypothalamic cells through a heavy isotope pulse/trace proteomic approach. *Cells, 9*(5), 1320. https://doi.org/10.3390/cells9051320
37. Kugeratski, F. G., et al. (2021). Quantitative proteomics identifies the core proteome of exosomes with syntenin-1 as the highest abundant protein and a putative universal biomarker. *Nature Cell Biology, 23*(6), 631–641. https://doi.org/10.1038/s41556-021-00693-y
38. Zhang, J., et al. (2017). Motile hepatocellular carcinoma cells preferentially secret sugar metabolism regulatory proteins via exosomes. *Proteomics*, 17(13–14). https://doi.org/10.1002/pmic.201700103
39. Abhange, K., et al. (2024). In-depth proteome profiling of small extracellular vesicles isolated from cancer cell lines and patient serum. *Journal of Proteome Research, 23*(1), 386–396. https://doi.org/10.1021/acs.jproteome.3c00614
40. Chang, M., Chang, Y. J., Chao, P. Y., & Yu, Q. (2018). Exosome purification based on PEG-coated Fe3O4 nanoparticles. *PLoS ONE, 13*(6), Article e0199438. https://doi.org/10.1371/journal.pone.0199438
41. Lee, K., Shao, H., Weissleder, R., & Lee, H. (2015). Acoustic purification of extracellular microvesicles. *ACS Nano, 9*(3), 2321–2327. https://doi.org/10.1021/nn506538f
42. Jeppesen, D. K., et al. (2014). Comparative analysis of discrete exosome fractions obtained by differential centrifugation. *Journal of Extracellular Vesicles, 3*, 25011. https://doi.org/10.3402/jev.v3.25011
43. Linares, R., et al. (2015). High-speed centrifugation induces aggregation of extracellular vesicles. *Journal of Extracellular Vesicles, 4*, 29509. https://doi.org/10.3402/jev.v4.29509
44. Takov, K., Yellon, D. M., & Davidson, S. M. (2019). Comparison of small extracellular vesicles isolated from plasma by ultracentrifugation or size-exclusion chromatography: Yield, purity and functional potential. *Journal of Extracellular Vesicles, 8*(1), 1560809. https://doi.org/10.1080/20013078.2018.1560809
45. Lozano-Ramos, I., et al. (2015). Size-exclusion chromatography-based enrichment of extracellular vesicles from urine samples. *Journal of Extracellular Vesicles, 4*, 27369. https://doi.org/10.3402/jev.v4.27369
46. Baranyai, T., et al. (2015). Isolation of exosomes from blood plasma: qualitative and quantitative comparison of ultracentrifugation and size exclusion chromatography methods. *PLoS ONE, 10*(12), Article e0145686. https://doi.org/10.1371/journal.pone.0145686
47. Clayton, A., et al. (2001). Analysis of antigen presenting cell derived exosomes, based on immuno-magnetic isolation and flow cytometry. *Journal of Immunological Methods, 247*(1–2), 163–174. https://doi.org/10.1016/s0022-1759(00)00321-5
48. Ueda, K., et al. (2014). Antibody-coupled monolithic silica microtips for highthroughput molecular profiling of circulating exosomes. *Science and Reports, 4*, 6232. https://doi.org/10.1038/srep06232
49. Multia, E., et al. (2019). Fast isolation of highly specific population of platelet-derived extracellular vesicles from blood plasma by affinity monolithic column, immobilized with anti-human CD61 antibody. *Analytica Chimica Acta, 1091*, 160–168. https://doi.org/10.1016/j.aca.2019.09.022
50. Zarovni, N., et al. (2015). Integrated isolation and quantitative analysis of exosome shuttled proteins and nucleic acids using immunocapture approaches. *Methods, 87*, 46–58. https://doi.org/10.1016/j.ymeth.2015.05.028
51. Kowal, J., et al. (2016). Proteomic comparison defines novel markers to characterize heterogeneous populations of extracellular vesicle subtypes. *Proceedings of the National Academy of Sciences, 113*(8), E968-977. https://doi.org/10.1073/pnas.1521230113

52. Wang, L., Bruce, T. F., Huang, S., & Marcus, R. K. (2019). Isolation and quantitation of exosomes isolated from human plasma via hydrophobic interaction chromatography using a polyester, capillary-channeled polymer fiber phase. *Analytica Chimica Acta, 1082*, 186–193. https://doi.org/10.1016/j.aca.2019.07.035
53. Bruce, T. F., et al. (2019). Exosome isolation and purification via hydrophobic interaction chromatography using a polyester, capillary-channeled polymer fiber phase. *Electrophoresis, 40*(4), 571–581. https://doi.org/10.1002/elps.201800417
54. Ji, X., et al. (2021). A novel method of high-purity extracellular vesicle enrichment from microliter-scale human serum for proteomic analysis. *Electrophoresis, 42*(3), 245–256. https://doi.org/10.1002/elps.202000223
55. Cheruvanky, A., et al. (2007). Rapid isolation of urinary exosomal biomarkers using a nanomembrane ultrafiltration concentrator. *American Journal of Physiology Renal Physiology, 292*(5), F1657-1661. https://doi.org/10.1152/ajprenal.00434.2006
56. Lobb, R. J., et al. (2015). Optimized exosome isolation protocol for cell culture supernatant and human plasma. *Journal of Extracellular Vesicles, 4*, 27031. https://doi.org/10.3402/jev.v4.27031
57. Vergauwen, G., et al. (2017). Confounding factors of ultrafiltration and protein analysis in extracellular vesicle research. *Science and Reports, 7*(1), 2704. https://doi.org/10.1038/s41598-017-02599-y
58. Busatto, S., et al. (2018). Tangential flow filtration for highly efficient concentration of extracellular vesicles from large volumes of fluid. *Cells, 7*(12), 273. https://doi.org/10.3390/cells7120273
59. Nordin, J. Z., et al. (2015). Ultrafiltration with size-exclusion liquid chromatography for high yield isolation of extracellular vesicles preserving intact biophysical and functional properties. *Nanomedicine, 11*(4), 879–883. https://doi.org/10.1016/j.nano.2015.01.003
60. Park, J., et al. (2021). Multiple-cycle polymeric extracellular vesicle precipitation and its evaluation by targeted mass spectrometry. *International Journal of Molecular Sciences, 22*(9), 4311. https://doi.org/10.3390/ijms22094311
61. Ludwig, A. K., et al. (2018). Precipitation with polyethylene glycol followed by washing and pelleting by ultracentrifugation enriches extracellular vesicles from tissue culture supernatants in small and large scales. *Journal of Extracellular Vesicles, 7*(1), 1528109. https://doi.org/10.1080/20013078.2018.1528109
62. Weng, Y., et al. (2016). Effective isolation of exosomes with polyethylene glycol from cell culture supernatant for in-depth proteome profiling. *The Analyst, 141*(15), 4640–4646. https://doi.org/10.1039/c6an00892e
63. Zhang, H., & Lyden, D. (2019). Asymmetric-flow field-flow fractionation technology for exomere and small extracellular vesicle separation and characterization. *Nature Protocols, 14*(4), 1027–1053. https://doi.org/10.1038/s41596-019-0126-x
64. Zhang, H., et al. (2018). Identification of distinct nanoparticles and subsets of extracellular vesicles by asymmetric flow field-flow fractionation. *Nature Cell Biology, 20*(3), 332–343. https://doi.org/10.1038/s41556-018-0040-4
65. Sitar, S., et al. (2015). Size characterization and quantification of exosomes by asymmetrical-flow field-flow fractionation. *Analytical Chemistry, 87*(18), 9225–9233. https://doi.org/10.1021/acs.analchem.5b01636
66. Chen, C., et al. (2010). Microfluidic isolation and transcriptome analysis of serum microvesicles. *Lab on a Chip, 10*(4), 505–511. https://doi.org/10.1039/b916199f
67. Guo, S. C., Tao, S. C., & Dawn, H. (2018). Microfluidics-based on-a-chip systems for isolating and analysing extracellular vesicles. *Journal of Extracellular Vesicles, 7*(1), 1508271. https://doi.org/10.1080/20013078.2018.1508271
68. Davies, R. T., et al. (2012). Microfluidic filtration system to isolate extracellular vesicles from blood. *Lab on a Chip, 12*(24), 5202–5210. https://doi.org/10.1039/c2lc41006k
69. Kanwar, S. S., Dunlay, C. J., Simeone, D. M., & Nagrath, S. (2014). Microfluidic device (ExoChip) for on-chip isolation, quantification and characterization of circulating exosomes. *Lab on a Chip, 14*(11), 1891–1900. https://doi.org/10.1039/c4lc00136b

70. Kim, J., Tan, Z., & Lubman, D. M. (2015). Exosome enrichment of human serum using multiple cycles of centrifugation. *Electrophoresis, 36*(17), 2017–2026. https://doi.org/10.1002/elps.201500131
71. Sodar, B. W., et al. (2016). Low-density lipoprotein mimics blood plasma-derived exosomes and microvesicles during isolation and detection. *Science and Reports, 6*, 24316. https://doi.org/10.1038/srep24316
72. An, M., Wu, J., Zhu, J., & Lubman, D. M. (2018). Comparison of an optimized ultracentrifugation method versus size-exclusion chromatography for isolation of exosomes from human serum. *Journal of Proteome Research, 17*(10), 3599–3605. https://doi.org/10.1021/acs.jproteome.8b00479
73. Blans, K., et al. (2017). Pellet-free isolation of human and bovine milk extracellular vesicles by size-exclusion chromatography. *Journal of Extracellular Vesicles, 6*(1), 1294340. https://doi.org/10.1080/20013078.2017.1294340
74. Wei, R., et al. (2020). Combination of size-exclusion chromatography and ultracentrifugation improves the proteomic profiling of plasma-derived small extracellular vesicles. *Biological Procedures Online, 22*, 12. https://doi.org/10.1186/s12575-020-00125-5
75. Koh, Y. Q., et al. (2018). Exosome enrichment by ultracentrifugation and size exclusion chromatography. *Front Biosci (Landmark Ed), 23*(5), 865–874. https://doi.org/10.2741/4621
76. Zhu, J., et al. (2021). Column-based technology for CD9-HPLC immunoaffinity isolation of serum extracellular vesicles. *Journal of Proteome Research, 20*(10), 4901–4911. https://doi.org/10.1021/acs.jproteome.1c00549
77. Wu, C.C., et al. (2024). Mag-Net: Rapid enrichment of membrane-bound particles enables high coverage quantitative analysis of the plasma proteome. *bioRxiv*. https://doi.org/10.1101/2023.06.10.544439
78. Bellotti, C., et al. (2021). High-grade extracellular vesicles preparation by combined size-exclusion and affinity chromatography. *Science and Reports, 11*(1), 10550. https://doi.org/10.1038/s41598-021-90022-y
79. Shi, T., et al. (2012). Advancing the sensitivity of selected reaction monitoring-based targeted quantitative proteomics. *Proteomics, 12*(8), 1074–1092. https://doi.org/10.1002/pmic.201100436
80. Shi, T., et al. (2016). Advances in targeted proteomics and applications to biomedical research. *Proteomics, 16*(15–16), 2160–2182. https://doi.org/10.1002/pmic.201500449
81. Aebersold, R., Burlingame, A. L., & Bradshaw, R. A. (2013). Western blots versus selected reaction monitoring assays: Time to turn the tables? *Molecular and Cellular Proteomics, 12*(9), 2381–2382. https://doi.org/10.1074/mcp.E113.031658
82. Ibrahim, Y. M., et al. (2017). New frontiers for mass spectrometry based upon structures for lossless ion manipulations. *The Analyst, 142*(7), 1010–1021. https://doi.org/10.1039/c7an00031f
83. Hebert, A. S., et al. (2018). Comprehensive single-shot proteomics with FAIMS on a hybrid orbitrap mass spectrometer. *Analytical Chemistry, 90*(15), 9529–9537. https://doi.org/10.1021/acs.analchem.8b02233

MALDI-MS Analysis of Exosomes

Mohamed A. Gab-Allah and Jeongkwon Kim

Abstract Exosomes are nanoscale extracellular vesicles (typically 30–150 nm in diameter) originating by an endocytosis process from various cell types and extensively circulating into surrounding biofluids. Recently, the applications of exosomes in the biomedical field have grown significantly, primarily due to their cell-specific cargos (proteins, nucleic acids, metabolites, and lipids) that enable intercellular communication between neighboring cells. The presence of these components in bodily fluids serves as promising biomarker resources for the diagnosis and prognosis of various diseases. Before their analysis, various isolation platforms can be used to enrich and purify exosomes without compromising their biological activity or integrity. The development of rapid and high-throughput analytical approaches such as matrix-assisted laser desorption ionization-mass spectrometry (MALDI-MS) continues to play a leading role in the characterization of exosomal components, demonstrating its effectiveness for biomarker discovery, clinical diagnostics, and disease monitoring. This is particularly attributed to the multiple features of this technique, including easy operation, direct analytical capability, and high tolerance to contamination (i.e., detergents, salts, buffers). This chapter mainly highlights the major applications of MALDI-MS in the analysis and characterization of exosomes. The basic principle, process, and analytical strategy of MALDI-MS are summarized. This chapter also offers a comprehensive overview of exosome biogenesis, composition, and function, along with their enrichment and characterization using various techniques to support a deeper understanding of MALDI-MS applications and utilization in exosome research.

Keywords Exosomes · MALDI · Mass spectrometry · Cell cargos · Biofluids · Enrichment procedures · Characterization techniques

M. A. Gab-Allah · J. Kim (✉)
Department of Chemistry, Chungnam National University, Daejeon, Republic of Korea
e-mail: jkkim48105@cnu.ac.kr

M. A. Gab-Allah
Organic and Protein Analysis Laboratory, National Institute of Standards, Giza, Egypt

R. K. Marcus and D. Lubman (eds.), *Exosomes and Extracellular Vesicles*,
https://doi.org/10.1007/978-3-031-89270-7_11

1 Introduction

Exosomes are nanosized extracellular vesicles secreted by nearly all living cells and can circulate throughout the human body [1]. In living cells, multivesicular bodies (MVBs) carrying intraluminal vesicles fuse with the plasma membrane to release exosomes into the extracellular environment through an endocytosis process [2]. Enclosed in a lipid bilayer, exosomes carry bioactive substances such as proteins, metabolites, lipids, and a variety of genetic materials, such as DNA, messenger RNA (mRNA), microRNA (miRNA), and long non-coding RNA [3]. Exosomes have long been regarded as a cellular waste disposal tool, which helps preserve cellular homeostasis [4]. However, they have recently been recognized as key factors in cellular communication, signal transduction, and serving as potential biomarkers for various diseases [5]. These multiple functionalities have connected exosomes to various physiological and pathological processes, underscoring their significance in both health and disease contexts. Particularly, the potential role of exosomes as biomarkers for disease diagnosis has led researchers to study their presence in various biological samples, including plasma [3, 6–8], serum [1, 9–11], urine [12–15], saliva [16], breast milk [17], blood cells [18], and human stem cells [19–21]. However, analyzing exosome contents from such complex biological samples is challenging and requires careful preparation since these samples often contain a complex mixture of other constituents that may interfere with the analytical process. Therefore, isolating pure exosomes is a critical step in studying their biological functions and potential therapeutic applications. To date, the most common methods to achieve efficient isolation and enrichment of exosomes include differential ultracentrifugation [22, 23], ultrafiltration, size-exclusion chromatography [24–26], and commercial kits [27, 28]. Additionally, more advanced techniques such as microfluidics [29–31] and precipitation-based methods [23, 32] have been developed to improve the purity and yield of isolated exosomes. Each method has its advantages and limitations, and the choice of technique often depends on the specific downstream analysis and the type of biological sample being studied. Combining two isolation methods can further increase the specificity and purity of the extracted exosomes [33–37].

Characterization of exosomes is essential for understanding their biological roles and potential applications in diagnostics and therapeutics. Various techniques have been employed to analyze exosome size, morphology, surface markers, and molecular composition [38]. Common methods include nanoparticle tracking analysis (NTA), dynamic light scattering (DLS) for size determination, as well as scanning electron microscopy (SEM), and transmission electron microscopy (TEM) for morphological analysis [39, 40]. Western blotting is the primary method for identifying specific surface markers, such as CD63, CD81, and CD9 on the surface of exosomes [5]. Enzyme-linked immunosorbent assay (ELISA) and nano-flow cytometry are also included for subgroup characterization [3]. Moreover, liquid chromatography-mass spectrometry has become one of the most sensitive analytical methods to provide information on the molecular mass and structural features of molecules [41–50]. While these methods can reveal the phenotype and biological characteristics of exosomes, they are often hindered by complex procedures, time-consuming preparation, high cost, and the need for sophisticated equipment, thereby limiting their practicality for clinical applications. Thus, it has become evident that developing reliable methods for the rapid and cost-effective analysis of exosomes in different biological matrices is essentially recommended.

A leading role in the field of exosome analysis is certainly played by matrix-assisted laser desorption ionization-mass spectrometry (MALDI-MS), a technique that has demonstrated its potential to provide detailed molecular profiling of exosomal proteins, lipids, and metabolites for diagnostic purposes while proceeding in a relatively straightforward manner. Indeed, MALDI-MS is distinguished by multiple features, including high throughput capacity, easy operation, cost-effectiveness, rapid analysis, direct analytical capability, and high tolerance to contamination [4, 28, 51]; thus, making it a more suitable candidate for extending methods to clinical practice, particularly when it comes to cancer diseases [3, 4, 8, 52].

The main purpose of this chapter is to comprehensively review the major benefits and applications of MALDI-MS in the analysis and characterization of exosomes, offering a comprehensive overview of its methodologies and clinical relevance. Further explanation on exosomes, their enrichment methods, and characterization techniques is also provided herein to enhance understanding of these topics within their respective contexts.

2 Biological Composition of Exosomes

Exosomes are often considered to be miniature replicas of the parent cell [53]. Despite being released by diverse cell types, they exhibit similarities in their biochemical composition due to their common endosomal origin [54]. The structure and composition of exosomes are depicted in Fig. 1. A large variety of substances have been identified in exosomes from different cell types, including approximately 9769 proteins, 1116 lipids, 3408 mRNAs, and 2838 miRNAs, as documented in the most recent updates provided by the exosome database (http://www.exocarta.org) [55]. This demonstrates their intricate complexity and potential functional diversity.

Generally, exosomes are highly enriched in proteins with diverse functions. Exosomal proteins have been identified into two groups; proteins that are common among all exosomes, such as those participating in membrane transport and vesicle fusion (annexins, flotillin, and Rab27a); MVB biogenesis, and exosome release (ALIX, TSG101, and flotillin); protein-folding process (heat shock proteins: HSP60, HSP70, and HSP90); cell penetration, membrane fusion, and protein trafficking (Tetraspanins: CD9, CD63, CD81, CD82, CD106); vesicle growth, release, and exosome biogenesis (programmed cell death 6 interacting protein: PDCD6IP or ALIX) [56, 57]. Other proteins commonly identified in exosomes include cytoskeletal proteins (actin, cofilin, tubulin, and moesin), metabolic enzymes, cell adhesion-related proteins (integrins), and proteins containing post-translational modifications, such as glycosylation and phosphorylation [58, 59]. Collectively, Alix, TSG101,

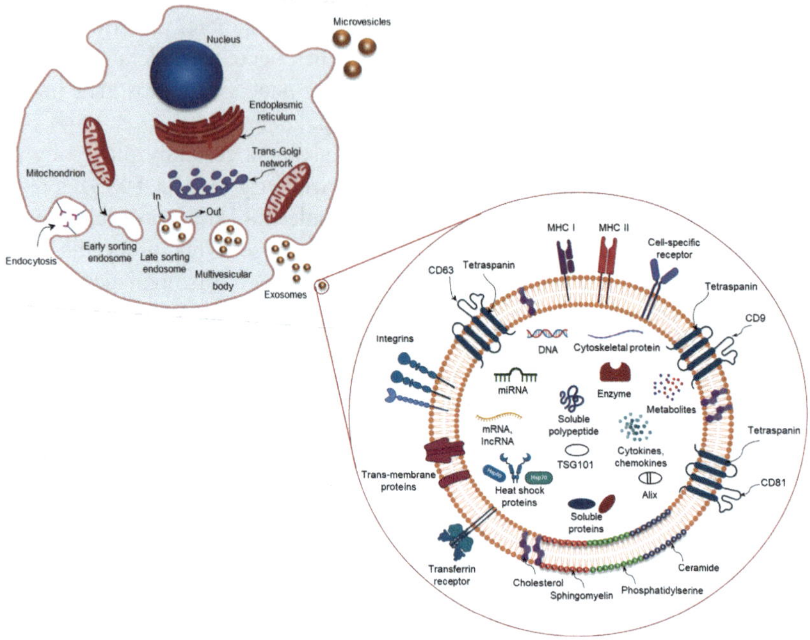

Fig. 1 Schematic illustration of exosome biogenesis, structure, and biological components. The main mechanism by which exosomes (30–150 nm) are released involves the fusion of MVBs carrying exosomes with the plasma membrane. In contrast, apoptotic bodies and microvesicles, which can be as large as 1 μm, are secreted through direct budding from the plasma membrane. Exosomes are rich in various biological molecules, including proteins, nucleic acids, lipids, and metabolites. Created with BioRender.com

HSP70, CD81, CD9, and CD63 are widely used as exosomal marker proteins, distinguishing exosomes from other extracellular vesicles and facilitating their isolation and detection through flow cytometric analysis, western blotting, and ELISA [54, 60, 61]. Another group belongs to cell-specific protein components that are intricately associated with the antigen-presenting process, including CD45 and MHC-I, II [2, 62–64].

In addition, exosomes are enriched in lipids, such as diglycerides, cholesterol, sphingolipids (including ceramide and sphingomyelin), lysobisphosphatidic acid (LBPA), disaturated lipids, phospholipids, glycerophospholipids including phosphatidylinositol (PI), phosphatidylcholine (PC), phosphatidylethanolamine (PE), phosphatidylserine (PS), and polyglycerophospholipids (i.e., bisphosphate) [2, 65]. Lipids play a crucial role not only in maintaining the structural integrity of exosomes but also in contributing to the process of exosome biogenesis and the regulation of homeostasis in recipient cells [65]. The presence of high-density lipids such as lyosbisphosphatidic acid (LBPA) in the inner membrane of MVBs leads to the generation of ILVs and subsequently the creation of exosomes.

Exosomes also contain a variety of genetic materials: DNA and RNAs, including messenger RNA (mRNA), microRNA (miRNA), and noncoding RNAs [66]. Some exosomal RNAs are specific to certain tissues, while others are found in exosomes regardless of their cellular origin [67]. This suggests the existence of different mechanisms involved in the sorting of RNA cargo. An important breakthrough in the field was the manifestation that exosome cargos include substantial quantities of mRNA, miRNA, and short noncoding RNA within the exosomes' lumen [68, 69]. Exosomal mRNA primarily plays roles in the cell cycle, angiogenesis, progression, migration, and histone modification [70]. Several RNAs found in exosomes have demonstrated the ability to generate proteins through an in vitro translation system [71, 72]. Exosomes transport nucleic acids to specific cells, where they undergo transcription and translation, thereby influencing the behavior of cells and exerting special functional roles [73].

3 Biogenesis of Exosomes

Exosomes are generated as intraluminal vesicles, inside multivesicular bodies (MVBs), by the endosome system, which plays a crucial role in the sorting and transporting of cellular vesicles to their designated locations [74]. A schematic illustration of exosome biogenesis is shown in Fig. 1. The main mechanism by which exosomes are released involves the fusion of MVBs carrying exosomes with the plasma membrane [2]. The process of exosome biogenesis initiates with the inward budding of the plasma membrane through an endocytosis process that creates a cup-shaped pocket holding extracellular proteins and the cell surface [65, 67]. This process subsequently leads to the formation of early endosomes, which in certain cases may merge with other endosomes generated by constituents of mitochondria, endoplasmic reticulum, and the trans-Golgi network. The early endosomes undergo invagination into late endosomes or MVBs, which entails the generation of intraluminal vesicles (ILVs) in the lumen of the organelles through a second inward budding of the endosomal membrane [2, 57]. Exosomes are characterized by their size, which ranges from 30 to 150 nm, distinguishing them from microvesicles with sizes typically ranging from 100 to 1000 nm. Additionally, B cell-derived exosomes have a density of 1.13 g/mL, whereas exosomes derived from epithelial cells have a slightly higher density of 1.19 g/mL [75].

It has been recognized that the formation and biogenesis of exosomes are significantly influenced by the endosomal sorting complex required for the transports (ESCRTs) mechanism, as reviewed recently [67, 76]. The components of ESCRTs comprise four separate proteins, i.e., ESCRT-0, ESCRT-I, ESCRT-II, and ESCRT-III, with other associated proteins, including apoptosis-linked gene 2-interacting protein X (ALIX), tumor susceptibility gene 101 protein (TSG101), vacuolar protein sorting-associated protein 4 (VPS4), and vesicle trafficking 1 (4VTA1), which cooperate in a stepwise manner [53, 70, 74, 75]. During the biogenesis of MVBs, the ubiquitin-binding subunits of ESCRT-0 are recruited by ubiquitinated cargo, the major pathway-specific signal in MVB biogenesis, to specific domains located on the endosomal membrane [65, 76]. The ESCRT-I and -II components facilitate the development of membrane buds and cargo encapsulation. In contrast, the ESCRT-III complex is dissociated from the MVB membrane by the vacuolar protein sorting 4 (Vps4) AAA ATPase and facilitates the separation of the vesicles from the cytoplasmic membrane [74, 77]. The excretion of exosomes into the external environment is facilitated by proteins belonging to the Rab-GTPase family, such as RAB2B, 5A, 7, 9A, 11, 27, and 35 [77–79]. The involvement of SNARE family proteins VAMP7 and YKT6 in the release process has also been suggested [78, 80]. Nonetheless, the exosomal protein "ALIX" is involved in processes related to endosomal membrane budding, abscission, and the selection of exosomal cargo, particularly through its interaction with syndecan and tetraspanin proteins (e.g., CD9, CD81, and CD63) [77, 79].

4 Biological Functions of Exosomes

The most unique function of exosomes may be their roles in intercellular communication, as they transport biomolecules, including proteins, lipids, and nucleic acids, between cells [28, 54, 81]. This exchange influences numerous biological processes and disease mechanisms. Via intercellular communication, these biologically active components can regulate gene expression, alter the fate and function of recipient cells, and potentially improve physiological functions or alleviate pathological processes [82, 83]. The contents of DNA and RNA within exosomes reflect the physiological or pathological conditions of the parent cell. For instance, exosomes derived from virus-infected cells carry viral RNA, while those from cancer cells exhibit distinct miRNA profiles compared to exosomes from normal cells [84]. In the immune system, exosomes can present antigens to immune cells, either activating or suppressing immune responses. For instance, dendritic cell-derived exosomes can stimulate T cells, enhancing immune activity, while cancer cell-derived exosomes may carry molecules that help tumors evade the immune system [85].

The contents of exosomes can also serve as prognostic markers and/or a graded basis for assessing cancer progression [28, 55, 74]. Additionally, they play a role in regulating tumor growth, metastasis, and angiogenesis, and contribute to mediating drug resistance in tumor cells [2, 86]. In cancer, exosomes support tumor growth and metastasis by transporting oncogenic factors such as mutant proteins, microRNAs, and long non-coding RNAs, which can modulate the tumor microenvironment to favor cancer cell survival and dissemination the tumor microenvironment, promoting cancer cell survival and spread [87, 88]. They also induce angiogenesis by delivering pro-angiogenic factors like VEGF to endothelial cells, facilitating tumor growth and providing pathways for metastasis [89].

Exosomes are involved in neurodegenerative diseases as well, spreading misfolded proteins and inflammatory signals throughout the nervous system [90, 91]. They also play a vital role in tissue repair and maintaining cellular homeostasis [92].

Exosomes from mesenchymal stem cells (MSCs) promote tissue regeneration by delivering growth factors and cytokines to damaged areas, encouraging cell proliferation and differentiation [54, 81]. This regenerative capability has been explored in treatments for myocardial infarction, wound healing, and neural injuries [75, 78]. Additionally, exosomes help cells adapt to environmental changes by removing unwanted components and transferring stress-related molecules [93]. The RNA content of exosomes shows significant specificity and diversity. Some RNAs are unique to specific tissues, while others are common across various cell types, indicating complex sorting mechanisms. These RNAs can modulate gene expression in recipient cells, influencing processes, such as cell cycle regulation and histone modification [94]. For instance, exosomal microRNAs can suppress target genes, affecting cell proliferation and differentiation, making them potential biomarkers for diseases [95].

Exosomes also play a role in metabolic regulation by carrying metabolic enzymes and substrates, impacting recipient cells' metabolic activities [96]. This is particularly relevant in metabolic diseases like diabetes, where exosome-mediated transfer of insulin-related molecules can affect glucose regulation [97]. Additionally, exosomal lipids can influence membrane fluidity and signaling pathways in recipient cells, further highlighting their role in metabolism [98].

5 Exosome Isolation Methods

The isolation of pure exosomes is a critical step in studying their biological functions and potential therapeutic applications. To date, various isolation platforms have been developed to enrich and isolate exosomes without compromising their biological activity or integrity. Common approaches include differential ultracentrifugation, ultrafiltration, size-exclusion chromatography, polymer-based precipitation, immunoaffinity capture, and commercial kits [27, 28]. Finding a standard method to extract intact exosomes with high repeatability and purity is a major challenge, while understanding the principles of each method is essential for selecting the most appropriate technique for ensuring the accuracy of downstream analyses [27].

5.1 Ultracentrifugation

Ultracentrifugation is the earliest and most commonly used method for exosome isolation due to its effectiveness in separating vesicles based on their size and density [99]. This technique is regarded as the gold standard for exosome isolation and is utilized in almost 80% of exosome processing [100]. Typically, the process involves a series of centrifugation steps at progressively higher speeds. Initially, low-speed centrifugation (300 × g and 2000 × g) removes cells and large debris. Subsequent mild-speed centrifugation (10,000 × g) helps to eliminate smaller cellular fragments and larger vesicles, leaving the exosomes in the supernatant. Finally, ultracentrifugation at very high speeds (typically over 100,000 × g) pellets the exosomes, which can then be resuspended in an appropriate buffer for further analysis [28]. The ultracentrifugation process for enriching exosomes from biological fluids is illustrated in Fig. 2a. One of the primary advantages of ultracentrifugation is its ability to process large volumes of samples, making it suitable for clinical and research applications where substantial quantities of exosomes are required [56, 74]. Moreover, this method is relatively simple and does not require expensive reagents or specialized equipment except for the initial instrument cost [62]. However, the technique has some limitations, including the potential for co-pelleting of non-exosomal particles, such as protein aggregates and other extracellular vesicles [12, 101]. This can affect the purity of the isolated exosomes, necessitating additional purification steps, such as density gradient ultracentrifugation, which further refines the separation based on buoyant density differences [102]. A lower centrifugation speed of 40,000 × g was employed to enrich exosomes, offering several benefits such as lower costs, the widespread availability of centrifuge instruments, simplicity of operation, and the ability to secure the process with a lid [103].

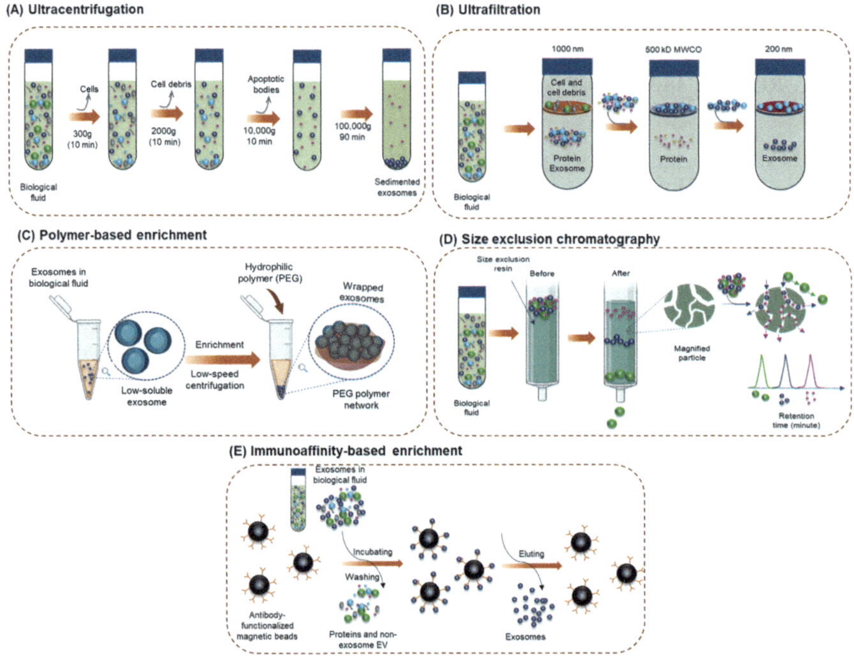

Fig. 2 Commonly used approaches for exosome enrichment and purification: **a** ultracentrifugation, **b** ultrafiltration, **c** polymer-based enrichment, **d** size exclusion chromatography, and **e** Immunoaffinity-based enrichment. Among these, ultracentrifugation is the most extensively studied method for isolating exosomes

5.2 *Ultrafiltration*

Ultrafiltration is a membrane-based technique used to isolate exosomes by leveraging size exclusion to separate these small vesicles from other components in biological fluids [28, 36]. The utilization rate of this technique is around 20% and it employs membrane filters with specific molecular weight cut-offs to selectively separate larger particles while allowing smaller molecules to pass through [104]. This method can be performed using various devices, including centrifugal filters and tangential flow filtration systems, depending on the sample volume and throughput requirements [105, 106]. Figure 2b depicts a diagrammatic illustration of the sequential ultrafiltration-based method for enriching and isolating exosomes from biological fluids. Biological fluids are initially passed through a 1000-nm filter to eliminate larger particles such as cells or cell debris, followed by filtration using a 500-kDa cut-off filter to get rid of smaller particles like free proteins. Eventually, exosomes smaller than 200 nm are collected using a 200 nm filter [107]. A key advantage of ultrafiltration is its ability to rapidly process large sample volumes efficiently, making it particularly useful for clinical applications where quick and scalable exosome isolation is needed [36]. Additionally, it is a gentle method that minimizes the risk of exosome damage, preserving their integrity and biological activity for downstream studies and therapeutic applications [56, 62, 108]. However, ultrafiltration can co-isolate similarly sized particles, such as other extracellular vesicles and protein aggregates, affecting the purity of exosome preparations [109]. Moreover, the ultrafiltration process may experience issues with clogging and vesicle entrapment, which can reduce the membrane's lifespan and isolation efficiency. Additionally, exosomes may adhere to the membranes, making them inaccessible for further downstream analyses. This can lead to low yields and potentially erroneous interpretations of test results [102]. To enhance the specificity and efficiency of exosome enrichment, ultrafiltration is often combined with other purification techniques like size-exclusion chromatography [33, 35, 36], density gradient centrifugation [22], or other precipitation-based enrichment [34, 37].

5.3 *Size Exclusion Chromatography*

Size exclusion chromatography (SEC) is one of the gentlest chromatography techniques commonly used for the purification of proteins, polysaccharides, and other macromolecules. Recently, exosomes from different biological fluids including serum, plasma, saliva, cerebrospinal fluid, urine, tears, and milk, have been isolated and purified using SEC [110]. Similar to ultrafiltration, SEC employs size-based separation to isolate exosomes from other macromolecules and particulate matter [100]. This method employs a porous stationary phase through which biological samples are passed, separating components according to their molecular size [62]. Components in a sample with large hydrodynamic radii, such as exosomes, traverse the column more quickly, as they are excluded from entering the pores of the matrix and eluted earlier. In contrast, components with small hydrodynamic radii can pass through the pores, leading to delayed elution [28, 111]. The process of SEC is shown in Fig. 2d. One of the primary advantages of SEC is its ability to maintain the structural integrity and biological activity of exosomes, as the process is gentle and does not involve harsh conditions [25, 35]. Additionally, this approach ensures low pollutant levels and high exosome homogeneity [110]. This is particularly important for downstream applications where functional exosomes are required, such as in therapeutic applications and biomarker discovery [24, 28]. Additionally, SEC can be performed relatively quickly and can be easily scaled up or down depending on the sample volume, making it a versatile tool in both research and clinical settings. Despite its benefits, there are some challenges associated with the SEC. The resolution of separation depends on the characteristics of the column matrix and the flow rate, which may need to be optimized for different sample types [102]. Additionally, while SEC effectively removes small contaminants, it may not fully eliminate similarly sized extracellular vesicles or protein aggregates, potentially affecting the purity of the exosome preparation [62, 111]. Therefore, SEC is often used in combination with other purification techniques, such as ultracentrifugation or immunoaffinity capture, to enhance the specificity and purity of the isolated exosomes [33–37]. More recently, the SEC was investigated by Park et al. [25] as an efficient characterization tool to evaluate the performance of exosome isolation from human urine.

5.4 Polymer-Based Precipitation

Polymer-based precipitation is a widely used method for exosome isolation, offering a straightforward and scalable approach. Originally developed for EV virus isolation [112], this method is frequently employed to isolate and purify small-sized particles such as exosomes due to the similar biophysical characteristics between exosomes and viruses [32, 113]. The method utilizes super-hydrophilic polymers such as polyethylene glycol (PEG), in concentrations varying from 8 to 15%, as a medium to precipitate exosomes from biological fluids [111]. The mechanism involves adding a polymer solution to the sample, which alters the solubility and dispersibility of exosomes, causing them to aggregate and precipitate out of the solution [100]. The precipitated exosomes are then collected by low-speed centrifugation, washed, and resuspended in a suitable buffer for further analysis. The process of polymer-based precipitation is shown in Fig. 2c. Currently, PEG-based precipitation kits, such as ExoQuick, Total Exosome Isolation, and ExtraPEG, have been developed to simplify and standardize the isolation process [32, 100, 114]. ExoQuick, for example, is a commercially available reagent that enables the efficient precipitation of exosomes from various biological fluids that are processed in only a few steps. The process involves adding a polymer solution to the samples, allowing the mixture to incubate for a set period, and then collecting the exosomes through centrifugation. Similarly, ExtraPEG offers a simple protocol for isolating exosomes with high yield and purity [32]. One of the main features of polymer-based precipitation is its flexibility, cost-effectiveness, simplicity, rapidity, and the ability to handle numerous samples at once [62]. Moreover, it does not require specialized equipment and can be performed with basic laboratory tools, making it accessible to a wide range of researchers. The gentle nature of the precipitation process further helps preserve the integrity and biological activity of the isolated exosomes [28].

While the polymer-based precipitation method yields a higher product, its limitation lies in the lower purity of the final product. The presence of polymer can hinder subsequent analysis of exosomes with -omics-based assays, as it introduces positively charged molecules and contaminants into the final exosome pellet [9, 35]. Selectivity is a challenge for PEG-based exosome enrichment because the PEG polymer decreases the solubility of both exosomes and proteins [115], leading to significant contamination, particularly in complex matrices like plasma or serum. To address this, multiple-cycle PEG-based enrichment [116] or combining PEG-based enrichment with other methods such as ultracentrifugation, SEC, filtration, or gel filtration, can be commonly used to enhance the purity of exosome products [13, 23, 100, 117]. Employing precipitation as a preliminary step before SEC resulted in the elimination of soluble protein and increased the resolution of the SEC method [117]. Another investigation demonstrated a scalable and reliable method by integrating PEG precipitation with ultracentrifugation [23].

5.5 *Immunoaffinity Techniques*

Immunoaffinity chromatography is a powerful and specific technique for isolating exosomes from complex mixtures based on the specific binding between their surface proteins and antibodies [22, 118]. This method employs antibodies that specifically bind to protein markers adhered to the exosome membrane's surface [62]. Exosomes can be easily isolated by immobilizing them on a solid support, allowing for their selective capture of complex biological samples [119]. Magnetic beads [120] and gold-loaded ferric oxide nanocubes [121] are commonly used in immunoaffinity protocols due to their ease of use and efficiency. These beads are coated with antibodies and mixed with the sample, allowing the exosomes to bind specifically. Figure 2e graphically represents the process of immunoaffinity-based enrichment. Several commercially available products utilize immunoaffinity-based enrichment, including ExoCap Streptavidin Kit (MBL International), Exosome Isolation Kit Pan (Miltenyi Biotec), and ExoQuant (Biovision). This process is relatively quick and can be performed under mild conditions, preserving the exosomes' functionality and viability for downstream applications. The ability of this approach to selectively capture exosomes based on surface markers makes it a valuable tool in both basic research and clinical applications, including biomarker discovery and therapeutic development [118, 119]. While this method can yield high-purity extracts, exosomes without specific surface markers recognized by the magnetic-labeled antibodies will not be isolated, potentially resulting in a reduced exosome yield [111]. Non-specific binding is another concern, where other particles or proteins in the sample may adhere to the antibodies, potentially contaminating the exosome preparation. Moreover, the method is relatively expensive due to the cost of antibodies and specialized reagents. To address some of these challenges, immunoaffinity chromatography is often used in conjunction with preprocessing methods such as SEC or centrifugation [122, 123]. These preliminary steps help remove protein aggregates and other large particles, enhancing the purity and specificity of the exosome isolation.

5.6 *Nano-Filtration*

Nano-filtration is an effective technique for isolating exosomes, utilizing successive filtration steps to achieve high purity and concentration [110]. It employs membranes with specific pore sizes to filter out impurities and cellular debris, initially removing particles smaller than 1000 nm. The filtrate is then repeatedly passed through exclusion filters to concentrate and extract exosomes. The final isolation uses a membrane with pores sized 50–250 nm to capture exosomes, while smaller particles pass through [26, 124]. This method is advantageous for handling large sample volumes and maintaining exosome integrity. However, it can suffer from contamination by non-exosomal peptides, such as alpha-1-antitrypsin and albumin, affecting purity and yield, particularly for exosome cargoes like RNA and miRNA [13, 26]. Additionally, interactions between vesicles and the filtration membranes can cause aggregates, clogging pores, and reduce efficiency. Despite these challenges, nano-filtration remains a robust technique, especially when combined with other purification methods like SEC or immunoaffinity chromatography, for enhanced exosome isolation [26].

5.7 *Microfluidics*

Microfluidics has emerged as a cutting-edge technology for the isolation of exosomes, leveraging factors such as size, density, and immunoaffinity to achieve high-throughput separation [29]. The fundamental principle behind this approach involves the use of microfluidic chips, which can capture and isolate exosomes based on their unique properties. This immuno-microfluidic technique allows for precise and efficient exosome isolation, offering significant advantages in terms of speed and purity [125]. Microfluidic devices utilize a variety of separation techniques, including size-based filtration, density gradient separation, and immunoaffinity capture. These methods enable the rapid and efficient isolation of exosomes from small sample volumes, significantly reducing the consumption of reagents and the time required for the process [29]. For instance, the ExoChip, equipped with CD63 antibodies, has been widely used to isolate exosomes with high specificity and efficiency [126]. This method not only preserves the integrity of the exosomes but also allows for the subsequent analysis of exosomal RNA and proteins [125]. The versatility of microfluidic-based exosome isolation is further enhanced by the incorporation of novel manipulations such as acoustic, electrophoretic, and electromagnetic techniques [127, 128]. Moreover, microfluidics can be combined with immunoaffinity techniques to identify specific subtypes of exosomes, facilitating detailed molecular profiling and biomarker discovery [125]. Despite the promising advantages, microfluidic exosome isolation faces challenges related to high cost and the complexity of the devices which can limit their widespread adoption in clinical settings [31, 102]. Nevertheless, the ongoing development of these methods may hold great potential for improving exosome isolation and is expected to become more accessible and reliable, paving the way for their integration into routine clinical practice and expanding their applications in diagnostics and therapeutics [30].

6 Exosome Characterization Techniques

Characterizing exosomes is essential for understanding their biological roles and potential applications in diagnostics and therapeutics. Various techniques are employed to analyze exosome size, morphology, surface markers, and molecular composition [38]. Mass spectrometry has become one of the most sensitive analytical methods to provide information on the molecular mass and structural features of molecules [41–50, 129, 130]. Particularly, MALDI-MS and liquid chromatography-tandem mass spectrometry (LC–MS/MS) offer detailed molecular profiling of exosomal proteins and lipids. In this chapter, our focus will be directed to MALDI-MS due to its efficacy in providing high-resolution molecular fingerprints of exosomal components, facilitating their use in research and clinical applications. Furthermore, MALDI-MS enables label-free and direct analysis, while also being a cost-effective and time-efficient method [28]. The main techniques commonly employed for exosome characterization are described below.

6.1 Western Blot Analysis

Western blot analysis, commonly known as immunoblotting, is a widely used technique for identifying specific proteins on the surface of exosomes isolated from biological fluids such as blood, urine, and saliva [22, 72, 100, 114]. This method provides valuable insights into the protein composition of exosomes and helps in verifying the presence of exosome-specific markers, such as CD63, CD9, CD81, TSG101, HSP70, and Alix, which are commonly used to confirm the successful isolation of exosomes [131, 132]. The procedure involves isolating exosomes, lysing them to release proteins, and separating these proteins according to their molecular weight using gel electrophoresis [25]. The separated proteins are then transferred onto a polyvinylidene fluoride (PVDF) membrane and probed with primary antibodies that specifically recognize the antigen on the surface of the membrane. The membrane is then treated with a secondary antibody that targets the species of the primary antibody used to identify the antigen. This secondary antibody is detected either through its fluorescent tag or via a horseradish peroxidase or alkaline phosphatase enzyme linked to the antibody to confirm the presence and purity of exosomes [56]. Western blotting offers good specificity and sensitivity for detecting specific proteins in a sample using antibodies [5, 38]. Furthermore, it allows the detection of multiple proteins simultaneously, making it valuable for validating exosome isolation and studying their role in diseases by comparing exosomes from healthy and diseased individuals [133, 134]. Despite being time-consuming and requiring substantial starting material, it remains a cornerstone in exosome research when combined with other methods.

6.2 Nanoparticle Tracking Analysis

Nanoparticle Tracking Analysis (NTA) is an optical technique used for sizing and counting vesicles and particles in a sample solution [5]. This technique involves illuminating particles with a laser light beam, which can be scattered by the interaction with the particles (under Brownian motion) [135]. By tracking these movements using a dark-field microscope with a camera mounted on it, the particle concentration and size distribution are estimated based on the Stokes–Einstein equation through the mean velocity and diffusivity of the particles [136]. Unlike DLS, which assesses bulk scattering, NTA focuses on the scattering of individual particles [137]. NTA is particularly valuable for exosome characterization because it allows for real-time analysis and provides high-resolution size distribution profiles. One of the primary strengths of NTA is its sensitivity to detect particles in the range of 10 nm to 1000 nm, which encompasses the typical size range of exosomes (50–150 nm) [138]. This sensitivity makes it an indispensable tool for distinguishing exosomes from other extracellular vesicles and particles that might be present in a biological sample. For clinical diagnostics, NTA is advantageous because it performs analyses in just a few minutes. Additionally, NTA can be used in conjunction with fluorescent labeling to provide further specificity, allowing researchers to characterize the surface immunophenotype and identify subpopulations of exosomes carrying particular biomarkers [139, 140]. However, the accuracy of size distribution measurements by NTA varies depending on surface properties and exosome diffusion behavior [40]. Another challenge with NTA is that it necessitates sample volumes of approximately 0.5 mL and requires careful optimization of data collection and analysis parameters [140].

6.3 Direct Light Scattering

Dynamic Light Scattering (DLS) is a widely utilized technique for characterizing exosomes, providing crucial information on their size distribution. Analogous to NTA, DLS relies on measuring light scattering caused by the Brownian motion of particles in suspension to determine particle size and concentration [56]. However, instead of measuring the diffusion coefficient directly, DLS analyzes the intensity fluctuations of the scattered light to determine the diffusion coefficients of the particles, allowing for the calculation of their size distribution [135]. One key advantage of DLS is its non-invasive nature, allowing for the characterization of exosomes in their native state, preserving their integrity and biological activity. Unlike NTA, DLS uses only a small sample and is also user-friendly, with few parameters required for optimization, thereby making it suitable for routine laboratory analysis [141]. However, DLS struggles to distinguish between particles of similar sizes and tends to provide less accurate results when characterizing vesicles with heterogeneous size distributions, as it does not analyze individual particles [5].

6.4 Electron Microscopy

Two primary types of EM are commonly used for the characterization of exosomes: transmission electron microscopy (TEM) and scanning electron microscopy (SEM) [142]. Because of the small size of exosomes and the simplicity of sample preparation, TEM is regarded as the gold standard for studying exosome morphology and structural integrity [142]. TEM offers a resolution of approximately 1 nm, and the negative staining procedure is both straightforward and quick, taking only 2–3 h [56, 142]. It involves passing a beam of electrons through a thinly prepared sample, resulting in high-resolution images that reveal detailed internal structures. By staining exosome samples with heavy metals like uranyl acetate, researchers can enhance contrast and visualize the characteristic cup-shaped morphology of exosomes [134, 143]. On the other hand, SEM provides detailed surface images of exosomes by scanning the surface with a focused beam of electrons. To achieve better resolution, samples are typically fixed, dehydrated, and sputter-coated before SEM analysis [5, 25]. SEM's ability to produce high-resolution images makes it an excellent tool for confirming the size distribution and surface characteristics of exosomes isolated from complex biological samples [144, 145]. While these techniques require specialized equipment and expertise, their ability to produce detailed images of exosomes makes them indispensable for confirming exosome presence and structural integrity [146, 147]. Therefore, EM remains a cornerstone of exosome research, complementing other analytical methods like nanoparticle tracking analysis and dynamic light scattering [134, 144].

6.5 *Flow Cytometry*

Flow cytometry is one of the most commonly used techniques for characterizing cells or larger micrometer-sized entities within a mixed population, using light scattering and fluorescence activation [56]. While advancements in flow cytometry technology are rapidly progressing with newer instruments achieving detection limits as low as 100–200 nm, the majority of devices still have a detection limit of 300–500 nm which is significantly larger than typical exosomal vesicles (30–150 nm), thereby making its application to exosomes challenging [56, 135]. Consequently, traditional flow cytometry analysis often employs exosome-coated beads rather than intact exosomes [5, 111]. Using fluorescently labeled antibodies targeting specific exosome surface markers, this method can identify and quantify various exosome subpopulations with enhanced sensitivity, providing detailed insights into their composition and function [135]. The fluorescent antibodies are visible under an epifluorescence microscope (EPI), where the sample generates a fluorescent signal detected as it passes the flow cytometer's laser [39].

6.6 *Atomic Force Microscopy*

Atomic force microscopy (AFM) is another valuable technique for characterizing exosomes, providing high-resolution imaging and detailed structural information [148]. Characterization of exosomes with this technique takes place by probing their surfaces using a probing tip mounted to the free end of a mechanical cantilever [39]. As the tip interacts with the exosome surface, deflections in the cantilever are measured by a laser beam reflected onto a photodetector [148]. These deflections are used to create a topographical map of the exosome, revealing detailed surface features that may not be accessible by some other microscopy methods. Exosome suspension is placed onto a mica holder and photographed after gentle drying using AFM. Topographic AFM images reveal the round shape of isolated exosomes, while phase images show substructures, revealing different components such as lipids and proteins [135]. However, the necessity for a flat and clean substrate like mica for precise imaging, combined with a relatively slow scanning process, makes it less suitable for high-throughput analysis [5]. Additionally, due to electrostatic attraction to the positively charged mica surface, exosomes may experience shape modification and flattening upon attachment [149].

6.7 *MALDI-MS*

Matrix-assisted laser desorption/ionization mass spectrometry (MALDI-MS) has emerged as the preferred method for the analysis of biological samples and has been used for various purposes in clinics. Its advantages, such as rapid analysis, ease of use, high throughput, and high tolerance to contamination, make it invaluable for detecting and identifying a wide range of biomolecules in exosomes without necessitating exosome lysis or labeling [3, 150].

6.7.1 Instrumental Basis and Process of MALDI-MS

Various ionization techniques are available for mass spectrometry, including chemical ionization, electrospray ionization (ESI), and MALDI. The selection of an ionization method is primarily determined by the objectives of the MS analysis and the characteristics of the samples. MALDI and ESI are particularly advantageous for the ionization and vaporization of large biomolecules due to their soft ionization capabilities [151]. These techniques minimize extensive fragmentation of the molecules, thereby preserving the integrity of the analyte molecules. This is particularly crucial in determining the correct mass of a biomolecule, particularly proteins, during MS analysis. The basic components of a MALDI system are depicted in Fig. 3. Upon irradiating the sample with a laser beam, usually in the UV range, the matrix absorbs the laser energy and converts it to electronically excited energy, facilitating the desorption and ion creation from large molecules [150]. Currently, Nd: YAG laser (wavelength of 355 nm) and N_2 pulsed laser (wavelength of 337 nm) are the most commonly used UV lasers in the MALDI system [28].

MALDI is usually coupled to TOF mass analyzers, creating MALDI-TOF/MS, due to its high throughput characteristics and its ability to handle the pulsed nature of the MALDI-generated ions efficiently [151]. In MALDI-TOF/MS, generated analyte ions are extracted into a field-free region (the flight tube) through a delayed potential pulse and finally detected. The ions travel through the TOF tube at velocities inversely proportional to the square root of their *m/z* ratios [152]. By measuring the time it takes for the ions to travel this distance, the *m/z* values can be accurately determined [153]. Importantly, the TOF mass analyzer is not constrained by mass range, making it an ideal match for MALDI, which produces singly charged ions with high *m/z* values [150, 151]. These features make MALDI-TOF/MS an attractive platform for many applications, including exosomes, which contain a wide array of biomolecules including proteins, lipids, nucleic acids, and metabolites. The resolution of a TOF mass analyzer depends on the length of the flight path and can be enhanced through techniques like delayed ion extraction and the use of a reflection. When a single TOF mass analyzer cannot adequately analyze target analytes in a complex mixture, coupling it with a second TOF mass analyzer enables tandem MS analysis (TOF/TOF mass analyzer) to allow the fragmentation of MALDI-generated precursor ions [151]. The first TOF mass analyzer selects the precursor ions, which are then fragmented in a collision cell. The second TOF mass analyzer subsequently analyzes these fragments to produce a tandem mass spectrum [154].

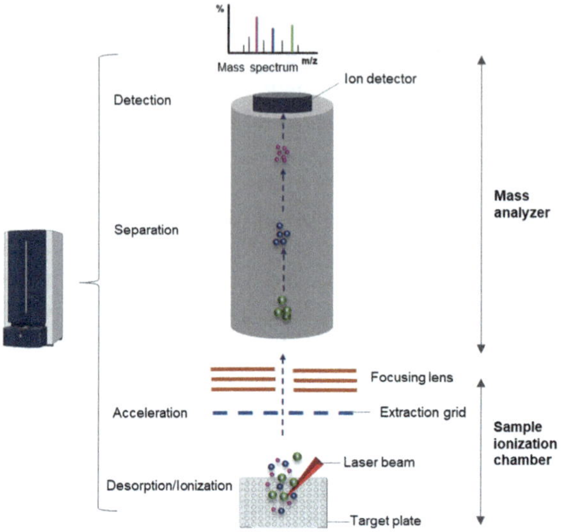

Fig. 3 Schematic illustration of the MALDI-TOF/MS principle. The sample is co-crystallized with the matrix compound on a clean MALDI sample plate that is placed inside the vacuum chamber of a mass spectrometer. A laser is irradiated at specific spots on the plate, causing the sample and matrix molecules to desorb and ionize. The resulting cloud of ionized molecules is then accelerated in an electric field into the TOF/MS for mass separation and detection. Lighter ions reach the detector faster, followed by heavier analytes. The generated mass spectrum represents the number of ions striking the detector over time. The separation of molecules occurs based on their *m/z* ratio; however, since the charge is typically single, the separation effectively reflects their molecular weight

6.7.2 The Role and Selection of Matrix

As indicated by its name, MALDI requires a matrix to assist the desorption and ionization of an organic molecule under UV irradiation. Matrices in MALDI are organic compounds of low molecular weights that co-crystallize with analyte molecules [155]. They serve multiple critical roles, such as providing a medium that can absorb the energy of the laser pulse, transferring it to the analyte molecules for desorption and ionization, protecting the analytes from direct laser damage, and preventing aggregation of the analyte molecules [156]. The crystallization process is influenced by the type of solvent used, the incubation period, and the matrix concentration [157]. The most commonly employed matrix compounds in MALDI are α-cyano-4-hydroxytrans-cinnamic acid (CHCA), 2,5-dihydroxybenzoic acid (DHB), and sinapinic acid (SA) [151]. These matrices are compatible with nitrogen and Nd: YAG lasers [158]. When exposed to laser irradiation, the analyte ions are incorporated into the matrix crystals and desorbed into the gas phase as matrix-analyte ion pairs. During this process, active species (radicals and electrons) are generated, leading to secondary reactions between the analyte and matrix in the plume, which ultimately result in the formation of analyte ions [159]. The ideal matrix should achieve (i) a good signal-to-noise (S/N) ratio, (ii) minimal fragmentation of the analyte, and (iii) a moderate background signal to reduce interference with peaks of interest, which is particularly a difficult task for small molecules, such as lipids [160, 161]. Another crucial factor is improving the uniformity of the co-crystals formed between the analyte and the matrix, as this significantly impacts the reproducibility of spectra and is therefore essential for accurate quantitative analysis [162].

When analyzing exosomes, which are complex biological particles consisting of proteins, lipids, and nucleic acids, the matrix must be chosen to optimize the ionization of these diverse molecules while minimizing fragmentation and background noise. One of the primary considerations in matrix selection is the nature of the analytes within the exosomes. For protein analysis, common matrices include CHCA and SA, both of which have shown efficacy in generating strong, reproducible signals for a broad range of proteins [10, 52]. CHCA is particularly favored for peptides and smaller proteins due to its high absorption coefficient at the laser wavelength and its ability to produce cleaner spectra with less fragmentation [163]. Conversely, SA is often preferred for larger proteins because it generates fewer multiply charged ions and more intact protein signals [164]. Super-DHB, a blend of 2-hydroxy-5-methoxybenzoic acid and DHB, is effective for glycan analysis, while for analysis of DNA molecules, 3-hydroxypicolinic acid is frequently employed [165].

6.7.3 Common Analytical Workflow of MALDI-MS

Another critical factor in MALDI-MS analysis is the sample preparation method, as it can significantly influence the quality of the analytical results. The MALDI process begins with co-crystallizing the analyte with a suitable matrix compound on a clean MALDI sample plate. The matrix must facilitate effective co-crystallization with the target analyte components to ensure uniform desorption and ionization [51, 156, 166]. This often involves optimizing the ratio of matrix to analyte and choosing solvents that can effectively dissolve both the matrix and the exosomal components [1, 3]. After drying and irradiation using the MALDI laser, the analyte molecules are desorbed and ionized, and the ions are accelerated in an electric field toward the mass analyzer for mass separation and detection [166]. This workflow is shown in Fig. 4 [167]. An alternative to MALDI, surface-assisted laser desorption/ionization (SALDI) does not require an organic matrix and uses solid nanomaterials as the matrix (see Fig. 4). Therefore, this technique promotes a more uniform sample distribution, and provides clear background spectra without observable interference peaks [168]. In some cases, pretreatment of biological samples is unnecessary; however, it is a common practice to conduct one or more clean-up or enrichment steps to concentrate analytes and/or remove interferents. These steps may include desalting, filtration, solid-phase extraction, liquid–liquid extraction, immunoaffinity isolation, or protein precipitation [150]. The "offline" relationship between sample preparation and analysis in MALDI allows for high throughput, making it practical for routine clinical analysis of large sample sets, such as those in 96- and 384-well plates. MALDI-MS can rapidly analyze large numbers of samples with minimal preparation time. Each sample spot can be processed in less than a minute, enabling the

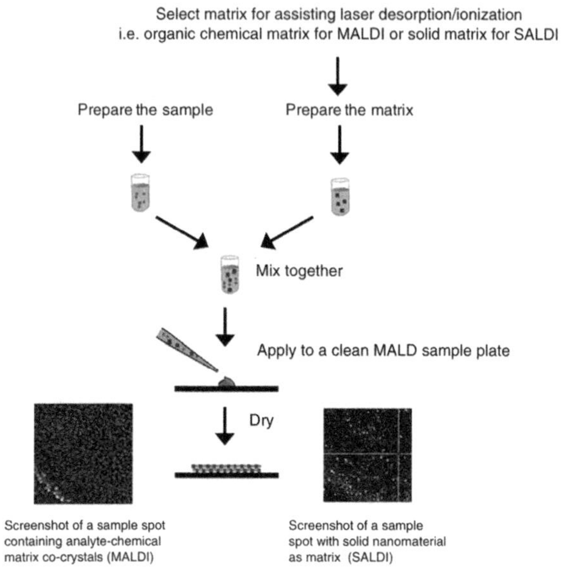

Fig. 4 Typical workflow for sample preparation in MALDI-TOF/MS and SALDI-TOF/MS. In traditional MALDI-TOF/MS, analytes are combined with chemical matrices and dried to form co-crystals. Conversely, in SALDI-TOF/MS, analytes are uniformly coated and evenly distributed on a solid matrix layer. Reprinted with permission from ref. [167] Copyright 2014, Springer

automated analysis of hundreds of samples within a few hours. Comparatively, ESI typically requires an additional step, "LC separation", with a turnaround time of 20 min to 1 h per sample. As a result, analyzing 100 samples using ESI MS takes between 30 and 100 h. Surface-assisted laser desorption/ionization-time-of-flight mass spectrometry (SALDI-TOF/MS) extends the capabilities of traditional MALDI-TOF/MS by incorporating surface affinity technology, allowing for the selective enrichment and analysis of target biomolecules from complex samples [169]. For instance, proteins of interest can be isolated and purified before laser ionization by binding with pre-modified materials on the ProteinChip surface utilizing biological or chemical affinities, while non-target components and contaminants can be washed away [169]. SALDI allows for the direct spotting of materials such as serum and urine without the need for further treatment. This technique also offers significant advantages in terms of selectivity, sensitivity, and throughput, making it a powerful tool for clinical diagnostics and biomolecular research [151].

6.7.4 MALDI-MS Analysis of Exosomal Proteins

The applications of MALDI-MS in biomarker discovery, disease diagnostics, and the elucidation of disease mechanisms continue to expand, underscoring the importance of this technology in advancing exosome research [11, 14, 52, 170]. When analyzed using MALDI-TOF/MS, peptides and proteins are mainly observed as singly charged protonated molecules with high sensitivity [3, 151]. This rapid exosome detection approach is proposed as a promising tool, especially for cancer research.

Lung cancer, one of the leading causes of cancer-related deaths worldwide, is characterized by uncontrolled cell growth in lung tissues, often resulting in metastasis to other parts of the body [171, 172]. Molecular analysis of lung cancer, including the study of exosomes, has become crucial for understanding its pathology. In a recent research, Choi et al. [11] investigated the profiling of extracellular vesicle (EV)-derived proteins to identify potential lung cancer biomarkers using serum samples from 80 individuals, both healthy and cancer patients. EVs were isolated via PEG-based precipitation and immunoaffinity separation, followed by proteomic analysis using 2-D gel electrophoresis and MALDI-TOF/MS. The expression of differentially upregulated proteins was validated by Western blotting. Among 55 upregulated protein spots, seven were highly expressed in patient-derived EVs, with CD5L showing the highest predictability for lung cancer (AUC value of 0.943) and correlating with cancer tissue expression. The findings suggest that EV-derived CD5L may serve as a potential biomarker for the non-invasive diagnosis of lung cancer.

Melanoma is among the most perilous forms of skin cancer, originating from mutated dermal melanocytes. While it usually appears on the skin, it can also develop in the mouth, eyes, or intestines. Recently, Zhu et al. isolated exosomes from different cell culture media using a differential centrifugation approach and used MALDI-TOF/MS to profile exosome proteins without the need for exosome lysis and labeling [52]. The protein-assigned fingerprints aided in the investigation of melanoma development via exosome-mediated intercellular communication and provided a qualitative and semiquantitative means of identifying several potential melanoma biomarkers.

Cancer of the urinary tract, also known as urothelial carcinoma (UC), is ranked as the ninth most common form of cancer globally [15]. UC is now detected by intravenous or computed tomography urography, urine cytology, and biopsy-assisted cystoscopy [173]. Chen and co-workers introduced an exosome proteome profiling approach by analyzing protein extracts using a simple and rapid MALDI-TOF/MS platform, demonstrating the potential of this MS technique in exosome analysis [14]. From 129 patients, they successfully identified urinary exosome proteins as potential biomarkers, such as histone H2B1K and alpha 1-antitrypsin, which could aid in the rapid and cost-effective diagnosis and prognosis of urothelial carcinoma.

Breast cancer is the primary cause of cancer-related mortality among women globally. It is classified into six molecular subtypes, each with distinct clinical significance, treatment strategies, therapeutic responses, and prognoses. Klinke II et al. [170] employed MALDI-TOF/MS and two-dimensional gel electrophoresis (2DE) to compare the secreted proteins by two different breast cancer cell lines (BT474 and SKBR3) to those of a normal human mammary epithelial cell line (184A1). The study revealed distinct secretome profiles between the cancerous and normal cell lines. The cancer cell exosomes are enriched in proteins that were found to be involved in antigen processing and glycolytic metabolism, which are pathways that are linked with tumor immunosurveillance evasion and deregulated cellular energetics. More recently, Zheng et al. [8] developed an integrated strategy for isolating and analyzing exosomes from human plasma collected from breast cancer (79), pancreatic cancer (57), and healthy (84) individuals, effectively discriminating between these different cancers using deep-learning fingerprints. Their method, sequential size-exclusion chromatography (SSEC), efficiently isolated exosomes with high purity within 2 h. Utilizing MALDI-TOF/MS, they profiled the exosome contents and constructed an artificial neural network model, Exo-ANN, achieving 80.0% diagnostic accuracy and an area under the curve of 0.91 in an independent test group. This approach shows promise for clinical liquid biopsy applications based on the analysis of plasma exosomes. In another research, Tutanov et al. [18] conducted a comprehensive analysis of proteomes from isolated plasma exosomes and exosomes associated with blood cells in both healthy individuals and those with breast cancer. They used ultrafiltration and ultracentrifugation methods for exosome isolation, and MALDI-MS was utilized for protein analysis, identifying 146 and 111 proteins (48 newly discovered proteins) in pathological and healthy samples, respectively. Their study, using a database of differently expressed proteins in human cancer, revealed that 64% of blood cell-associated exosomal proteins in cancer patients originated from tumors. Functional enrichment analysis showed distinct compositions: plasma exosomes in breast cancer exhibited increased transport and catalytic proteins, while blood cell-associated exosomes showed a shift toward transcription-regulating proteins. These findings underscore the significant role of exosomes in tumor dissemination and highlight their potential in cancer diagnostics and therapeutics.

Osteosarcoma is an aggressive bone cancer that primarily affects children and adolescents, typically arising in the long bones such as the femur, tibia, and humerus [4]. Recent research on osteosarcoma has utilized exosome profiling to identify potential biomarkers and therapeutic targets. Shen's group utilized MALDI-TOF/MS along with multivariate statistical analyses to profile plasma exosomes for evaluating lung metastasis in osteosarcoma patients [3]. Forty osteosarcoma patients (20 with and 20 without lung metastasis) and 12 healthy controls were included. Exosomes were isolated and analyzed, allowing for efficient differentiation between osteosarcoma patients and healthy controls, as well as between metastatic and non-metastatic cases. Moreover, seven exosomal proteins were identified as potential biomarkers for osteosarcoma lung metastasis, highlighting the method's promise for clinical diagnosis and monitoring of the disease.

Brain tumors encompass various types, with gliomas being the most aggressive and lethal primary brain cancer [19]. These tumors are heterogeneous and can be categorized based on molecular properties, histological characteristics, differentiation patterns, location, and anaplasia features [174]. Sequential centrifugal ultrafiltration was successfully used by Di Giuseppe et al. [19] for the efficient isolation of EVs from the culture medium of glioblastoma-derived stem-like cells (GSCs). These EVs were then categorized into two subtypes: microvesicles and exosomes using electron microscopy and Western blot analysis. Subsequent 2-DE and MALDI-TOF analysis enabled the identification of distinct protein sets with various functions within each EV subpopulation. These findings suggest new prognostic markers and potential drug targets, although further validation with a larger sample size is needed. A significant role of EVs in pathological conditions was also highlighted by the Filatov group, who developed methods for the isolation and purification of exosomes derived from human cancer glioblastoma cell lines in vitro, along with characterizing their protein content [20]. More specifically, the combination of MALDI-TOF with 2-DE was used to gather more comprehensive data on the proteome of exosomes released by brain tumor cells. In total, over 30 proteins, including specific tumor markers, were identified using this approach.

Globally, hepatocellular carcinoma (HCC) is the second-largest cause of cancer-related mortality and the fifth most prevalent type of cancer [175]. An essential step for epithelial cancer cells to become mesenchymal and metastasis is an epithelial-to-mesenchymal transition. During this transition, cells lose their cell–cell adhesions and become more motile, enabling them to migrate to distant anatomical regions and contribute to the development of metastatic tumors. The study by Karaosmanoglu et al. [175] was designed to explore the differences between complete and partial epithelial to mesenchymal transition (EMT) in hepatocellular carcinoma (HCC) cells by over-expressing the EMT-associated transcription factor Slug. Using MALDI-TOF/MS and ELISA assays, the research revealed that Slug over-expression induces complete and partial EMT in HCC-derived HepG2 and Huh7 cells. Complete EMT showed downregulation of E-cadherin and upregulation of ZEB2, while partial EMT was marked by upregulation of E-cadherin and downregulation of vimentin and ZEB2. Additionally, Slug was found to enhance chemo-resistance through modulation of ABC transporters ABCB1 and ABCG2, and the stemness marker CD133. Furthermore, Slug-mediated partial EMT was linked to increased exosomal secretion of modified fibronectin 1 (FN1), collagen type II alpha 1 (COL2A1), and native fibrinogen gamma chain (FGG), which were suggested as biomarkers. In another research, Kim's group explored the utility of MALDI-FTICR-MS for exosome analysis, demonstrating its effectiveness for non-invasive clinical diagnostics [1]. They isolated exosomes from human serum via centrifugation and analyzed them using MALDI-FTICR-MS, identifying a distinctive protein peak at *m/z* 7766 as platelet factor 4 (PLF4). This approach provided high mass accuracy and resolution, facilitating reliable protein identification. Additionally, analysis of exosomes from 27 liver disease patients revealed PLF4 as the most abundant peak, along with minor peaks. These findings highlight MALDI-FTICR-MS as a valuable and efficient method for exosome detection.

Malignant mesothelioma (MM), a cancer originating from the mesodermal tissue that lines the coelomic cavities, currently lacks an effective curative therapy for this condition [176]. Due to the limited data available on tumor antigens in MM, this particular tumor was selected as a model system for studying the properties of tumor-derived exosomes. Hegmans et al. [63] conducted a study on the protein composition of exosomes derived from mesothelioma cells, utilizing MALDI-TOF for analysis and characterization. They identified a diverse range of proteins, including MHC class I, heat shock proteins HSC70 and HSP90, annexins, PV-1, and several cytoskeletal proteins such as ezrin, moesin, actinin-4, desmoplakin, and fascin. Additionally, the study detected the developmental endothelial locus-1 (DEL-1), a potent angiogenic factor. These findings suggest that mesothelioma cell-derived exosomes express proteins involved in signal transduction, migration, adhesion, and antigen presentation, indicating their significant role in tumor-environment interactions.

MALDI-MS is often integrated with other techniques to enhance its analytical capabilities. For example, coupling MALDI-MS with liquid chromatography (LC) allows for the separation of complex protein mixtures before mass spectrometric analysis, improving the resolution and accuracy of protein identification [177]. This combined approach has been utilized in exosome research to achieve more comprehensive proteomic profiles. Welton et al. [21] conducted the first proteomics analysis of exosomes derived from bladder cancer cells, highlighting the utility of the LC-MALDI-TOF/TOF MS approach. They purified exosomes from HT1376 bladder cancer cells using ultracentrifugation and validated their purity via Western blotting and flow cytometry. This analysis identified 353 high-quality proteins, with 72 previously unreported in human exosome studies. The identified proteome showed significant overlap with carcinoma-origin exosomes, including proteins like basigin and galectin-3, trophoblast glycoprotein (5T4), and others. These findings pave the way for developing urine exosome-based diagnostic tools for monitoring bladder cancer. In another study, Ye et al. [7] combined MALDI-TOF/MS with LC–MS/MS to achieve detailed molecular profiling of exosomes. This integration provided comprehensive insights into the protein and metabolite composition of exosomes. The researchers also employed the EXODUS ultrafast-isolation system, which facilitated the high-purity separation of exosomes from plasma, enhancing the accuracy and sensitivity of subsequent biomarker detection and analysis.

6.7.5 MALDI-MS Analysis of Exosomal Lipids

As explained earlier, exosomes are rich in lipids such as diglycerides, cholesterol, sphingolipids (including ceramide and sphingomyelin), lyosbisphosphatidic acid (LBPA), disaturated lipids, phospholipids, including phosphatidylinositol, phosphatidylcholine, phosphatidylethanolamine, phosphatidylserine. These lipids are essential for maintaining the structural integrity of exosomes and play significant roles in exosome biogenesis and regulating homeostasis in recipient cells [65]. One of the primary applications of MALDI-MS in exosome lipidomics is the profiling of lipid species, which can reveal unique lipid signatures associated with different cell types, disease states, or treatment conditions [178].

Total lipids are often extracted using organic solvents and then directly deposited onto the metal target. This can be done either with or without combining them with the matrix. The instrument can operate in negative ion mode, covering a mass range of *m/z* 400–2000, and providing high mass resolution [179]. For instance, Lobasso et al. [178] investigated the lipid profiles of exosomes derived from melanoma cell lines with different metastatic propensities (LCP and SK-Mel28) using MALDI-TOF/MS. Ultracentrifugation was used to enrich exosomes, and their lipid composition was analyzed to find possible lipid biomarkers for various melanoma cell migration and invasiveness. Specifically, for certain phosphatidylinositol species, they discovered that weakly metastatic (LCP) cells compared to highly metastatic (SK-Mel28) cells had shorter and more saturated fatty acid tails. Exosome membranes exhibited higher levels of phosphatidic acid, lysophosphatidylcholine, and sphingomyelin relative to the parental cells. Additionally, bis(monoacylglycero) phosphate was identified as a specific lipid marker of exosomes.

The phospholipids in EVs released by non-small-cell lung cancer cell lines were effectively characterized by Jung et al. [180] to identify potential marker peaks that could predict resistance to the signal transduction inhibitor gefitinib. The Bligh-Dyer extraction procedure was used to separate phospholipids from EVs generated by drug-resistant (PC9R) and drug-responsive (PC9) cell lines. Lipid analysis was performed using MALDI-TOF/MS. By comparing the lipidomes of EVs from the two sets of samples, they found that the phospholipid composition of the vesicles was significantly distinct between the two groups. Statistical analyses identified 35 differentially regulated lipids, suggesting that EV phospholipid profiles may serve as predictive biomarkers for drug resistance in NSCLC patients treated with gefitinib.

More recently, Jalaludin et al. [51] explored the lipidomic profiling of exosomes from human serum using MALDI-FTICR-MS. Multiple cycles of centrifugation were employed to isolate exosomes from human serum. They compared four matrices and three sample preparation methods, identifying 172 lipid species in human serum and 89 in exosomes. The CHCA matrix detected the highest number of exosomal lipids, particularly phosphatidylcholine, likely due to the positive ion mode used. This study highlights the potential of MALDI-FTICR-MS for detailed lipidomic analysis of exosomes, offering valuable insights for future biomarker research.

MALDI-MS can provide complementary data on anionic lipid classes, such as sulfatides and phosphatidylinositols, which are less accessible with other techniques. Peterka et al. [6] utilized MALDI-MS alongside ultrahigh-performance supercritical fluid chromatography-mass spectrometry (UHPSFC/MS) and ultrahigh-performance liquid chromatography-mass spectrometry (UHPLC/MS) to characterize the lipid profiles of exosomes isolated from human plasma. They identified 244 lipids in exosomes and 191 in plasma, revealing significant differences in the levels of phosphatidylcholines, diacylglycerols, triacylglycerols, and lysophosphatidylcholines between exosomes and plasma. In contrast, sphingomyelins, phosphatidylinositols, and sulfatides showed similar profiles in both biological matrices. This study highlights the value of combining MALDI-MS with other mass spectrometry techniques for comprehensive lipidomic profiling.

Recently, extracellular vesicles from the oviductal fluid, known as oviductosomes or oEVs, have been suggested to play a crucial role in the early communication between the embryo and the maternal environment [181]. A study using intact cell MALDI-TOF/MS was established to analyze the phospholipid content of bovine blastocysts developed in vitro in the presence or absence of oEVs [182]. While oEV exposure did not alter blastocyst yield or cryotolerance, it significantly changed their phospholipid profiles, increasing levels of phosphatidylcholines, phosphatidylethanolamines, and sphingomyelins with long-chain fatty acids. The lipidomic analysis revealed that many of the altered lipid species in embryos were also present in oEVs, suggesting that oEVs may modify embryonic lipid composition by incorporating their lipids into embryo membranes and affecting lipid metabolism. This study offers the first insights into how oEVs regulate the lipidome of early embryos.

6.7.6 MALDI-MS Analysis of Exosomal Metabolites

The application of MALDI-MS for the analysis of exosome-derived metabolites is a relatively new and emerging field, with limited research available in the current literature. Despite the growing interest in exosome research, the specific application of MALDI-MS to study exosomal metabolites has not been extensively explored. One pioneering study explored the application of MALDI-TOF/MS in analyzing metabolic fingerprints of urinary exosomes from patients with systemic lupus erythematosus (SLE) compared to healthy controls (HCs) [183]. The study employed high-performance liquid size-exclusion chromatography (HPL-SEC) to isolate exosomes, followed by MALDI-TOF/MS for high-throughput and efficient metabolomic profiling. The results demonstrated the method's effectiveness in distinguishing SLE patients from HCs, achieving an area under the curve (AUC) of 1.00 and an accuracy of 90.32% in unsupervised clustering. This study also highlights the potential of MALDI-MS in identifying disease-specific metabolic biomarkers in exosomes, paving the way for exosome-driven precision medicine and large-scale clinical screening.

The relative scarcity of research in this domain emphasizes both the novelty and the opportunity for further exploration. As the technology and methodologies continue to advance, there is great potential for MALDI-MS to become a standard tool for exosome metabolomics. Future research could focus on expanding the range of detectable metabolites, improving quantification techniques, and integrating MALDI-MS with other analytical methods to provide a more comprehensive understanding of exosomal functions and their roles in various diseases.

7 Conclusion and Future Outlook

Currently, exosomes are recognized as key elements in intercellular communication and their specific cargos have emerged as potential biomarkers in various physiological and pathological conditions. Over the years, MALDI-MS has been increasingly used in the characterization of exosomes, demonstrating its potential in a wide range of applications. The majority of research work has been devoted to proteomic studies, where MALDI-MS has proven invaluable in identifying and characterizing proteins within exosomes. Frequently studied topics in this domain include the identification of new biomarkers for early diagnosis, tumor differentiation, follow-up of disease progression, metastasis monitoring, prediction of drug resistance, and the process of epithelial-to-mesenchymal transition. In addition to proteomics, MALDI-MS has also been applied to lipidomic and metabolomic analyses of exosomes, further expanding our understanding of their roles in cellular communication and disease mechanisms.

A comprehensive characterization of exosomes requires multiple analytical techniques, as they contain diverse components linked to various biological functions and diseases. To achieve this, some studies use combined LC–MS and MALDI approaches to capture a more complete profile of the exosomes. While MS techniques offer significant promise for addressing complex bioanalytical challenges, there are notable limitations, especially in translating these strategies from research to clinical practice. Key challenges include developing more efficient and cost-effective isolation protocols for exosomes that can be automated and standardized for integration with analytical instrumentation. Additionally, the complexity of biological matrices and the heterogeneous composition of exosome contents complicate the task. Although LC–MS is often used for analyzing exosome components due to its ability to manage sample complexity, MALDI-MS offers distinct advantages, including high-throughput capacity, rapid processing, cost-effectiveness, and ease of use, making it a more suitable candidate for extending methods to clinical practice.

The future of exosome analysis using MALDI-MS holds significant promise. As technology advances, enhanced sensitivity and specificity are expected, enabling the detection of low-abundance biomarkers crucial for early disease diagnosis. Integrating MALDI-MS with other omics approaches will provide a comprehensive understanding of exosomal content and its role in disease. Furthermore, standardizing protocols and improving data analysis techniques will be essential for ensuring reproducibility and translating findings into clinical applications.

Acknowledgements This research was supported by Basic Science Research Program through the National Research Foundation of Korea (NRF), funded by the Ministry of Education (RS-2023-00219710314820926400001 and 2016R1D1A1B02008854).

Conflict of Interest The authors declare no conflict of interest.

References

1. Nguyen, H.-Q., Lee, D., Kim, Y., Paek, M., Kim, M., Jang, K.-S., Oh, J., Lee, Y.-S., Yeon, J. E., Lubman, D. M., & Kim, J. (2019). Platelet factor 4 as a novel exosome marker in MALDI-MS analysis of exosomes from human serum. *Analytical Chemistry, 91*(20), 13297–13305. https://doi.org/10.1021/acs.analchem.9b04198
2. Moeinzadeh, L., Razeghian-Jahromi, I., Zarei-Behjani, Z., Bagheri, Z., & Razmkhah, M. (2022). Composition, biogenesis, and role of exosomes in tumor development. *Stem Cells International, 2022*(1), 8392509. https://doi.org/10.1155/2022/8392509
3. Han, Z., Peng, C., Yi, J., Wang, Y., Liu, Q., Yang, Y., Long, S., Qiao, L., & Shen, Y. (2021). Matrix-assisted laser desorption ionization mass spectrometry profiling of plasma exosomes evaluates osteosarcoma metastasis. *iScience, 24*(8). https://doi.org/10.1016/j.isci.2021.102906
4. Zambonin, C. (2022). MALDI-TOF/MS analysis of extracellular vesicles released by cancer cells. *Applied Sciences*, 12. https://doi.org/10.3390/app12126149
5. Jalaludin, I., Lubman, D. M., & Kim, J. (2023). A guide to mass spectrometric analysis of extracellular vesicle proteins for biomarker discovery. *Mass Spectrometry Reviews, 42*(2), Article e21749. https://doi.org/10.1002/mas.21749
6. Peterka, O., Jirásko, R., Chocholoušková, M., Kuchař, L., Wolrab, D., Hájek, R., Vrána, D., Strouhal, O., Melichar, B., & Holčapek, M. (2020) Lipidomic characterization of exosomes isolated from human plasma using various mass spectrometry techniques. *Biochimica et*

Biophysica Acta (BBA) - Molecular and Cell Biology of Lipids, 1865(5), 158634 https://doi.org/10.1016/j.bbalip.2020.158634

7. Ye, W., Pan, R., Shi, K.-Q., Li, H.-P., Lee, L. P., & Liu, F. (2022). Isolation of small extracellular vesicles from a drop of plasma via EXODUS and their fingerprint proteomics profiling by MALDI-TOF MS. *Biosensors and Bioelectronics: X, 10*, Article 100099. https://doi.org/10.1016/j.biosx.2021.100099
8. Zheng, H., Zhao, J., Wang, X., Yan, S., Chu, H., Gao, M., & Zhang, X. (2022). Integrated pipeline of rapid isolation and analysis of human plasma exosomes for cancer discrimination based on deep learning of MALDI-TOF MS fingerprints. *Analytical Chemistry, 94*(3), 1831–1839. https://doi.org/10.1021/acs.analchem.1c04762
9. Wang, J., Yang, K., Yuan, W., & Gao, Z. (2018). Determination of serum exosomal H19 as a noninvasive biomarker for bladder cancer diagnosis and prognosis. *Medical Science Monitor: International Medical Journal of Experimental and Clinical Research*, 24, 9307–9316. https://doi.org/10.12659/msm.912018
10. Xie, X.-F., Chu, H.-J., Xu, Y.-F., Hua, L., Wang, Z.-P., Huang, P., Jia, H.-L., & Zhang, L. (2019). Proteomics study of serum exosomes in Kawasaki disease patients with coronary artery aneurysms. *Cardiology Journal, 26*(5), 584–593. https://doi.org/10.5603/CJ.a2018.0032
11. Choi, E. S., Faruque, H. A., Kim, J. H., Kim, K. J., Choi, J. E., Kim, B. A., Kim, B., Kim, Y. J., Woo, M. H., Park, J. Y., Hur, K., Lee, M. Y., Kim, D. S., Lee, S. Y., & Kim, E. (2021). CD5L as an extracellular vesicle-derived biomarker for liquid biopsy of lung cancer. *Diagnostics (Basel)*, **11**(4). https://doi.org/10.3390/diagnostics11040620
12. Vago, R., Radano, G., Zocco, D., & Zarovni, N. (2022). Urine stabilization and normalization strategies favor unbiased analysis of urinary EV content. *Scientific Reports, 12*(1), 17663. https://doi.org/10.1038/s41598-022-22577-3
13. Alvarez, M. L., Khosroheidari, M., Kanchi Ravi, R., & DiStefano, J. K. (2012). Comparison of protein, microRNA, and mRNA yields using different methods of urinary exosome isolation for the discovery of kidney disease biomarkers. *Kidney International, 82*(9), 1024–1032. https://doi.org/10.1038/ki.2012.256
14. Lin, S.-Y., Chang, C.-H., Wu, H.-C., Lin, C.-C., Chang, K.-P., Yang, C.-R., Huang, C.-P., Hsu, W.-H., Chang, C.-T., & Chen, C.-J. (2016). Proteome profiling of urinary exosomes identifies Alpha 1-Antitrypsin and H2B1K as diagnostic and prognostic biomarkers for urothelial carcinoma. *Scientific Reports, 6*(1), 34446. https://doi.org/10.1038/srep34446
15. Ploeg, M., Aben, K. K. H., & Kiemeney, L. A. (2009). The present and future burden of urinary bladder cancer in the world. *World Journal of Urology, 27*(3), 289–293. https://doi.org/10.1007/s00345-009-0383-3
16. Michael, A., Bajracharya, S. D., Yuen, P. S. T., Zhou, H., Star, R. A., Illei, G. G., & Alevizos, I. (2010). Exosomes from human saliva as a source of microRNA biomarkers. *Oral Diseases, 16*(1), 34–38. https://doi.org/10.1111/j.1601-0825.2009.01604.x
17. Miyake, H., Lee, C., Chusilp, S., Bhalla, M., Li, B., Pitino, M., Seo, S., O'Connor, D. L., & Pierro, A. (2020). Human breast milk exosomes attenuate intestinal damage. *Pediatric Surgery International, 36*(2), 155–163. https://doi.org/10.1007/s00383-019-04599-7
18. Tutanov, O. S., Proskura, K. V., Grigor'eva, A. E., Tsentalovich, Y. P., & Tamkovich, S. N. (2020). Identification of tumor dissemination facilitating proteins in exosomes associated with blood cells of breast cancer patients. *Russian Journal of Bioorganic Chemistry, 46*(6), 1018–1033. https://doi.org/10.1134/S1068162020060357
19. Di Giuseppe, F., Carluccio, M., Zuccarini, M., Giuliani, P., Ricci-Vitiani, L., Pallini, R., De Sanctis, P., Di Pietro, R., Ciccarelli, R., & Angelucci, S. (2021). Proteomic characterization of two extracellular vesicle subtypes isolated from human glioblastoma stem cell secretome by sequential centrifugal ultrafiltration. *Biomedicines, 9*(2). https://doi.org/10.3390/biomedicines9020146
20. Shtam, T. A., Naryzhny, S. N., Landa, S. B., Burdakov, V. S., Artamonova, T. O., & Filatov, M. V. (2012). Purification and in vitro analysis of exosomes secreted by malignantly transformed human cells. *Cell and Tissue Biology, 6*(4), 317–325. https://doi.org/10.1134/S1990519X12040116

21. Welton, J. L., Khanna, S., Giles, P. J., Brennan, P., Brewis, I. A., Staffurth, J., Mason, M. D., & Clayton, A. (2010). Proteomics analysis of bladder cancer exosomes. *Molecular & Cellular Proteomics, 9*(6), 1324–1338. https://doi.org/10.1074/mcp.M000063-MCP201
22. Greening, D. W., Xu, R., Ji, H., Tauro, B.J., & Simpson, R. J. (2015). A protocol for exosome isolation and characterization: evaluation of ultracentrifugation, density-gradient separation, and immunoaffinity capture methods. In A. Posch (Eds.), *Proteomic profiling: Methods and protocols* (pp. 179–209). Springer New York. https://doi.org/10.1007/978-1-4939-2550-6_15
23. Ludwig, A.-K., De Miroschedji, K., Doeppner, T. R., Börger, V., Ruesing, J., Rebmann, V., Durst, S., Jansen, S., Bremer, M., Behrmann, E., Singer, B. B., Jastrow, H., Kuhlmann, J. D., El Magraoui, F., Meyer, H. E., Hermann, D. M., Opalka, B., Raunser, S., Epple, M., Horn, P. A., & Giebel, B. (2018). Precipitation with polyethylene glycol followed by washing and pelleting by ultracentrifugation enriches extracellular vesicles from tissue culture supernatants in small and large scales. *Journal of Extracellular Vesicles, 7*(1), 1528109. https://doi.org/10.1080/20013078.2018.1528109
24. Gámez-Valero, A., Monguió-Tortajada, M., Carreras-Planella, L., Franquesa, M., Beyer, K., & Borràs, F. E. (2016). Size-exclusion chromatography-based isolation minimally alters extracellular vesicles' characteristics compared to precipitating agents. *Science and Reports, 6*, 33641. https://doi.org/10.1038/srep33641
25. Park, S., Jalaludin, I., Hwang, H., Ko, M., Adelipour, M., Hwan, M., Cho, N., Kim, K. K., Lubman, D. M., & Kim, J. (2023). Size-exclusion chromatography for the characterization of urinary extracellular vesicles. *Journal of Chromatography B, 1228*, Article 123828. https://doi.org/10.1016/j.jchromb.2023.123828
26. Sidhom, K., Obi, P. O., & Saleem, A. (2020). A review of exosomal isolation methods: is size exclusion chromatography the best option? *International Journal of Molecular Sciences*, 21. https://doi.org/10.3390/ijms21186466
27. Burtenshaw, D., Regan, B., Owen, K., Collins, D., McEneaney, D., Megson, I. L., Redmond, E. M., & Cahill, P. A. (2022). Exosomal composition, biogenesis and profiling using point-of-care diagnostics-implications for cardiovascular disease. *Frontiers in Cell and Developmental Biology, 10*, Article 853451. https://doi.org/10.3389/fcell.2022.853451
28. Jalaludin, I., Lubman, D. M., & Kim, J. (2021). MALDI-MS: a powerful but underutilized mass spectrometric technique for exosome research. *Mass Spectrometry Letters, 12*(3), 93–105. https://doi.org/10.5478/MSL.2021.12.3.93
29. Gholizadeh, S., Shehata Draz, M., Zarghooni, M., Sanati-Nezhad, A., Ghavami, S., Shafiee, H., & Akbari, M. (2017). Microfluidic approaches for isolation, detection, and characterization of extracellular vesicles: Current status and future directions. *Biosensors and Bioelectronics, 91*, 588–605. https://doi.org/10.1016/j.bios.2016.12.062
30. Guo, S.-C., Tao, S.-C., & Dawn, H. (2018). Microfluidics-based on-a-chip systems for isolating and analysing extracellular vesicles. *Journal of Extracellular Vesicles, 7*(1), 1508271. https://doi.org/10.1080/20013078.2018.1508271
31. Iliescu, F. S., Vrtačnik, D., Neuzil, P., & Iliescu, C. (2019). Microfluidic technology for clinical applications of exosomes. *Micromachines, 10*. https://doi.org/10.3390/mi10060392
32. Rider, M. A., Hurwitz, S. N., & Meckes, D. G. (2016). ExtraPEG: A polyethylene glycol-based method for enrichment of extracellular vesicles. *Scientific Reports, 6*(1), 23978. https://doi.org/10.1038/srep23978
33. Benedikter, B. J., Bouwman, F. G., Vajen, T., Heinzmann, A. C. A., Grauls, G., Mariman, E. C., Wouters, E. F. M., Savelkoul, P. H., Lopez-Iglesias, C., Koenen, R. R., Rohde, G. G. U., & Stassen, F. R. M. (2017). Ultrafiltration combined with size exclusion chromatography efficiently isolates extracellular vesicles from cell culture media for compositional and functional studies. *Scientific Reports, 7*(1), 15297. https://doi.org/10.1038/s41598-017-15717-7
34. Gao, M., Cai, J., Zitkovsky, H. S., Chen, B., & Guo, L. (2022). *Comparison of yield, purity, and functional properties of large-volume exosome isolation using ultrafiltration and polymer-based precipitation. Plastic and Reconstructive Surgery*, 149(3). https://doi.org/10.1097/PRS.0000000000008830

35. Nordin, J. Z., Lee, Y., Vader, P., Mäger, I., Johansson, H. J., Heusermann, W., Wiklander, O. P. B., Hällbrink, M., Seow, Y., Bultema, J. J., Gilthorpe, J., Davies, T., Fairchild, P. J., Gabrielsson, S., Meisner-Kober, N. C., Lehtiö, J., Smith, C. I. E., Wood, M. J. A., & Andaloussi, S. E. L. (2015). Ultrafiltration with size-exclusion liquid chromatography for high yield isolation of extracellular vesicles preserving intact biophysical and functional properties. *Nanomedicine: Nanotechnology, Biology and Medicine, 11*(4), 879–883 https://doi.org/10.1016/j.nano.2015.01.003
36. Shu, S. L., Allen, C. L., Benjamin-Davalos, S., Koroleva, M., MacFarland, D., Minderman, H., & Ernstoff, M. S. (2021). A rapid exosome isolation using ultrafiltration and size exclusion chromatography (REIUS) method for exosome isolation from melanoma cell lines. In K. M. Hargadon (Ed.), *Melanoma: Methods and protocols* (pp. 289–304). Springer US. https://doi.org/10.1007/978-1-0716-1205-7_22
37. Stam, J., Bartel, S., Bischoff, R., & Wolters, J. C. (2021). Isolation of extracellular vesicles with combined enrichment methods. *Journal of Chromatography B, 1169*, Article 122604. https://doi.org/10.1016/j.jchromb.2021.122604
38. Liu, X., Wang, C., Meng, H., Liao, S., Zhang, J., Guan, Y., Tian, H., & Peng, J. (2022). Research progress on exosomes in osteonecrosis of the femoral head. *Orthopaedic Surgery, 14*(9), 1951–1957. https://doi.org/10.1111/os.13393
39. Szatanek, R., Baj-Krzyworzeka, M., Zimoch, J., Lekka, M., Siedlar, M., & Baran, J. (2017). The methods of choice for extracellular vesicles (EVs) characterization. *International Journal of Molecular Sciences, 18*(6), 1153. https://doi.org/10.3390/ijms18061153
40. Dragovic, R. A., Gardiner, C., Brooks, A. S., Tannetta, D. S., Ferguson, D. J. P., Hole, P., Carr, B., Redman, C. W. G., Harris, A. L., Dobson, P. J., Harrison, P., & Sargent, I. L. (2011). Sizing and phenotyping of cellular vesicles using nanoparticle tracking analysis. *Nanomedicine: Nanotechnology, Biology and Medicine*, **7**(6), 780–788 https://doi.org/10.1016/j.nano.2011.04.003
41. Tahoun, I. F., Gab-Allah, M. A., Yamani, R. N., & Shehata, A. B. (2021). Development and validation of a reliable LC-MS/MS method for simultaneous determination of deoxynivalenol and T-2 toxin in maize and oats. *Microchemical Journal, 169*, Article 106599. https://doi.org/10.1016/j.microc.2021.106599
42. Gab-Allah, M. A., Choi, K., & Kim, B. (2021). Development of isotope dilution-liquid chromatography/tandem mass spectrometry for the accurate determination of type-A trichothecenes in grains. *Food Chemistry, 344*, Article 128698. https://doi.org/10.1016/j.foodchem.2020.128698
43. Gab-Allah, M. A., Choi, K., & Kim, B. (2021). Accurate determination of type B trichothecenes and conjugated deoxynivalenol in grains by isotope dilution–liquid chromatography tandem mass spectrometry. *Food Control, 121*, Article 107557. https://doi.org/10.1016/j.foodcont.2020.107557
44. Gab-Allah, M. A., Mekete, K. G., Choi, K., & Kim, B. (2021). Occurrence of major type-B trichothecenes and deoxynivalenol-3-glucoside in cereal-based products from Korea. *Journal of Food Composition and Analysis, 99*, Article 103851. https://doi.org/10.1016/j.jfca.2021.103851
45. Gab-Allah, M. A., Tahoun, I. F., Yamani, R. N., Rend, E. A., & Shehata, A. B. (2022). Natural occurrence of deoxynivalenol, nivalenol and deoxynivalenol-3-glucoside in cereal-derived products from Egypt. *Food Control, 137*, Article 108974. https://doi.org/10.1016/j.foodcont.2022.108974
46. Lijalem, Y. G., Gab-Allah, M. A., Choi, K., & Kim, B. (2022). Development of isotope dilution-liquid chromatography/tandem mass spectrometry for the accurate determination of zearalenone and its metabolites in corn. *Food Chemistry, 384*, Article 132483. https://doi.org/10.1016/j.foodchem.2022.132483
47. Gab-Allah, M. A., Choi, K., & Kim, B. (2023). Type B trichothecenes in cereal grains and their products: recent advances on occurrence, toxicology, analysis and post-harvest decontamination strategies. *Toxins, 15*. https://doi.org/10.3390/toxins15020085

48. Gab-Allah, M. A., Choi, K., & Kim, B. (2022). Development of isotope dilution–liquid chromatography/tandem mass spectrometry as a candidate reference method for the accurate determination of patulin in apple products. *Analytical and Bioanalytical Chemistry, 414*(5), 1867–1879. https://doi.org/10.1007/s00216-021-03817-3
49. Gab-Allah, M. A., Tahoun, I. F., Yamani, R. N., Rend, E. A., & Shehata, A. B. (2022). Eco-friendly and sensitive analytical method for determination of T-2 toxin and HT-2 toxin in cereal products using UPLC-MS/MS. *Journal of Food Composition and Analysis, 107*, Article 104395. https://doi.org/10.1016/j.jfca.2022.104395
50. Lijalem, Y. G., Gab-Allah, M. A., Yu, H., Choi, K., & Kim, B. (2024). Development of isotope dilution-ultrahigh-performance liquid chromatography-tandem mass spectrometry for the accurate determination of aflatoxins in grains. *Journal of Food Composition and Analysis, 126*, Article 105896. https://doi.org/10.1016/j.jfca.2023.105896
51. Jalaludin, I., Nguyen, H.-Q., Jang, K.-S., Lee, J., Lubman, D. M., & Kim, J. (2023). Matrix-assisted laser desorption/ionization-Fourier-transform ion cyclotron resonance-mass spectrometry analysis of exosomal lipids from human serum. *Rapid Communications in Mass Spectrometry, 37*(2), Article e9427. https://doi.org/10.1002/rcm.9427
52. Zhu, Y., Pick, H., Gasilova, N., Li, X., Lin, T.-E., Laeubli, H. P., Zippelius, A., Ho, P.-C., & Girault, H. H. (2019). MALDI detection of exosomes: A potential tool for cancer studies. *Chem, 5*(5), 1318–1336. https://doi.org/10.1016/j.chempr.2019.04.007
53. Xu, M., Ji, J., Jin, D., Wu, Y., Wu, T., Lin, R., Zhu, S., Jiang, F., Ji, Y., Bao, B., Li, M., Xu, W., & Xiao, M. (2023). The biogenesis and secretion of exosomes and multivesicular bodies (MVBs): Intercellular shuttles and implications in human diseases. *Genes & Diseases, 10*(5), 1894–1907. https://doi.org/10.1016/j.gendis.2022.03.021
54. Yu, B., Zhang, X., & Li, X. (2014). Exosomes derived from mesenchymal stem cells. *International Journal of Molecular Sciences, 15*(3), 4142–4157. https://doi.org/10.3390/ijms15034142
55. Dai, J., Su, Y., Zhong, S., Cong, L., Liu, B., Yang, J., Tao, Y., He, Z., Chen, C., & Jiang, Y. (2020). Exosomes: Key players in cancer and potential therapeutic strategy. *Signal Transduction and Targeted Therapy, 5*(1), 145. https://doi.org/10.1038/s41392-020-00261-0
56. Doyle, L. M., & Wang, M. Z. (2019). Overview of extracellular vesicles, their origin, composition, purpose, and methods for exosome isolation and analysis. *Cells*, 8. https://doi.org/10.3390/cells8070727
57. Vitha, A. E., Kollefrath, A. W., Huang, C.-Y.C., & Garcia-Godoy, F. (2018). Characterization and therapeutic uses of exosomes: A new potential tool in orthopedics. *Stem Cells and Development, 28*(2), 141–150. https://doi.org/10.1089/scd.2018.0205
58. Di Vizio, D., Morello, M., Dudley, A. C., Schow, P. W., Adam, R. M., Morley, S., Mulholland, D., Rotinen, M., Hager, M. H., Insabato, L., Moses, M. A., Demichelis, F., Lisanti, M. P., Wu, H., Klagsbrun, M., Bhowmick, N. A., Rubin, M. A., D'Souza-Schorey, C., & Freeman, M. R. (2012). Large oncosomes in human prostate cancer tissues and in the circulation of mice with metastatic disease. *The American Journal of Pathology, 181*(5), 1573–1584. https://doi.org/10.1016/j.ajpath.2012.07.030
59. Morello, M., Minciacchi, V., de Candia, P., Yang, J., Posadas, E., Kim, H., Griffiths, D., Bhowmick, N., Chung, L., & Gandellini, P. (2013). Large oncosomes mediate intercellular transfer of functional microRNA. *Cell Cycle, 12*(22), 3526–3536. https://doi.org/10.4161/cc.26539
60. Jalalian, S. H., Ramezani, M., Jalalian, S. A., Abnous, K., & Taghdisi, S. M. (2019). Exosomes, new biomarkers in early cancer detection. *Analytical Biochemistry, 571*, 1–13. https://doi.org/10.1016/j.ab.2019.02.013
61. Kim, J.-H., Kim, E., & Lee, M. Y. (2018). Exosomes as diagnostic biomarkers in cancer. *Molecular & Cellular Toxicology, 14*(2), 113–122. https://doi.org/10.1007/s13273-018-0014-4
62. Zhang, Y., Bi, J., Huang, J., Tang, Y., Du, S., & Li, P. (2020). Exosome: A review of its classification, isolation techniques, storage, diagnostic and targeted therapy applications. *International Journal of Nanomedicine, 15*(null), 6917–6934. https://doi.org/10.2147/IJN.S264498

63. Hegmans, J. P. J. J., Bard, M. P. L., Hemmes, A., Luider, T. M., Kleijmeer, M. J., Prins, J.-B., Zitvogel, L., Burgers, S. A., Hoogsteden, H. C., & Lambrecht, B. N. (2004). Proteomic analysis of exosomes secreted by human mesothelioma cells. *The American Journal of Pathology, 164*(5), 1807–1815. https://doi.org/10.1016/S0002-9440(10)63739-X
64. Yin, K., Wang, S., & Zhao, R. C. (2019). Exosomes from mesenchymal stem/stromal cells: A new therapeutic paradigm. *Biomarker Research, 7*(1), 8. https://doi.org/10.1186/s40364-019-0159-x
65. Théry, C., Zitvogel, L., & Amigorena, S. (2002). Exosomes: Composition, biogenesis and function. *Nature Reviews Immunology, 2*(8), 569–579. https://doi.org/10.1038/nri855
66. Zhang, J., Li, S., Li, L., Li, M., Guo, C., Yao, J., & Mi, S. (2015). Exosome and exosomal microRNA: Trafficking, sorting, and function. *Genomics, Proteomics & Bioinformatics, 13*(1), 17–24. https://doi.org/10.1016/j.gpb.2015.02.001
67. Yue, B., Yang, H., Wang, J., Ru, W., Wu, J., Huang, Y., Lan, X., Lei, C., & Chen, H. (2020). Exosome biogenesis, secretion and function of exosomal miRNAs in skeletal muscle myogenesis. *Cell Proliferation, 53*(7), Article e12857. https://doi.org/10.1111/cpr.12857
68. van den Boorn, J. G., Daßler, J., Coch, C., Schlee, M., & Hartmann, G. (2013). Exosomes as nucleic acid nanocarriers. *Advanced Drug Delivery Reviews, 65*(3), 331–335. https://doi.org/10.1016/j.addr.2012.06.011
69. Xie, Y., Dang, W., Zhang, S., Yue, W., Yang, L., Zhai, X., Yan, Q., & Lu, J. (2019). The role of exosomal noncoding RNAs in cancer. *Molecular Cancer, 18*(1), 37. https://doi.org/10.1186/s12943-019-0984-4
70. Frydrychowicz, M., Kolecka-Bednarczyk, A., Madejczyk, M., Yasar, S., & Dworacki, G. (2015). Exosomes—structure, biogenesis and biological role in non-small-cell lung cancer. *Scandinavian Journal of Immunology, 81*(1), 2–10. https://doi.org/10.1111/sji.12247
71. Xue, X.-Y., Liu, Y.-X., Wang, C., Gu, X.-J., Xue, Z.-Q., Zang, X.-L., Ma, X.-D., Deng, H., Liu, R., Pan, L., & Liu, S.-H. (2020). Identification of exosomal miRNAs as diagnostic biomarkers for cholangiocarcinoma and gallbladder carcinoma. *Signal Transduction and Targeted Therapy, 5*(1), 77. https://doi.org/10.1038/s41392-020-0162-6
72. Barceló, M., Castells, M., Bassas, L., Vigués, F., & Larriba, S. (2019). Semen miRNAs contained in exosomes as non-invasive biomarkers for prostate cancer diagnosis. *Scientific Reports, 9*(1), 13772. https://doi.org/10.1038/s41598-019-50172-6
73. Kalluri, R., & LeBleu, V. S. (2020). The biology, function, and biomedical applications of exosomes. *Science, 367*(6478). https://doi.org/10.1126/science.aau6977
74. Burkova, E. E., Sedykh, S. E., & Nevinsky, G. A. (2021). Human placenta exosomes: biogenesis, isolation, composition, and prospects for use in diagnostics. *International Journal of Molecular Sciences, 22*(4), 2158. https://doi.org/10.3390/ijms22042158
75. Zhang, Y., Liu, Y., Liu, H., & Tang, W. H. (2019). Exosomes: Biogenesis, biologic function and clinical potential. *Cell & Bioscience, 9*(1), 19. https://doi.org/10.1186/s13578-019-0282-2
76. Krylova, S. V., & Feng, D. (2023). The machinery of exosomes: Biogenesis, release, and uptake. *International Journal of Molecular Sciences, 24*(2), 1337. https://doi.org/10.3390/ijms24021337
77. Hessvik, N. P., & Llorente, A. (2018). Current knowledge on exosome biogenesis and release. *Cellular and Molecular Life Sciences, 75*(2), 193–208. https://doi.org/10.1007/s00018-017-2595-9
78. Gurunathan, S., Kang, M. H., & Kim, J. H. (2021). A comprehensive review on factors influences biogenesis, functions, therapeutic and clinical implications of exosomes. *International Journal of Nanomedicine, 16*, 1281–1312. https://doi.org/10.2147/ijn.S291956
79. Kowal, J., Tkach, M., & Théry, C. (2014). Biogenesis and secretion of exosomes. *Current Opinion in Cell Biology, 29*, 116–125. https://doi.org/10.1016/j.ceb.2014.05.004
80. Villarroya-Beltri, C., Baixauli, F., Gutiérrez-Vázquez, C., Sánchez-Madrid, F., & Mittelbrunn, M. (2014). Sorting it out: Regulation of exosome loading. *Seminars in Cancer Biology, 28*, 3–13. https://doi.org/10.1016/j.semcancer.2014.04.009
81. Lou, G., Chen, Z., Zheng, M., & Liu, Y. (2017). Mesenchymal stem cell-derived exosomes as a new therapeutic strategy for liver diseases. *Experimental & Molecular Medicine, 49*(6), e346–e346. https://doi.org/10.1038/emm.2017.63

82. Mondal, J., Pillarisetti, S., Junnuthula, V., Saha, M., Hwang, S. R., Park, I.-K., & Lee, Y.-K. (2023). Hybrid exosomes, exosome-like nanovesicles and engineered exosomes for therapeutic applications. *Journal of Controlled Release, 353*, 1127–1149. https://doi.org/10.1016/j.jconrel.2022.12.027
83. Tkach, M., & Théry, C. (2016). Communication by extracellular vesicles: Where we are and where we need to go. *Cell, 164*(6), 1226–1232. https://doi.org/10.1016/j.cell.2016.01.043
84. Crenshaw, B. J., Gu, L., Sims, B., & Matthews, Q. L. (2018). Exosome biogenesis and biological function in response to viral infections. *Open Virol J, 12*, 134–148. https://doi.org/10.2174/1874357901812010134
85. Théry, C., Ostrowski, M., & Segura, E. (2009). Membrane vesicles as conveyors of immune responses. *Nature Reviews Immunology, 9*(8), 581–593. https://doi.org/10.1038/nri2567
86. Aheget, H., Mazini, L., Martin, F., Belqat, B., Marchal, J. A., & Benabdellah, K. (2021). Exosomes: Their role in pathogenesis, diagnosis and treatment of diseases. *Cancers, 13*(1), 84. https://doi.org/10.3390/cancers13010084
87. Sun, Z., Yang, S., Zhou, Q., Wang, G., Song, J., Li, Z., Zhang, Z., Xu, J., Xia, K., Chang, Y., Liu, J., & Yuan, W. (2018). Emerging role of exosome-derived long non-coding RNAs in tumor microenvironment. *Molecular Cancer, 17*(1), 82. https://doi.org/10.1186/s12943-018-0831-z
88. Wang, J. P., Tang, Y. Y., Fan, C. M., Guo, C., Zhou, Y. H., Li, Z., Li, X. L., Li, Y., Li, G. Y., Xiong, W., Zeng, Z. Y., & Xiong, F. (2018). The role of exosomal non-coding RNAs in cancer metastasis. *Oncotarget, 9*(15), 12487–12502. https://doi.org/10.18632/oncotarget.23552
89. Olejarz, W., Kubiak-Tomaszewska, G., Chrzanowska, A., & Lorenc, T. (2020). Exosomes in angiogenesis and anti-angiogenic therapy in cancers. *International Journal of Molecular Sciences, 21*(16), 5840. https://doi.org/10.3390/ijms21165840
90. Coleman, B. M., & Hill, A. F. (2015). Extracellular vesicles—their role in the packaging and spread of misfolded proteins associated with neurodegenerative diseases. *Seminars in Cell & Developmental Biology, 40*, 89–96. https://doi.org/10.1016/j.semcdb.2015.02.007
91. Janas, A. M., Sapoń, K., Janas, T., Stowell, M. H. B., & Janas, T. (2016). Exosomes and other extracellular vesicles in neural cells and neurodegenerative diseases. *Biochimica et Biophysica Acta (BBA) - Biomembranes*, 1858(6), 1139–1151. https://doi.org/10.1016/j.bbamem.2016.02.011
92. Wan, R., Hussain, A., Behfar, A., Moran, S. L., & Zhao, C. (2022). The therapeutic potential of exosomes in soft tissue repair and regeneration. *International Journal of Molecular Sciences, 23*(7), 3869. https://doi.org/10.3390/ijms23073869
93. Javdani-Mallak, A., & Salahshoori, I. (2024). Environmental pollutants and exosomes: A new paradigm in environmental health and disease. *Science of The Total Environment, 925*, Article 171774. https://doi.org/10.1016/j.scitotenv.2024.171774
94. Ferguson, S. W., & Nguyen, J. (2016). Exosomes as therapeutics: The implications of molecular composition and exosomal heterogeneity. *Journal of Controlled Release, 228*, 179–190. https://doi.org/10.1016/j.jconrel.2016.02.037
95. Valadi, H., Ekström, K., Bossios, A., Sjöstrand, M., Lee, J. J., & Lötvall, J. O. (2007). Exosome-mediated transfer of mRNAs and microRNAs is a novel mechanism of genetic exchange between cells. *Nature Cell Biology, 9*(6), 654–659. https://doi.org/10.1038/ncb1596
96. Yang, E., Wang, X., Gong, Z., Yu, M., Wu, H., & Zhang, D. (2020). Exosome-mediated metabolic reprogramming: The emerging role in tumor microenvironment remodeling and its influence on cancer progression. *Signal Transduction and Targeted Therapy, 5*(1), 242. https://doi.org/10.1038/s41392-020-00359-5
97. Zhang, W., Song, M., Fang, Z., Chen, F., Yuan, H., Gao, X., & Liu, K. (2024). Role of extracellular vesicles in insulin resistance: Signaling pathways, bioactive substances, miRNAs, and therapeutic potential. *Cell Biochemistry and Function, 42*(3), Article e4013. https://doi.org/10.1002/cbf.4013
98. Record, M., Carayon, K., Poirot, M., & Silvente-Poirot, S. (2014). Exosomes as new vesicular lipid transporters involved in cell–cell communication and various pathophysiologies. *Biochimica et Biophysica Acta (BBA) - Molecular and Cell Biology of Lipids, 1841*(1), 108–120. https://doi.org/10.1016/j.bbalip.2013.10.004

99. Meng, F., Xue, X., Yin, Z., Gao, F., Wang, X., & Geng, Z. (2022). Research progress of exosomes in bone diseases: Mechanism, diagnosis and therapy. *Frontiers in Bioengineering and Biotechnology, 10*, Article 866627. https://doi.org/10.3389/fbioe.2022.866627
100. Lai, J. J., Chau, Z. L., Chen, S.-Y., Hill, J. J., Korpany, K. V., Liang, N.-W., Lin, L.-H., Lin, Y.-H., Liu, J. K., Liu, Y.-C., Lunde, R., & Shen, W.-T. (2022). Exosome processing and characterization approaches for research and technology development. *Advanced Science, 9*(15), 2103222. https://doi.org/10.1002/advs.202103222
101. Wang, L., Yu, X., Zhou, J., & Su, C. (2023). Extracellular vesicles for drug delivery in cancer treatment. *Biological Procedures Online, 25*(1), 28. https://doi.org/10.1186/s12575-023-00220-3
102. Li, P., Kaslan, M., Lee, S. H., Yao, J., & Gao, Z. (2017). Progress in exosome isolation techniques. *Theranostics, 7*(3), 789–804. https://doi.org/10.7150/thno.18133
103. Kim, J., Tan, Z., & Lubman, D. M. (2015). Exosome enrichment of human serum using multiple cycles of centrifugation. *Electrophoresis, 36*(17), 2017–2026. https://doi.org/10.1002/elps.201500131
104. Van Deun, J., Mestdagh, P., Agostinis, P., Akay, Ö., Anand, S., Anckaert, J., Martinez, Z. A., Baetens, T., Beghein, E., Bertier, L., Berx, G., Boere, J., Boukouris, S., Bremer, M., Buschmann, D., Byrd, J. B., Casert, C., Cheng, L., Cmoch, A., … Hendrix, A. (2017). EV-TRACK: Transparent reporting and centralizing knowledge in extracellular vesicle research. *Nature Methods, 14*(3), 228–232. https://doi.org/10.1038/nmeth.4185
105. Hua, X., Zhu, Q., Liu, Y., Zhou, S., Huang, P., Li, Q., & Liu, S. (2023). A double tangential flow filtration-based microfluidic device for highly efficient separation and enrichment of exosomes. *Analytica Chimica Acta, 1258*, Article 341160. https://doi.org/10.1016/j.aca.2023.341160
106. Heinemann, M. L., Ilmer, M., Silva, L. P., Hawke, D. H., Recio, A., Vorontsova, M. A., Alt, E., & Vykoukal, J. (2014). Benchtop isolation and characterization of functional exosomes by sequential filtration. *Journal of Chromatography A, 1371*, 125–135. https://doi.org/10.1016/j.chroma.2014.10.026
107. Yang, D., Zhang, W., Zhang, H., Zhang, F., Chen, L., Ma, L., Larcher, L. M., Chen, S., Liu, N., Zhao, Q., Tran, P. H. L., Chen, C., Veedu, R. N., & Wang, T. (2020). Progress, opportunity, and perspective on exosome isolation–efforts for efficient exosome-based theranostics. *Theranostics, 10*(8), 3684–3707. https://doi.org/10.7150/thno.41580
108. Heinemann, M. L., & Vykoukal, J. (2017). Sequential filtration: A gentle method for the isolation of functional extracellular vesicles. In W. P. Kuo, & S. Jia (Eds.), Extracellular vesicles: Methods and protocols. Springer New York, pp 33–41. https://doi.org/10.1007/978-1-4939-7253-1_4
109. Vergauwen, G., Dhondt, B., Van Deun, J., De Smedt, E., Berx, G., Timmerman, E., Gevaert, K., Miinalainen, I., Cocquyt, V., Braems, G., Van den Broecke, R., Denys, H., De Wever, O., & Hendrix, A. (2017). Confounding factors of ultrafiltration and protein analysis in extracellular vesicle research. *Scientific Reports, 7*(1), 2704. https://doi.org/10.1038/s41598-017-02599-y
110. Li, J., Zhang, Y., Dong, P.-Y., Yang, G.-M., & Gurunathan, S. (2023). A comprehensive review on the composition, biogenesis, purification, and multifunctional role of exosome as delivery vehicles for cancer therapy. *Biomedicine & Pharmacotherapy, 165*, Article 115087. https://doi.org/10.1016/j.biopha.2023.115087
111. Rech, J., Getinger-Panek, A., Gałka, S., & Bednarek, I. (2022). Origin and composition of exosomes as crucial factors in designing drug delivery systems. *Applied Sciences, 12*. https://doi.org/10.3390/app122312259
112. Gardiner, C., Vizio, D. D., Sahoo, S., Théry, C., Witwer, K. W., Wauben, M., & Hill, A. F. (2016). Techniques used for the isolation and characterization of extracellular vesicles: Results of a worldwide survey. *Journal of Extracellular Vesicles, 5*(1), 32945. https://doi.org/10.3402/jev.v5.32945
113. Lewis, G. D., & Metcalf, T. G. (1988). Polyethylene glycol precipitation for recovery of pathogenic viruses, including hepatitis A virus and human rotavirus, from oyster, water, and sediment samples. *Applied and Environmental Microbiology, 54*(8), 1983–1988. https://doi.org/10.1128/aem.54.8.1983-1988.1988

114. Ding, M., Wang, C., Lu, X., Zhang, C., Zhou, Z., Chen, X., Zhang, C.-Y., Zen, K., & Zhang, C. (2018). Comparison of commercial exosome isolation kits for circulating exosomal microRNA profiling. *Analytical and Bioanalytical Chemistry, 410*(16), 3805–3814. https://doi.org/10.1007/s00216-018-1052-4
115. Kim, J., Lee, H., Park, K., & Shin, S. (2020). Rapid and efficient isolation of exosomes by clustering and scattering. *Journal of Clinical Medicine, 9*. https://doi.org/10.3390/jcm9030650
116. Park, J., Go, E.-B., Oh, J. S., Lee, J. K., & Lee, S.-Y. (2012). Multiple-cycle polymeric extracellular vesicle precipitation and its evaluation by targeted mass spectrometry. *International Journal of Molecular Sciences*, 22. https://doi.org/10.3390/ijms22094311
117. Martínez-Greene, J. A., Hernández-Ortega, K., Quiroz-Baez, R., Resendis-Antonio, O., Pichardo-Casas, I., Sinclair, D. A., Budnik, B., Hidalgo-Miranda, A., Uribe-Querol, E., Ramos-Godínez, M. D. P., & Martínez-Martínez, E. (2021). Quantitative proteomic analysis of extracellular vesicle subgroups isolated by an optimized method combining polymer-based precipitation and size exclusion chromatography. *Journal of Extracellular Vesicles, 10*(6), e12087. https://doi.org/10.1002/jev2.12087
118. Yousif, G., Qadri, S., Parray, A., Akhthar, N., Shuaib, A., & Haik, Y. (2022). Exosomes derived neuronal markers: Immunoaffinity isolation and characterization. *NeuroMolecular Medicine, 24*(3), 339–351. https://doi.org/10.1007/s12017-021-08696-6
119. Doldán, X., Fagúndez, P., Cayota, A., Laíz, J., & Tosar, J. P. (2016). Electrochemical sandwich immunosensor for determination of exosomes based on surface marker-mediated signal amplification. *Analytical Chemistry, 88*(21), 10466–10473. https://doi.org/10.1021/acs.analchem.6b02421
120. Koliha, N., Wiencek, Y., Heider, U., Jüngst, C., Kladt, N., Krauthäuser, S., Johnston, I. C. D., Bosio, A., Schauss, A., & Wild, S. (2016). A novel multiplex bead-based platform highlights the diversity of extracellular vesicles. *Journal of Extracellular Vesicles, 5*(1), 29975. https://doi.org/10.3402/jev.v5.29975
121. Boriachek, K., Masud, M. K., Palma, C., Phan, H.-P., Yamauchi, Y., Hossain, M. S. A., Nguyen, N.-T., Salomon, C., & Shiddiky, M. J. A. (2019). Avoiding pre-isolation step in exosome analysis: direct isolation and sensitive detection of exosomes using gold-loaded nanoporous ferric oxide nanozymes. *Analytical Chemistry, 91*(6), 3827–3834. https://doi.org/10.1021/acs.analchem.8b03619
122. Sharma, S., LeClaire, M., & Gimzewski, J. K. (2018). Ascent of atomic force microscopy as a nanoanalytical tool for exosomes and other extracellular vesicles. *Nanotechnology, 29*(13), Article 132001. https://doi.org/10.1088/1361-6528/aaab06
123. Song, E., Zhang, C., Israelow, B., Lu-Culligan, A., Prado, A. V., Skriabine, S., Lu, P., Weizman, O.-E., Liu, F., Dai, Y., Szigeti-Buck, K., Yasumoto, Y., Wang, G., Castaldi, C., Heltke, J., Ng, E., Wheeler, J., Alfajaro, M. M., Levavasseur, E., … Iwasaki, A. (2021). Neuroinvasion of SARS-CoV-2 in human and mouse brain. *Journal of Experimental Medicine, 218*(3), Article e20202135. https://doi.org/10.1084/jem.20202135
124. McNamara, R. P., Caro-Vegas, C. P., Costantini, L. M., Landis, J. T., Griffith, J. D., Damania, B.A., & Dittmer, D. P. (2018). Large-scale, cross-flow based isolation of highly pure and endocytosis-competent extracellular vesicles. *Journal of Extracellular Vesicles, 7*(1), 1541396. https://doi.org/10.1080/20013078.2018.1541396
125. Chen, C., Skog, J., Hsu, C.-H., Lessard, R. T., Balaj, L., Wurdinger, T., Carter, B. S., Breakefield, X. O., Toner, M., & Irimia, D. (2010). Microfluidic isolation and transcriptome analysis of serum microvesicles. *Lab on a Chip, 10*(4), 505–511. https://doi.org/10.1039/B916199F
126. Chen, F. W., Li, C., & Ioannou, Y. A. (2010). Cyclodextrin induces calcium-dependent lysosomal exocytosis. *PLoS ONE, 5*(11), Article e15054. https://doi.org/10.1371/journal.pone.0015054
127. Lee, K., Shao, H., Weissleder, R., & Lee, H. (2015). Acoustic purification of extracellular microvesicles. *ACS Nano, 9*(3), 2321–2327. https://doi.org/10.1021/nn506538f
128. Davies, R. T., Kim, J., Jang, S. C., Choi, E.-J., Gho, Y. S., & Park, J. (2012). Microfluidic filtration system to isolate extracellular vesicles from blood. *Lab on a Chip, 12*(24), 5202–5210. https://doi.org/10.1039/C2LC41006K

129. Gab-Allah, M. A., Getachew Lijalem, Y., Yu, H., Lee, S., Baek, S.-Y., Han, J., Choi, K., & Kim, B. (2023). Development of a certified reference material for the accurate determination of type B trichothecenes in corn. *Food Chemistry, 404*, Article 134542. https://doi.org/10.1016/j.foodchem.2022.134542
130. Lijalem, Y. G., Gab-Allah, M. A., Yu, H., Choi, K., & Kim, B. (2024). Development of a corn flour certified reference material for the accurate determination of zearalenone. *Analytical and Bioanalytical Chemistry, 416*(13), 3173–3183. https://doi.org/10.1007/s00216-024-05265-1
131. Mahmood, T., & Yang, P. C. (2012). Western blot: Technique, theory, and trouble shooting. *North American Journal of Medical Sciences, 4*(9), 429–434. https://doi.org/10.4103/1947-2714.100998
132. Lee, M., Ban, J.-J., Im, W., & Kim, M. (2016). Influence of storage condition on exosome recovery. *Biotechnology and Bioprocess Engineering, 21*(2), 299–304. https://doi.org/10.1007/s12257-015-0781-x
133. Ghosh, R., Gilda, J. E., & Gomes, A. V. (2014). The necessity of and strategies for improving confidence in the accuracy of western blots. *Expert Review of Proteomics, 11*(5), 549–560. https://doi.org/10.1586/14789450.2014.939635
134. Théry, C., Amigorena, S., Raposo, G., & Clayton, A. (2006). Isolation and characterization of exosomes from cell culture supernatants and biological fluids. *Current Protocols in Cell Biology*, 30(1), 3.22.1–3.22.29. https://doi.org/10.1002/0471143030.cb0322s30
135. Ramesh, D., Shaju, M. C., Kumar, A., Pandey, S., Nayak, R., Alka, V., Munjal, S., Salimi, A., Pai, K. S. R., & Bakkannavar, S. M. (2023). Biological, pathological, and multifaceted therapeutic functions of exosomes to target cancer. *Oncology Research, 32*(1), 73–94. https://doi.org/10.32604/or.2023.030401
136. Dragovic, R. A., Gardiner, C., Brooks, A. S., Tannetta, D. S., Ferguson, D. J., Hole, P., Carr, B., Redman, C. W., Harris, A. L., Dobson, P. J., Harrison, P., & Sargent, I. L. (2011). Sizing and phenotyping of cellular vesicles using nanoparticle tracking analysis. *Nanomedicine, 7*(6), 780–788. https://doi.org/10.1016/j.nano.2011.04.003
137. Shao, H., Chung, J., Balaj, L., Charest, A., Bigner, D. D., Carter, B. S., Hochberg, F. H., Breakefield, X. O., Weissleder, R., & Lee, H. (2012). Protein typing of circulating microvesicles allows real-time monitoring of glioblastoma therapy. *Nature Medicine, 18*(12), 1835–1840. https://doi.org/10.1038/nm.2994
138. Soo, C. Y., Song, Y., Zheng, Y., Campbell, E. C., Riches, A. C., Gunn-Moore, F., & Powis, S. J. (2012). Nanoparticle tracking analysis monitors microvesicle and exosome secretion from immune cells. *Immunology, 136*(2), 192–197. https://doi.org/10.1111/j.1365-2567.2012.03569.x
139. Shimomura, T., Seino, R., Umezaki, K., Shimoda, A., Ezoe, T., Ishiyama, M., & Akiyoshi, K. (2020). New lipophilic fluorescent dyes for exosome labeling: Monitoring of cellular uptake of exosomes. *bioRxiv*, 2020.02.02.931295 https://doi.org/10.1101/2020.02.02.931295
140. Thane, K. E., Davis, A. M., & Hoffman, A. M. (2019). Improved methods for fluorescent labeling and detection of single extracellular vesicles using nanoparticle tracking analysis. *Scientific Reports, 9*(1), 12295. https://doi.org/10.1038/s41598-019-48181-6
141. Palmieri, V., Lucchetti, D., Gatto, I., Maiorana, A., Marcantoni, M., Maulucci, G., Papi, M., Pola, R., De Spirito, M., & Sgambato, A. (2014). Dynamic light scattering for the characterization and counting of extracellular vesicles: A powerful noninvasive tool. *Journal of Nanoparticle Research, 16*(9), 2583. https://doi.org/10.1007/s11051-014-2583-z
142. Au-Jung, M. K., & Au-Mun, J. Y. (2018). *Sample Preparation and Imaging of Exosomes by Transmission Electron Microscopy. JoVE*, 131, e56482. https://doi.org/10.3791/56482
143. Matsumoto, A., Takahashi, Y., Chang, H.-Y., Wu, Y.-W., Yamamoto, A., Ishihama, Y., & Takakura, Y. (2020). Blood concentrations of small extracellular vesicles are determined by a balance between abundant secretion and rapid clearance. *Journal of Extracellular Vesicles, 9*(1), 1696517. https://doi.org/10.1080/20013078.2019.1696517
144. Raposo, G., & Stoorvogel, W. (2013). Extracellular vesicles: Exosomes, microvesicles, and friends. *Journal of Cell Biology, 200*(4), 373–383. https://doi.org/10.1083/jcb.201211138

145. Sokolova, V., Ludwig, A.-K., Hornung, S., Rotan, O., Horn, P. A., Epple, M., & Giebel, B. (2011). Characterisation of exosomes derived from human cells by nanoparticle tracking analysis and scanning electron microscopy. *Colloids and Surfaces B: Biointerfaces, 87*(1), 146–150. https://doi.org/10.1016/j.colsurfb.2011.05.013
146. Wu, Y., Deng, W., & Klinke Ii, D. J. (2015). Exosomes: improved methods to characterize their morphology, RNA content, and surface protein biomarkers. *Analyst, 140*(19), 6631–6642. https://doi.org/10.1039/C5AN00688K
147. Stetefeld, J., McKenna, S. A., & Patel, T. R. (2016). Dynamic light scattering: A practical guide and applications in biomedical sciences. *Biophysical Reviews, 8*(4), 409–427. https://doi.org/10.1007/s12551-016-0218-6
148. Sbarigia, C., Tacconi, S., Mura, F., Rossi, M., Dinarelli, S., & Dini, L. (2022). High-resolution atomic force microscopy as a tool for topographical mapping of surface budding. *Frontiers in Cell and Developmental Biology, 10*. https://doi.org/10.3389/fcell.2022.975919
149. Au-Skliar, M., & Au-Chernyshev, V. S. (2019). Imaging of extracellular vesicles by atomic force microscopy. *JoVE*, (151), e59254. https://doi.org/10.3791/59254
150. Duncan, M. W., Nedelkov, D., Walsh, R., & Hattan, S. J. (2016). Applications of MALDI mass spectrometry in clinical chemistry. *Clinical Chemistry, 62*(1), 134–143. https://doi.org/10.1373/clinchem.2015.239491
151. Li, D., Yi, J., Han, G., & Qiao, L. (2022). MALDI-TOF mass spectrometry in clinical analysis and research. *ACS Measurement Science Au, 2*(5), 385–404. https://doi.org/10.1021/acsmeasuresciau.2c00019
152. Guilhaus, M. (1995). Special feature: Tutorial. Principles and instrumentation in time-of-flight mass spectrometry. Physical and instrumental concepts. *Journal of Mass Spectrometry, 30*(11), 1519–1532. https://doi.org/10.1002/jms.1190301102
153. Harland, P. W. (1995). 11–Ion mass analyzers. In F. B. Dunning, & R. G. Hulet (Eds.), *Methods in experimental physics* (pp. 321–348). Academic Press. https://doi.org/10.1016/S0076-695X(08)60662-6
154. Medzihradszky, K. F., Campbell, J. M., Baldwin, M. A., Falick, A. M., Juhasz, P., Vestal, M. L., & Burlingame, A. L. (2000). The characteristics of peptide collision-induced dissociation using a high-performance MALDI-TOF/TOF tandem mass spectrometer. *Analytical Chemistry, 72*(3), 552–558. https://doi.org/10.1021/ac990809y
155. Gab-Allah, M. A., & Kim, J. (2024). A comprehensive review of recent advances in the enrichment and mass spectrometric analysis of glycoproteins and glycopeptides in complex biological matrices. *Mass Spectrometry Letters, 15*(1), 1–25. https://doi.org/10.5478/MSL.2024.15.1.1
156. Darie-Ion, L., Whitham, D., Jayathirtha, M., Rai, Y., Neagu, A.-N., Darie, C. C., & Petre, B. A. (2022). Applications of MALDI-MS/MS-based proteomics in biomedical research. *Molecules, 27*. https://doi.org/10.3390/molecules27196196
157. Römpp, A., & Spengler, B. (2013). Mass spectrometry imaging with high resolution in mass and space. *Histochemistry and Cell Biology, 139*(6), 759–783. https://doi.org/10.1007/s00418-013-1097-6
158. Stump, M. J., Fleming, R. C., Gong, W.-H., Jaber, A. J., Jones, J. J., Surber, C. W., & Wilkins, C. L. (2002). Matrix-assisted laser desorption mass spectrometry. *Applied Spectroscopy Reviews, 37*(3), 275–303. https://doi.org/10.1081/ASR-120014277
159. Liu, B.-H., Lee, Y. T., & Wang, Y.-S. (2009). Incoherent production reactions of positive and negative ions in matrix-assisted laser desorption/ionization. *Journal of the American Society for Mass Spectrometry, 20*(6), 1078–1086. https://doi.org/10.1016/j.jasms.2009.02.012
160. Torres-Sangiao, E., Leal Rodriguez, C., & García-Riestra, C. (2021). Application and perspectives of MALDI–TOF mass spectrometry in clinical microbiology laboratories. *Microorganisms, 9*. https://doi.org/10.3390/microorganisms9071539
161. Qiao, Z., & Lissel, F. (2021). MALDI matrices for the analysis of low molecular weight compounds: Rational design, challenges and perspectives. *Chemistry – An Asian Journal, 16*(8), 868–878. https://doi.org/10.1002/asia.202100044

162. O'Rourke, M. B., Djordjevic, S. P., & Padula, M. P. (2018). The quest for improved reproducibility in MALDI mass spectrometry. *Mass Spectrometry Reviews, 37*(2), 217–228. https://doi.org/10.1002/mas.21515
163. Pang, R. T. K., Johnson, P. J., Chan, C. M. L., Kong, E. K. C., Chan, A. T. C., Sung, J. J. Y., & Poon, T. C. W. (2004). Technical evaluation of MALDI-TOF mass spectrometry for quantitative proteomic profiling matrix formulation and application. *Clinical Proteomics, 1*(3), 259–270. https://doi.org/10.1385/CP:1:3-4:259
164. Yang, J., & Caprioli, R. M. (2011). Matrix sublimation/recrystallization for imaging proteins by mass spectrometry at high spatial resolution. *Analytical Chemistry, 83*(14), 5728–5734. https://doi.org/10.1021/ac200998a
165. Roskey, M. T., Juhasz, P., Smirnov, I. P., Takach, E. J., Martin, S. A., & Haff, L. A. (1996). DNA sequencing by delayed extraction-matrix-assisted laser desorption/ionization time of flight mass spectrometry. *Proceedings of the National Academy of Sciences U S A, 93*(10), 4724–4729. https://doi.org/10.1073/pnas.93.10.4724
166. Patel, R. (2015). MALDI-TOF MS for the diagnosis of infectious diseases. *Clinical Chemistry, 61*(1), 100–111. https://doi.org/10.1373/clinchem.2014.221770
167. Ng, E. W. Y., Wong, M. Y. M., & Poon, T. C. W. (2014). Advances in MALDI mass spectrometry in clinical diagnostic applications. In N. L.S. Tang and T. Poon (Eds.), Chemical diagnostics: From bench to bedside. Springer Berlin Heidelberg, pp 139–175. https://doi.org/10.1007/128_2012_413
168. Song, K., & Cheng, Q. (2020). Desorption and ionization mechanisms and signal enhancement in surface assisted laser desorption ionization mass spectrometry (SALDI-MS). *Applied Spectroscopy Reviews, 55*(3), 220–242 https://doi.org/10.1080/05704928.2019.1570519
169. Seibert, V., Wiesner, A., Buschmann, T., & Meuer, J. (2004). Surface-enhanced laser desorption ionization time-of-flight mass spectrometry (SELDI TOF-MS) and ProteinChip® technology in proteomics research. *Pathology - Research and Practice, 200*(2), 83–94. https://doi.org/10.1016/j.prp.2004.01.010
170. Klinke Ii, D. J., Kulkarni, Y. M., Wu, Y., & Byrne-Hoffman, C. (2014). Inferring alterations in cell-to-cell communication in HER2+ breast cancer using secretome profiling of three cell models. *Biotechnology and Bioengineering, 111*(9), 1853–1863. https://doi.org/10.1002/bit.25238
171. Schabath, M. B., & Cote, M. L. (2019). Cancer progress and priorities: Lung cancer. *Cancer Epidemiology, Biomarkers & Prevention, 28*(10), 1563–1579. https://doi.org/10.1158/1055-9965.EPI-19-0221
172. Rudin, C. M., Brambilla, E., Faivre-Finn, C., & Sage, J. (2021). Small-cell lung cancer. *Nature Reviews Disease Primers, 7*(1), 3. https://doi.org/10.1038/s41572-020-00235-0
173. Babjuk, M., Burger, M., Zigeuner, R., Shariat, S.F., van Rhijn, B. W. G., Compérat, E., Sylvester, R. J., Kaasinen, E., Böhle, A., Palou Redorta, J., & Rouprêt, M. (2013). EAU guidelines on non–muscle-invasive urothelial carcinoma of the bladder: Update 2013. *European Urology, 64*(4), 639–653. https://doi.org/10.1016/j.eururo.2013.06.003
174. Sourani, A., Saghaei, S., Sabouri, M., Soleimani, M., & Dehghani, L. (2021). A systematic review of extracellular vesicles as non-invasive biomarkers in glioma diagnosis, prognosis, and treatment response monitoring. *Molecular Biology Reports, 48*(10), 6971–6985. https://doi.org/10.1007/s11033-021-06687-1
175. Karaosmanoğlu, O., Banerjee, S., & Sivas, H. (2018). Identification of biomarkers associated with partial epithelial to mesenchymal transition in the secretome of slug over-expressing hepatocellular carcinoma cells. *Cellular Oncology (Dordrecht), 41*(4), 439–453. https://doi.org/10.1007/s13402-018-0384-6
176. Hoogsteden, H. C., Langerak, A. W., van der Kwast, T. H., Versnel, M. A. R. A., & van Gelder, T. (1997). Malignant pleural mesothelioma. *Critical Reviews in Oncology/Hematology, 25*(2), 97–126. https://doi.org/10.1016/S1040-8428(96)00231-4
177. Aebersold, R., & Mann, M. (2003). Mass spectrometry-based proteomics. *Nature, 422*(6928), 198–207. https://doi.org/10.1038/nature01511

178. Lobasso, S., Tanzarella, P., Mannavola, F., Tucci, M., Silvestris, F., Felici, C., Ingrosso, C., Corcelli, A., & Lopalco, P. (2021). A lipidomic approach to identify potential biomarkers in exosomes from melanoma cells with different metastatic potential. *Frontiers in Physiology, 12.* https://doi.org/10.3389/fphys.2021.748895
179. Aresta, A. M., De Vietro, N., & Zambonin, C. Analysis and characterization of the extracellular vesicles released in non-cancer diseases using matrix-assisted laser desorption ionization/mass spectrometry. *International Journal of Molecular Sciences*, 25. https://doi.org/10.3390/ijms25084490
180. Jung, J. H., Lee, M. Y., Choi, D.-Y., Lee, J. W., You, S., Lee, K. Y., Kim, J., & Kim, K. P. (2015). Phospholipids of tumor extracellular vesicles stratify gefitinib-resistant nonsmall cell lung cancer cells from gefitinib-sensitive cells. *Proteomics, 15*(4), 824–835. https://doi.org/10.1002/pmic.201400243
181. Bridi, A., Perecin, F., & Silveira, J. C. (2020). Extracellular vesicles mediated early embryo–maternal interactions. *International Journal of Molecular Sciences*, 21. https://doi.org/10.3390/ijms21031163
182. Banliat, C., Le Bourhis, D., Bernardi, O., Tomas, D., Labas, V., Salvetti, P., Guyonnet, B., Mermillod, P., & Saint-Dizier, M. (2020). Oviduct fluid extracellular vesicles change the phospholipid composition of bovine embryos developed in vitro. *International Journal of Molecular Sciences, 21.* https://doi.org/10.3390/ijms21155326
183. Yan, S., Huang, Z., Chen, X., Chen, H., Yang, X., Gao, M., & Zhang, X. (2023). Metabolic profiling of urinary exosomes for systemic lupus erythematosus discrimination based on HPL-SEC/MALDI-TOF MS. *Analytical and Bioanalytical Chemistry, 415*(26), 6411–6420. https://doi.org/10.1007/s00216-023-04916-z

LC–MS/MS-Based Glycomic and Glycoproteomic Analysis of Exosomes Derived from Cells and Biofluids

Mojibola Fowowe, Judith Nwaiwu, David M. Lubman, and Yehia Mechref

Abstract Exosomes are small extracellular vesicles (EVs) released by most cell types. Their pivotal role in intracellular communication has led to a growing interest in their study as biomarkers for several diseases, including neurodegenerative diseases and cancer. Emerging evidence now demonstrates the presence and involvement of glycoconjugates in the biogenesis, cellular recognition, and uptake of exosomes by recipient cells. Studies on the glycan composition of exosomes in different diseased conditions have been conducted, highlighting the significance of exosomal glycans as diagnostic biomarkers. In this chapter, we describe two efficient protocols for the isolation of intact exosomes derived from various sources including cells and biofluids such as serum, plasma, and cerebrospinal fluid. Finally, we present a comprehensive liquid chromatography-tandem mass spectrometry (LC–MS/MS) workflow for the glycomic and glycoproteomic analysis of exosome content.

Keywords Exosomes · EV isolation · Glycans · Glycopeptides · HILIC enrichment · C18 desalting · LC–MS/MS

Abbreviations

EVs	Extracellular vesicles
TEM	Transmission electron microscopy

M. Fowowe · J. Nwaiwu · Y. Mechref (✉)
Department of Chemistry and Biochemistry, Texas Tech University, Lubbock, TX, USA
e-mail: yehia.mechref@ttu.edu

M. Fowowe
e-mail: mfowowe@ttu.edu

J. Nwaiwu
e-mail: jnwaiwu@ttu.edu

D. M. Lubman
Department of Surgery, University of Michigan Medical Center, Ann Arbor, MI, USA
e-mail: dmlubman@umich.edu

R. K. Marcus and D. Lubman (eds.), *Exosomes and Extracellular Vesicles*,
https://doi.org/10.1007/978-3-031-89270-7_12

CCM	Concentrated culture medium
LC–MS/MS	Liquid chromatography-tandem mass spectrometry
QTOF/HCD	Quadrupole time-of-flight/ higher energy collisional dissociation
HILIC	Hydrophilic interaction liquid chromatography
PBS	Phosphate-buffered saline
FBS	Fetal bovine serum
CSF	Cerebrospinal fluid

1 Introduction

Extracellular vesicles (EVs) are membrane-bound structures enclosed by a phospholipid bilayer that play a crucial role in intercellular communication and the exchange of biomolecules within multicellular organisms [1, 2]. Their ability to transport bioactive cargo, such as proteins, lipids, nucleic acids, and signaling molecules, has positioned them as crucial mediators of cell-to-cell communication, influencing neighboring and distant cells [3]. Virtually all cell types secrete them and can be found in various bodily fluids, including blood, urine, saliva, synovial fluid, amniotic fluid, breast milk, and cerebrospinal fluid [4–12]. EVs influence a wide range of physiological and pathological processes, including immune regulation, tissue repair, and cancer metastasis [13–17]. There are three major subtypes of EVs: microvesicles (MVs), exosomes, and apoptotic bodies, distinct in their biogenesis pathway, size, cargo, and biofunction [18]. Exosomes or small extracellular vesicles ranging in size from 30 to 150 nm have increasingly gained attention over the past decade in biomedical studies that explore their contents as disease biomarkers, and therapeutic studies that explore them as potential natural drug delivery vehicles and drug targets. Mass spectrometry-based proteomic characterization of exosome content has been widely employed in many diseases to unravel exosome-specific signatures of disease development, progression, and therapy [19]. A recent study by An et al. identified eight exosomal proteins whose expressions were altered in pancreatic cancer patients who underwent chemoradiotherapy [20]. Other studies have identified exosome-related protein biomarkers for neurodegenerative diseases like Alzheimer's disease, [21, 22] Parkinson's disease, [23, 24] and amyotrophic lateral sclerosis [25, 26]. While much interest in exosomes has been channeled toward proteomic studies, the possible role of glycans in exosome structure and function has only recently come to light [27–30].

Glycosylation is the process of attaching carbohydrate molecules (glycans) to proteins or lipids, and it is a ubiquitous post-translational modification in eukaryotic cells. The glycosylation pattern of exosomes is of particular interest because it can influence their stability, targeting, and recognition by receptors on recipient cells. Additionally, exosomal glycosylation has been implicated in various diseases, including cancer, [29, 31, 32] neurological disorders, [33–35] and immune dysregulation [36–38]. Several studies have investigated the glycosylation of exosomes derived from different cell types and under various physiological and pathological conditions. For instance, a survey by Batista et al. analyzed the glycosylation profiles of exosomes/microvesicles derived from a variety of biological sources and observed an enrichment and depletion of certain glycan epitopes when compared with their parent cell membranes. Affected structures included high-mannose, sialylated, polylactosamine, terminal blood group A and B antigens, and complex *N*-linked glycans.

Despite the prospects, the study of exosome glycosylation presents unique challenges, chiefly the inherent heterogeneity of exosomes, even within the same cell type or biological sample, which can lead to variability in their glycosylation patterns [39]. Additionally, the low abundance of exosomes in biological fluids, such as blood or urine, can make it difficult to obtain sufficient quantities for comprehensive glycosylation analysis. Isolating and purifying exosomes from complex biological matrices is another challenge, as various isolation techniques may introduce biases or lead to the co-isolation of non-exosomal contaminants, impacting the accuracy of the analysis. Furthermore, the structural complexity of glycans, with their diverse combinations of monosaccharides, branching patterns, and linkages, complicates the accurate characterization and identification of specific glycan structures on exosomes. To address these issues, standardized protocols and efficient analytical methods for analyzing exosome glycosylation are required.

In this chapter, we describe two efficient protocols for isolating intact exosomes from cell culture media and biofluids including serum, plasma, and cerebrospinal fluid. Finally, we present a comprehensive liquid chromatography-tandem mass spectrometry (LC–MS/MS) workflow for the glycomic and glycoproteomic analysis of exosome content.

2 Materials

2.1 Cell Culture Preparation

1. Cell line.
2. Culture media.
3. Penicillin–streptomycin solution.
4. Phosphate-buffered saline (PBS).
5. Fetal bovine serum (FBS).
6. EV-depleted fetal bovine serum (ED-FBS) (*see* Note 1).
7. Trypsin.

2.2 Exosome Isolation by Differential Centrifugation

1. Human serum or plasma, cerebrospinal fluid (CSF), or concentrated culture medium (CCM).
2. Sterile-filtered phosphate-buffered saline (PBS, 1X) for washing and resuspending exosomes.
3. Fixed-angle rotor microcentrifuge for differential centrifugation and removal of cell debris and microvesicles.
4. Preparative ultracentrifuge (Beckman Coulter or equivalent) for pelleting exosomes.
5. Type 50.2 Ti or Type 70.1 Ti fixed-angle rotor for the ultracentrifuge.
6. Polycarbonate ultracentrifuge tubes.

2.3 Exosome Isolation by Immunoaffinity Capture

Human serum or plasma, cerebrospinal fluid (CSF), or concentrated culture medium (CCM).

1. MagCapture Exosome Isolation Kit PS Ver. 2.
2. Tube shaker.
3. Magnetic stand.

2.4 Exosome Characterization by Transmission Electron Microscopy (TEM)

1. NERL™ reagent-grade water.
2. 1% uranyl acetate.
3. Carbon-coated grid.
4. Hitachi Model H-7650 transmission electron microscope.

2.5 Exosome Lysis and Protein Quantitation

1. Exosomes.
2. 50 mM ammonium bicarbonate (ABC) buffer solution.
3. 5% sodium deoxycholate (SDC).
4. 1.5 or 2.0 mL microcentrifuge tubes, conical, with screw cap.
5. BeadBug microtube homogenizer.
6. 400 μm molecular biology-grade zirconium beads.
7. Micro BCA protein assay kit.
8. 96 well plate.
9. Microplate reader.

2.6 Glycomics and Glycoproteomics Sample Preparation

Tryptic Digestion

1. 50 mM ammonium bicarbonate (ABC) buffer solution.
2. 200 mM dithiothreitol (DTT).
3. 200 mM iodoacetamide (IAA).
4. HPLC-grade acetonitrile (ACN).
5. HPLC-grade water.
6. Formic acid (FA).
7. Mass spectrometry-grade trypsin/Lys-C.

C18 Desalting

1. C18 TopTips.
2. HPLC-grade acetonitrile (ACN).
3. HPLC-grade water.
4. Formic acid (FA).
5. Three buffer solutions: Buffer A includes 100% H_2O, 0.1% FA; buffer B includes 60% ACN, 40% H_2O, 0.1% FA; buffer C contains 100% ACN, 0.1% FA.

Glycopeptide enrichment using PolyHYDROXYETHYL A (HILIC)

1. PolyHydroxyethyl A (HILIC) TopTips.
2. HPLC-grade acetonitrile (ACN).
3. HPLC-grade water.
4. Trifluoroacetic acid (TFA) 99%, spectrometric grade.
5. Two buffer solutions: Loading buffer containing 80% ACN, 1% TFA, and elution buffer containing 0.1% TFA.

2.7 LC–MS/MS Analysis

1. Mobile Phase A (MPA): 2% acetonitrile (ACN), 0.1% formic acid (FA) solution.
2. Mobile Phase B (MPB): 0.1% FA in ACN (*see* Note 2).
3. C18 Acclaim PepMap 100 trap column (particle size 3 μm, 100 Å pore size, 75 μm id, 20 mm length).
4. C18 Acclaim PepMap RSLC column (particle size 2 μm, 100 Å pore size, 75 μm id, 150 mm length).
5. Dionex 3000 UltiMate Nano LC system.
6. Orbitrap Fusion Lumos Tribrid mass spectrometer or Q Exactive HF Hybrid Quadrupole Orbitrap mass spectrometer.

3 Methods

The procedure below outlines detailed steps for the preparation and isolation of small EVs or exosomes from biological samples. The workflow is depicted in Fig. 1. The characterization of exosomes by TEM imaging and proteomics is also discussed. Lastly, a comprehensive workflow for the glycomic and glycoproteomic analysis of exosome cargo is presented and depicted in Fig. 2.

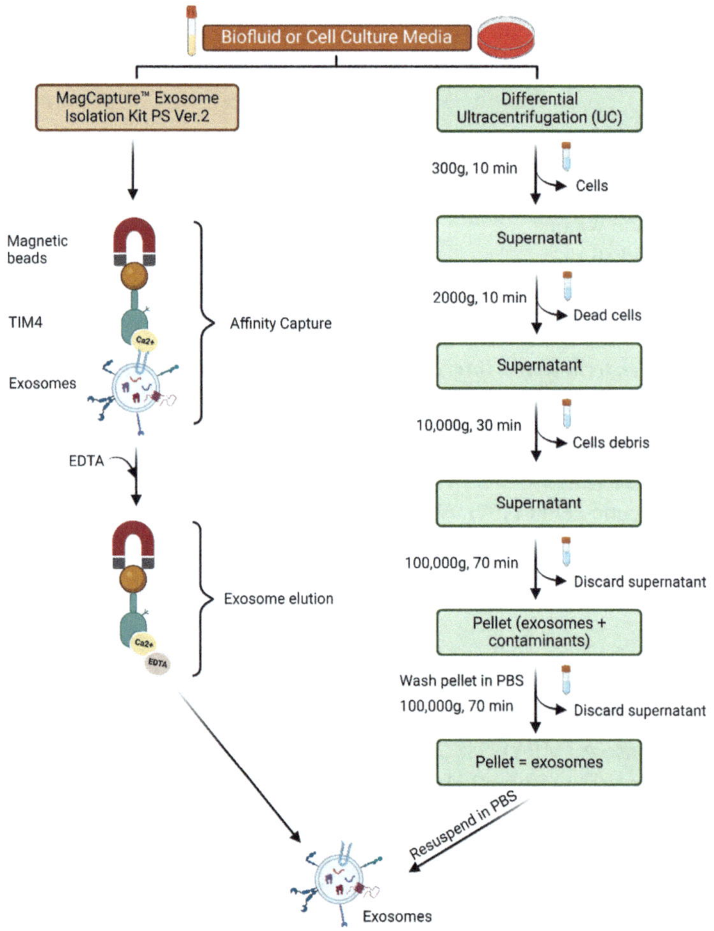

Fig. 1 Workflow for isolating exosomes from serum, plasma, cerebrospinal fluid, and cell culture media using differential ultracentrifugation and an affinity-based method (MagCapture Exosome Isolation Kit PS Ver. 2)

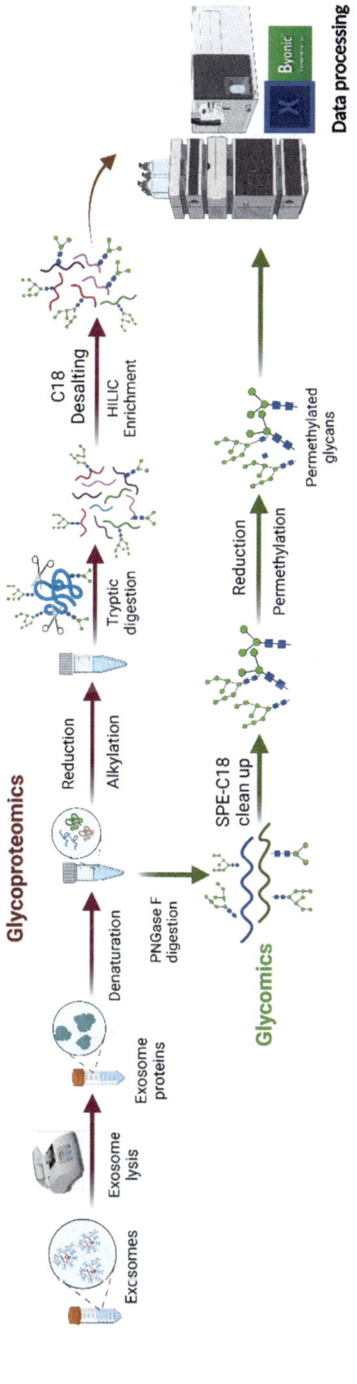

Fig. 2 Glycopeptide and glycan sample preparation workflow

3.1 Cell Culture and Concentrated Culture Medium (CCM) Preparation

The following steps describe the preparation of cell culture samples for the extraction of exosomes:

1. Culture cells in a 175 mL sterile culture flask in their appropriate media.
2. Prepare the medium by adding 10% FBS and 2% penicillin and mix properly (*see* Note 3).
3. Incubate at 37 °C and 5% CO_2.
4. At 80% confluency, detach cells with trypsin and collect cells in a centrifuge tube.
5. Prepare exosome medium using 10% ED-FBS and 2% penicillin (ED-culture media) (*see* Note 1).
6. Remove previous media and pellet cells by centrifuging at 3600 × g for 2 min.
7. Discard the supernatant containing media and add 2 mL of prepared culture media containing ED-FBS to cells, pipette severally until the cells are completely mixed in media.
8. Transfer 1 mL of cell mixture into a new culture flask and add 23 mL of ED-culture media.
9. Swirl the flask to distribute the cells evenly.
10. Incubate at 37 °C and 5% CO_2.
11. Examine cells every 2 to 3 days and change media, if necessary (*see* Note 4).
12. Collect media containing exosomes at 80% confluency (*see* Note 5).

3.2 Exosome Isolation by Differential Centrifugation

1. Centrifuge the freshly harvested cell culture media, serum, plasma, or cerebrospinal fluid at 300 × g, 4 °C for 10 min to pellet cells and large bodies. Carefully transfer the supernatant to a new tube (*see* Note 6).
2. Centrifuge the supernatant at 3000 × g, 4 °C for 10 min to pellet dead cells and cell debris. Carefully transfer the supernatant to a new tube (*see* Note 6).
3. Centrifuge the supernatant at 10,000 × g, 4 °C for 30 min to pellet microvesicles. Carefully transfer the supernatant to a new tube (*see* Note 6).
4. Centrifuge the supernatant at 100,000 × g, 4 °C for 70 min to pellet exosomes. Carefully discard the supernatant without disturbing the exosome pellet (*see* Note 7).
5. Wash the pelleted exosomes with sterile-filtered PBS and centrifuge again at 100,000 × g, 4 °C for 70 min to obtain purer exosomes (*see* Note 8).
6. Resuspend the exosome pellet in 100–200 μL PBS and store in −80 °C until further downstream processing.

3.3 Exosome Isolation by Immunoaffinity Capture

1. Follow steps 1–3 from Sect. 3.2 to prepare the sample for exosome isolation using the commercial MagCapture Exosome Isolation Kit PS Ver. 2.
2. Extract exosomes from the samples by following the manufacturer's instructions.

3.4 Exosome Characterization by Transmission Electron Microscopy (TEM)

TEM is a powerful tool for characterizing exosomes, offering high-resolution imaging of their size and morphology. Figure 3 shows the regular image of an exosome as viewed using the TEM.

1. Place a drop of the sample on the carbon-coated side of the grid (darker side) and leave on there for 1 min. Next, blot the excess sample off with a lint-free filter paper.
2. Place a drop of freshly prepared 1% uranyl acetate directly on top of the carbon-coated side of the grid where the sample was added. Stain for 30 s. Next, blot the excess 1% uranyl acetate off with a lint-free filter paper.
3. Dry grids for at least 3 h to remove any moisture in the sample.
4. Obtain images of exosomes using a transmission electron microscope operating at an accelerating voltage of 100 kV.

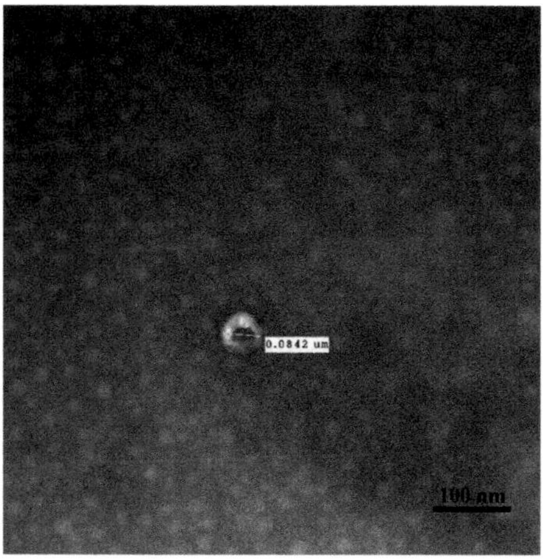

Fig. 3 Representative electron micrograph of whole-mount exosome preparation showing the size and round shape morphology of the vesicle. Exosomes were isolated from human blood serum using the MagCapture Exosome Isolation Kit PS Ver. 2, deposited onto carbon-coated EM grids, and negatively stained

3.5 Protein Extraction and Quantitation

1. Fill a 2.0 mL microcentrifuge tube (conical, screw cap) with zirconium beads (400 μm, molecular biology grade) up to 0.5 cm from the bottom of the tube.
2. Add 100 μL of 5% SDC solution to 100 μL of exosome suspension in PBS and vortex the mixture.
3. Gently transfer the mixture to the 2.0 mL microcentrifuge tube containing the zirconium beads.
4. Homogenize the mixture using a BeadBug microtube homogenizer at 4 °C and 4,000 rpm for 5 min (*see* Notes 9).
5. Sonicate the sample on ice for 1 h (*see* Note 10).
6. Centrifuge at 21,000 × g for 10 min.
7. Collect the supernatant.
8. Determine the protein concentration using a bicinchoninic acid (BCA) protein assay kit or a UV–Vis NanoDrop spectrophotometer (*see* Note 11).

3.6 Proteomic Characterization of Exosomes

Exosomes can be characterized based on their physical properties and their cargo content. To characterize exosomes by their protein cargo, an LC–MS/MS-based proteomics analysis is performed, and the identified proteins are compared with top exosome marker proteins reported in the literature and deposited into exosome study databases such as ExoCarta and Vesiclepedia. Proteomic analysis follows a similar procedure to glycoproteomic analysis, as discussed in Sect. 3.7, with the exception of the glycopeptide enrichment step. Figure 4 shows a representative extracted ion chromatogram (EIC), and mass spectra of a tryptic digested peptide identified in the 14–3-3 family of exosome marker proteins, confirming the presence of exosomes after isolation.

3.7 Glycoproteomics Sample Preparation

The following steps describe the preparation of glycopeptides from the extracted proteins for LC–MS/MS glycoproteomics analysis. Exosomes can also be characterized by examining glycopeptide structures from marker proteins as shown in Fig. 5.

Tryptic Digestion

1. Transfer 50 μg protein to a new Eppendorf tube and dry protein in a SpeedVac vacuum concentrator.
2. Resuspend the sample in 50 μL of 50 mM ABC buffer and vortex properly.
3. Denature the sample by heating at 90 °C for 15 min.

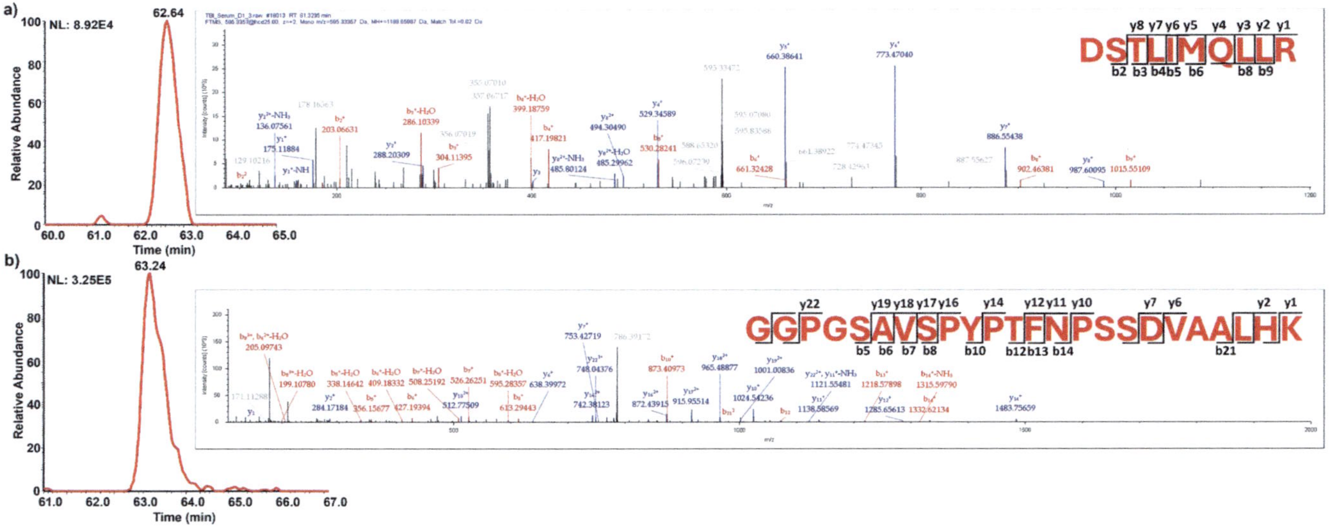

Fig. 4 Extracted ion chromatograms (EIC) of tryptic digested peptides from **a** 14–3-3 protein family and **b** Annexin A1. The insets are the HCD tandem mass spectra corresponding to the peptides. These proteins are among the commonly identified proteins in exosomes and confirm their presence after isolation

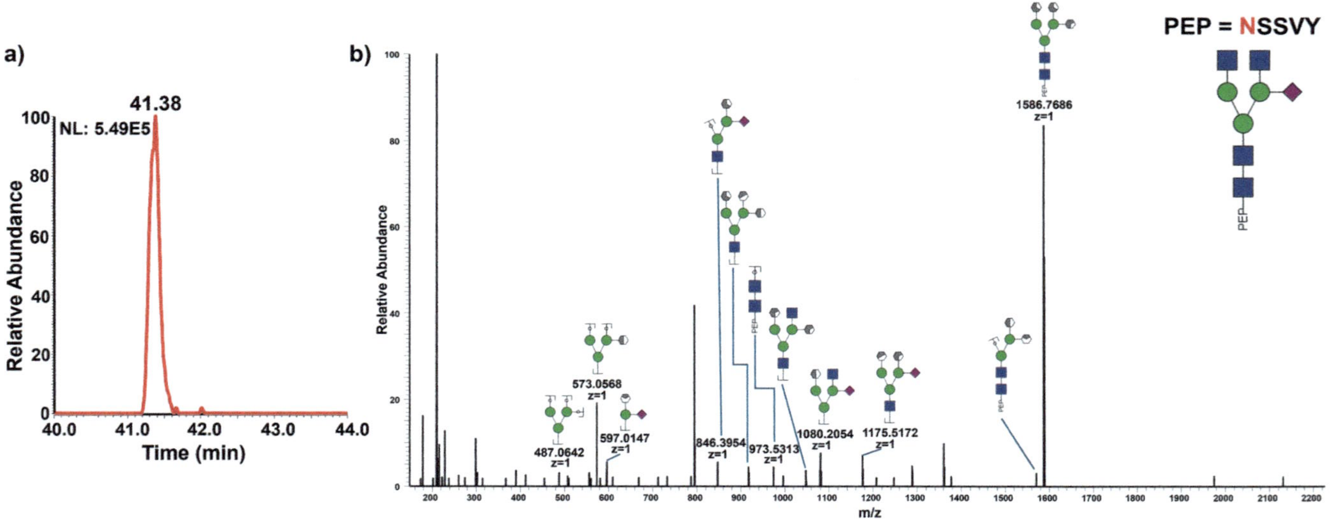

Fig. 5 **a** Extracted ion chromatogram (EIC) of a tryptic digested glycopeptide derived from CD81 protein, a well-known exosome marker, and its **b** HCD tandem mass spectra, confirming the presence of exosomes after isolation

4. Add 200 mM DTT (1/40 of sample volume) (*see* Note 12).
5. Incubate the sample at 60 °C for 45 min.
6. Add 200 mM IAA to the sample (four times DTT volume) (*see* Note 13).
7. Incubate at 37.5 °C for 45 min.
8. Add 200 mM DTT again to quench the alkylation.
9. Incubate at 37.5 °C for 30 min.
10. Check the pH of the samples (preferred pH = 8) (*see* Note 14).
11. Dissolve trypsin in commercial suspension buffer or 0.1 M HCl.
12. Add trypsin solution to the sample at a trypsin to sample protein mass ratio of 1:25.
13. Incubate at 37.5 °C for 18 h (overnight).
14. Quench trypsin activity by adding neat formic acid to a final concentration of 0.5% (v/v) or by heating the reaction mixture at 90 °C for 10 min.
15. Centrifuge the samples at 14.8 krpm for 10 min.
16. Carefully transfer the supernatant to a new Eppendorf tube and dry it using a SpeedVac vacuum concentrator.

C18 Desalting

C18 desalting procedure is essential to purify protein samples prior to LC–MS/MS analysis and HILIC enrichment. Here, TopTip C18 desalting procedure is described:

1. Fill one-third of the TopTip column with C18 material.
2. Wash the column with 50 μL of buffer B, spin down at 1,000 × g for 1 min, and repeat the step three times.
3. Wash the column with 50 μL of buffer A, spin down at 1,000 × g for 1 min, and repeat the step three times.
4. Load the sample into the column and spin down at 500 × g for 40 s.
5. Incubate for 2 min.
6. Centrifuge at 500 × g for 1 min and re-load the pass through into the tip. Repeat steps 4–6 three times.
7. Wash the column with 50 μL of buffer A, spin down at 1,000 × g for 1 min, and repeat the step three times.
8. Elute the sample with 50 μL of buffer B, spin down at 1,000 × g for 1 min, and repeat the step three times.
9. Elute the sample with 50 μL of buffer C, spin down at 2,000 × g for 1 min, and repeat the step three times.
10. Collect and dry the eluents in a SpeedVac vacuum concentrator.

Glycopeptide Enrichment Using PolyHYDROXYETHYL a (HILIC)

Glycopeptides are low in abundance compared to peptides, hence enrichment of glycopeptides prior to LC–MS/MS analysis is important. Here, HILIC enrichment protocol using PolyHydroxyethyl A is described:

1. Resuspend dried and purified peptide samples in 100 μL of loading buffer.
2. Wash a TopTip PolyHYDROXYETHYL A (HILIC) spin column with 100 μL of water, spin down at 1,000 × g for 1 min, and repeat the step two more times.

3. Wash the column with 100 μL of loading buffer, spin down at 1,000 × g for 1 min, and repeat three times.
4. Load the sample into the column and spin down at 1,000 × g for 1 min. Reload the flow-through two more times.
5. Wash the column three times with 100 μL of loading buffer, spin down at 1,000 × g for 1 min.
6. Elute the sample with 50 μL of elution buffer, spin down at 1,000 × g for 1 min, and repeat this step two more times.
7. Combine the eluted fractions and dry them.

3.8 Glycomics Sample Preparation

PNGase F digestion

1. Transfer 50–100 μg of proteins to a 1.5 mL Eppendorf tube and add 100 μL of 50 mM ABC buffer solution.
2. Denature proteins by heating in a water bath at 90 °C for 15 min.
3. Add PNGase F enzyme (1500 units to 20 μg protein) and incubate at 37 °C for 18 h.
4. Dry the sample using a SpeedVac vacuum concentrator.

SPE-C18 Cleanup

1. Resuspend the dried sample in 300 μL of 5% acetic acid, vortex, and spin down.
2. Wash the SPE-C18 cartridge three times with 1 mL of methanol.
3. Wash the cartridge three times with 1 mL of 5% acetic acid.
4. Load the sample onto the cartridge and gently elute with 5% acetic acid by applying external air pressure. Repeat this process three times.
5. Dry the collected sample in a SpeedVac vacuum concentrator.

Reduction

1. Add 10 μL of 10 μg/μL borane ammonia solution to the dried glycan samples, vortex, and spin down, then incubate at 60 °C for 1 h.
2. After incubation, add 1 mL methanol to the reduced sample to remove the residual reduction solution and dry in a SpeedVac vacuum concentrator, repeat this step four to five times or until there is no more visible white powder in the tube.

Permethylation

Permethylation of glycans is done to increase hydrophobicity and sensitivity in LC–MS/MS glycan analysis [40].

1. Resuspend reduced glycan samples in 30 μL DMSO and 1.2 μL HPLC-grade water. Vortex and spin down.
2. Place an empty clean top tip column into a 2 mL tube with spacers to allow flow-through.

3. In a 50 mL centrifuge tube, add NaOH beads and DMSO, ensure beads are completely immersed in the DMSO.
4. Fill one-third of the column with NaOH beads, then centrifuge at 1800 rpm for 2 min.
5. Wash the column with 200 μL DMSO and centrifuge at 1800 rpm for 2 min.
6. Now add 20 μL iodomethane (CH_3I) to the sample, vortex, and spin down.
7. Replace the collecting tube with a clean labeled tube and quickly load the sample into the column.
8. Incubate sample at room temperature in the dark for 25 min.
9. Add 20 μL of iodomethane to the column, then incubate for 15 min.
10. Centrifuge at 1800 rpm for 2 min.
11. Add 30 μL of ACN to the column, centrifuge at 1800 rpm again for 2 min.
12. Use a syringe to expel the residual sample in the column into the collection tube.
13. Dry the sample.
14. Resuspend the sample in 20% ACN, 0.1% FA solution prior to LC–MS/MS analysis.

3.9 LC–MS/MS Analysis

LC–MS/MS analysis will be performed after C18 desalting and HILIC enrichment for the peptide and glycopeptide samples, respectively, and after permethylation for the glycan samples. The analysis is performed using a Dionex 3000 UltiMate Nano LC system coupled to a Q Exactive HF Orbitrap mass spectrometer or a Fusion Lumos Tribid Orbitrap mass spectrometer. An Acclaim PepMap 100 C18 trap column is used for online desalting of the samples.

3.9.1 LC Condition

NanoLC Conditions (Peptides and Glycopeptides)

1. Resuspend the sample to a final concentration of 0.5–1 ug/uL.
2. The injection amount can be 1 ug or 2 ug depending on a good total ion count intensity.
3. Adjust column oven temperature to 29.5 °C.
4. Glycopeptides are separated on a reversed-phased C18 Acclaim PepMap 100 Å capillary column (150 mm × 75 μm).
5. A multistep gradient with total elution time of 120 min and a flow rate of 0.350 uL/min is applied starting at 5% MPB for 10 min, with a gradual increase to 30% over 75 min, and to 50% over 13 min. Then, ramp up to 90% MPB in 2 min and keep constant for 5 min to wash the column. Finally, decrease to 5% B in 1 min and keep it constant for 14 min to equilibrate the column.

NanoLC Conditions (Permethylated Glycans)

1. Resuspend the sample to a final concentration of 1 ug/uL.
2. Adjust column oven temperature to 55 °C.
3. Permethylated glycans are separated on a reversed-phased C18 Acclaim PepMap 100 Å capillary column (150 mm × 75 μm).
4. A multistep gradient with a total elution time of 60 min is applied for the glycan sample: starting at 20% MPB for 10 min, gradually increasing to 42% over 40 min, and to 95% over 60 min. Finally, decrease to 20% MPB in 1 min and keep it constant for 9 min to equilibrate the column.

3.9.2 MS Condition

1. The following data acquisition parameters can be set up on Orbitrap MS systems like the Fusion Lumos Tribrid MS or the Q Exactive HF Hybrid Quadrupole Orbitrap MS:
2. The nanoESI source voltage is set to 1.6 kV in positive ion mode.
3. The ion transfer tube temperature is set to 275 °C.
4. The data is acquired in the data-dependent acquisition (DDA) mode with the following scan parameters:
5. Full MS scan set to a resolution of 120 K.
6. Scan range set to 400–2000 m/z.
7. An AGC target value of $1e^6$.
8. Maximum injection time of 50 ms.
9. Detector type set to Orbitrap.
10. For the tandem MS scan, the top 20 most intense precursor ions are selected for higher energy collision dissociation (HCD) and the following scan parameters are used:
11. Resolution of 30 K.
12. An AGC target value of $1e^5$.
13. Maximum injection time of 50 ms.
14. Detector type set to Orbitrap.
15. Charge states 2–8 are included.
16. Isolation window of 1.6 m/z.
17. Dynamic exclusion set to 15 s.
18. Collision energy set to stepped NCEs 15, 30, 45.

3.9.3 Data Processing

The proteomics LC–MS/MS raw data files are initially processed using Proteome Discoverer™ 2.5 software (Thermo Scientific) for the identification and quantification of proteins present in the exosome fractions. The search is performed against a focused database of the top 100 exosome marker proteins curated from the ExoCarta and Vesiclepedia repository. The following search parameters are utilized with the SequestHT search engine:

1. Cleavage sites, RK; cleavage side, C-terminal; digestion specificity, fully specific; missed cleavages, 2; precursor mass tolerance ± 10 ppm; fragment mass tolerance ± 0.02 Da.
2. Carbamidomethylation of cysteine as fixed modification, oxidation of methionine, and acetylation of the protein N-terminal as variable modification.
3. The results are filtered to exclude proteins identified by a single peptide and to include peptides and proteins within an FDR cutoff of 1%.

For the glycoproteomics data processing, raw LC–MS/MS data are processed using Byonic™ software (Protein Metrics by Dotmatics). The software search parameters are set as follows:

1. Cleavage sites, RK; cleavage side, C-terminal; digestion specificity, fully specific; missed cleavages, 2; precursor mass tolerance ± 10 ppm; fragment mass tolerance ± 20 ppm; fragmentation type, QTOF/HCD; recalibration (lock mass), none.
2. Fixed modification—carbamidomethylation (C); variable modifications—oxidation (M) and acetylation (protein N-term).
3. Glycan modifications and protein database are chosen based on the sample type.

After completing the Byonic™ run, the result is analyzed and filtered using Byos Software™. The parameters used for the filtering are given thus: 2D FDR < 0.01; PEP 2D < 0.01; |LogProb| > 1; score > 300. Results without glycan modifications are filtered out. After the filtering step, a manual data processing using XCalibur™ Qual Browser is performed to verify the extracted ion chromatogram (EIC) and to quantify each identified glycopeptide. The check is conducted within a mass tolerance of ± 10 ppm and a Gaussian smoothing of 11 points. The peak area of the EIC is utilized as the abundance of the corresponding glycopeptide (*see* Note 15).

4 Notes

1. EV-depleted fetal bovine serum can be prepared by centrifuging FBS at 100,000 × g and 4 °C for 18 h, and then sterilizing the supernatant using a 0.22 μm filter. Commercial preparations of EV-depleted fetal bovine serum also exist as an alternative. It is important to use the EV-depleted fetal bovine serum when culturing the cells to harvest exosomes because fetal bovine serum (FBS) contains its own EVs along with other components that support cell growth. These FBS-derived EVs may produce artifacts in the experiment because they are difficult to distinguish from the EVs of interest [41]. EV-depleted media helps eliminate this contamination, allowing more focus on the exosomes secreted by the cell of interest.
2. All solvents must be HPLC grade. Mobile phases A and B should be degassed after preparation and stored at room temperature.

3. Cells are highly sensitive to their environment, thus growing cells in normal media composition prior to the exosome media would help adapt them to the growth conditions and minimize metabolic stress.
4. The introduction of the ED-FBS could impact the doubling time of cells, thus investigate cells frequently and change media to remove dead and floating cells.
5. Culture media and all devices used for exosome extraction must be kept on ice. Carrying out the protocol on ice would minimize protein degradation and ensure extraction of intact exosomes.
6. Care must be taken when transferring the supernatants from steps 1–3 to avoid disturbing the pellets which may increase the risk of contamination.
7. If a fixed-angle rotor is used in the ultracentrifugation step, it could help to make a tiny mark on the part of the ultracentrifuge tube facing outward to aid in the visualization of the exosome pellets. Also, ultracentrifuge tubes must be filled to sufficient volume and properly balanced in the ultracentrifuge to prevent tube collapse which may lead to low exosome recovery. Ice-cold PBS can be used to top up the tubes to a reasonable volume, usually beyond half of the total volume of the tube.
8. The pelleted exosomes can be washed multiple times as necessary to obtain a purer exosome fraction that is relatively free from protein and other contamination. However, caution must be exercised when deciding on the number of washing steps to perform. It is essential to optimize the washing steps to balance the need for purity with the preservation of exosome integrity. Excess washing presents some deleterious outcomes, including a reduction in the total exosome yield due to disruption or sample loss. Each centrifugation and resuspension cycle presents an opportunity for exosomes to adhere to surfaces, aggregate, or become damaged, leading to decreased yield and purity of the isolated exosomes.
9. The BeadBug microtube homogenizer should be maintained at 4 °C to prevent unwanted protein degradation in the sample. Also, the homogenization process should be conducted in at least five cycles with each cycle consisting of 30 s of beads beating with 30 s intervals.
10. Sonication enhances the efficiency of the overall lysis procedure and improves protein dissolution.
11. Total protein content can be approximated with varying degrees of sensitivity and accuracy using colorimetric assays, fluorometric assays, or by simply taking absorbance readings. Depending on the yield, the sensitivity of protein assays may be challenged. Therefore, it is recommended to not dilute the protein extract when carrying out the BCA assay. Also, the protein concentration should lie within the linear range of the calibration curve to ensure accuracy of measurement.
12. DTT solution should be prepared freshly and kept in the dark. The amount of DTT solution added should be 1/40 of the sample volume.
13. IAA solutions should also be prepared freshly and kept in the dark. The amount of IAA should be four times the volume of DTT.

14. The ideal pH for enzymatic digestion is around 8. If the pH is below 8, adjust it by adding a little amount of 100 mM ABC buffer solution.
15. Other software like Skyline (MacCoss Lab Software) can also be employed for the manual verification and quantification of glycopeptide structures.

Acknowledgements This work is supported by grants from the National Institutes of Health, NIH, including 1R01GM112490-08 (YM), 1R01GM130091-05 (YM), and 1U01CA225753-05 (YM/DML). The work is also supported by grants from the Robert A. Welch Foundation No. D-0005 and The CH Foundation (YM). The work was also supported by the National Cancer Institute (NCI) grant numbers 3R01CA160254 (DML), 3R01CA160254-S1 (DML), and 1R01CA258240 (DML).

Competing Interests The authors have no conflicts of interest to declare that are relevant to the content of this chapter.

References

1. Maas, S. L. N., Breakefield, X. O., & Weaver, A. M. (2017). Extracellular vesicles: Unique intercellular delivery vehicles. *Trends in Cell Biology, 27*(3), 172–188. https://doi.org/10.1016/j.tcb.2016.11.003 From NLM.
2. Raposo, G., & Stoorvogel, W. (2013). Extracellular vesicles: Exosomes, microvesicles, and friends. *Journal of Cell Biology, 200*(4), 373–383. https://doi.org/10.1083/jcb.201211138
3. Di Bella, M. A. (2022). Overview and update on extracellular vesicles: Considerations on exosomes and their application in modern medicine. *Biology (Basel), 11*(6). https://doi.org/10.3390/biology11060804 From NLM.
4. Gallo, A., Tandon, M., Alevizos, I., & Illei, G. G. (2012). The majority of microRNAs detectable in serum and saliva is concentrated in exosomes. *PLoS ONE, 7*(3), Article e30679. https://doi.org/10.1371/journal.pone.0030679 From NLM.
5. Blans, K., Hansen, M. S., Sørensen, L. V., Hvam, M. L., Howard, K. A., Möller, A., Wiking, L., Larsen, L. B., & Rasmussen, J. T. (2017). Pellet-free isolation of human and bovine milk extracellular vesicles by size-exclusion chromatography. *Journal of Extracellular Vesicles, 6*(1), 1294340. https://doi.org/10.1080/20013078.2017.1294340 From NLM.
6. Clayton, A., Boilard, E., Buzas, E. I., Cheng, L., Falcón-Perez, J. M., Gardiner, C., Gustafson, D., Gualerzi, A., Hendrix, A., Hoffman, A., et al. (2019). Considerations towards a roadmap for collection, handling and storage of blood extracellular vesicles. *Journal of Extracellular Vesicles, 8*(1), 1647027. https://doi.org/10.1080/20013078.2019.1647027 From NLM.
7. Iwai, K., Minamisawa, T., Suga, K., Yajima, Y., & Shiba, K. (2016). Isolation of human salivary extracellular vesicles by iodixanol density gradient ultracentrifugation and their characterizations. *Journal of Extracellular Vesicles, 5*, 30829. https://doi.org/10.3402/jev.v5.30829 From NLM.
8. Kalani, M. Y. S., Alsop, E., Meechoovet, B., Beecroft, T., Agrawal, K., Whitsett, T. G., Huentelman, M. J., Spetzler, R. F., Nakaji, P., Kim, S., et al. (2020). Extracellular microRNAs in blood differentiate between ischaemic and haemorrhagic stroke subtypes. *Journal of Extracellular Vesicles, 9*(1), 1713540. https://doi.org/10.1080/20013078.2020.1713540 From NLM.
9. Chen, P., Ruan, A., Zhou, J., Huang, L., Zhang, X., Ma, Y., & Wang, Q. (2020). Extraction and identification of synovial tissue-derived exosomes by different separation techniques. *Journal of Orthopaedic Surgery and Research, 15*(1), 97. https://doi.org/10.1186/s13018-020-01604-x
10. Sheller-Miller, S., & Menon, R. (2020). Chapter Ten–Isolation and characterization of human amniotic fluid-derived exosomes. In *Methods in Enzymology*, S. Spada, L. Galluzzi (Eds.), (vol. 645, pp 181–194). Academic Press

11. Brown, C. J., Gaunitz, S., Wang, Z., Strindelius, L., Jacobson, S. C., Clemmer, D. E., Trinidad, J. C., & Novotny, M. V. (2020). Glycoproteomic analysis of human urinary exosomes. *Analytical Chemistry, 92*(21), 14357–14365. https://doi.org/10.1021/acs.analchem.0c01952
12. Li, M., Huang, L., Chen, J., Ni, F., Zhang, Y., & Liu, F. (2021). Isolation of exosome nanoparticles from human cerebrospinal fluid for proteomic analysis. *ACS Applied Nano Materials, 4*(4), 3351–3359. https://doi.org/10.1021/acsanm.0c02622
13. Robbins, P. D., Dorronsoro, A., & Booker, C. N. (2016). Regulation of chronic inflammatory and immune processes by extracellular vesicles. *The Journal of Clinical Investigation, 126*(4), 1173–1180. https://doi.org/10.1172/jci81131 From NLM.
14. Caruso, S., & Poon, I. K. H. (2018). Apoptotic cell-derived extracellular vesicles: more than just debris. *Frontiers in Immunology, 9*, 1486. https://doi.org/10.3389/fimmu.2018.01486 From NLM.
15. Veerman, R. E., Güçlüler Akpinar, G., Eldh, M., & Gabrielsson, S. (2019). Immune cell-derived extracellular vesicles–functions and therapeutic applications. *Trends in Molecular Medicine, 25*(5), 382–394. https://doi.org/10.1016/j.molmed.2019.02.003 From NLM.
16. Lema, D. A., & Burlingham, W. J. (2019). Role of exosomes in tumour and transplant immune regulation. *Scandinavian Journal of Immunology, 90*(5), Article e12807. https://doi.org/10.1111/sji.12807 From NLM.
17. Luo, T., Kang, Y., Liu, Y., Li, J., & Li, J. (2023). Small extracellular vesicles in breast cancer brain metastasis and the prospect of clinical application. *Frontiers in Bioengineering and Biotechnology, 11*, 1162089. https://doi.org/10.3389/fbioe.2023.1162089 From NLM.
18. Doyle, L. M., & Wang, M. Z. (2019). Overview of extracellular vesicles, their origin, composition, purpose, and methods for exosome isolation and analysis. *Cells, 8*(7). https://doi.org/10.3390/cells8070727 From NLM.
19. Schey, K. L., Luther, J. M., & Rose, K. L. (2015). Proteomics characterization of exosome cargo. *Methods, 87*, 75–82. https://doi.org/10.1016/j.ymeth.2015.03.018 From NLM.
20. An, M., Lohse, I., Tan, Z., Zhu, J., Wu, J., Kurapati, H., Morgan, M. A., Lawrence, T. S., Cuneo, K. C., & Lubman, D. M. (2017). Quantitative proteomic analysis of serum exosomes from patients with locally advanced pancreatic cancer undergoing chemoradiotherapy. *Journal of Proteome Research, 16*(4), 1763–1772. https://doi.org/10.1021/acs.jproteome.7b00024 From NLM.
21. Sun, R., Wang, H., Shi, Y., Gao, D., Sun, Z., Chen, Z., Jiang, H., & Zhang, J. (2019). A pilot study of urinary exosomes in Alzheimer's disease. *Neuro-Degenerative Diseases, 19*(5–6), 184–191. https://doi.org/10.1159/000505851 From NLM.
22. Rajendran, L., Honsho, M., Zahn, T. R., Keller, P., Geiger, K. D., Verkade, P., & Simons, K. (2006). Alzheimer's disease beta-amyloid peptides are released in association with exosomes. *Proceedings of the National Academy of Sciences U S A, 103*(30), 11172–11177. https://doi.org/10.1073/pnas.0603838103 From NLM.
23. Han, C., Xiong, N., Guo, X., Huang, J., Ma, K., Liu, L., Xia, Y., Shen, Y., Li, J., Jiang, H., et al. (2019). Exosomes from patients with Parkinson's disease are pathological in mice. *Journal ofMolecular Medicine (Berlin, Germany), 97*(9), 1329–1344. https://doi.org/10.1007/s00109-019-01810-z From NLM.
24. Chang, C., Lang, H., Geng, N., Wang, J., Li, N., & Wang, X. (2013). Exosomes of BV-2 cells induced by alpha-synuclein: Important mediator of neurodegeneration in PD. *Neuroscience Letters, 548*, 190–195. https://doi.org/10.1016/j.neulet.2013.06.009 From NLM.
25. Chen, P. C., Wu, D., Hu, C. J., Chen, H. Y., Hsieh, Y. C., & Huang, C. C. (2020). Exosomal TAR DNA-binding protein-43 and neurofilaments in plasma of amyotrophic lateral sclerosis patients: A longitudinal follow-up study. *Journal of the Neurological Sciences, 418*, Article 117070. https://doi.org/10.1016/j.jns.2020.117070 From NLM.
26. Gomes, C., Keller, S., Altevogt, P., & Costa, J. (2007). Evidence for secretion of Cu, Zn superoxide dismutase via exosomes from a cell model of amyotrophic lateral sclerosis. *Neuroscience Letters, 428*(1), 43–46. https://doi.org/10.1016/j.neulet.2007.09.024 From NLM.
27. Williams, C., Pazos, R., Royo, F., González, E., Roura-Ferrer, M., Martinez, A., Gamiz, J., Reichardt, N. C., & Falcón-Pérez, J. M. (2019). Assessing the role of surface glycans of

extracellular vesicles on cellular uptake. *Science and Reports, 9*(1), 11920. https://doi.org/10.1038/s41598-019-48499-1 From NLM.

28. Shimoda, A., Sawada, S.-I., Sasaki, Y., & Akiyoshi, K. (2019). Exosome surface glycans reflect osteogenic differentiation of mesenchymal stem cells: Profiling by an evanescent field fluorescence-assisted lectin array system. *Scientific Reports, 9*(1), 11497. https://doi.org/10.1038/s41598-019-47760-x
29. Wu, L., & Gao, C. (2023). Comprehensive overview the role of glycosylation of extracellular vesicles in cancers. *ACS Omega, 8*(50), 47380–47392. https://doi.org/10.1021/acsomega.3c07441
30. Yamamoto, A., Yasue, Y., Takahashi, Y., & Takakura, Y. (2021). Determining the role of surface Glycans in the pharmacokinetics of small extracellular vesicles. *Journal of Pharmaceutical Sciences, 110*(9), 3261–3267. https://doi.org/10.1016/j.xphs.2021.05.021
31. Becker, A., Thakur, B. K., Weiss, J. M., Kim, H. S., Peinado, H., & Lyden, D. (2016). Extracellular vesicles in cancer: Cell-to-cell mediators of metastasis. *Cancer Cell, 30*(6), 836–848. https://doi.org/10.1016/j.ccell.2016.10.009 From NLM.
32. Lin, S., Zhou, S., & Yuan, T. (2020). The, "sugar-coated bullets" of cancer: Tumor-derived exosome surface glycosylation from basic knowledge to applications. *Clinical and Translational Medicine, 10*(6), Article e204. https://doi.org/10.1002/ctm2.204
33. Xiao, L., Hareendran, S., & Loh, Y. P. (2021). Function of exosomes in neurological disorders and brain tumors. *Extracellular Vesicles and Circulating Nucleic Acids, 2*, 55–79. https://doi.org/10.20517/evcna.2021.04 From NLM.
34. Saeedi, S., Israel, S., Nagy, C., & Turecki, G. (2019). The emerging role of exosomes in mental disorders. *Translational Psychiatry, 9*(1), 122. https://doi.org/10.1038/s41398-019-0459-9
35. Zhang, N., He, F., Li, T., Chen, J., Jiang, L., Ouyang, X.-P., & Zuo, L. (2021). Role of exosomes in brain diseases. *Frontiers in Cellular Neuroscience, 15*, Review. https://doi.org/10.3389/fncel.2021.743353
36. Jiang, Y., Chai, X., Chen, S., Chen, Z., Tian, H., Liu, M., & Wu, X. (2022). Exosomes from the uterine cavity mediate immune dysregulation via inhibiting the JNK signal pathway in endometriosis. *Biomedicines, 10*(12). https://doi.org/10.3390/biomedicines10123110 From NLM.
37. Schwarzenbach, H., & Gahan, P.B. (2021). Exosomes in immune regulation. *Noncoding RNA, 7*(1). https://doi.org/10.3390/ncrna7010004 From NLM.
38. Cao, H., Abd Aziz, N. H., Xavier, J. R., Shafiee, M. N., Kalok, A., Jee, B., Salker, M. S., & Singh, Y. (2022). Dysregulated exosomes result in suppression of the immune response of pregnant COVID-19 convalescent women. *Frontiers in Molecular Biosciences, 9*, Brief Research Report. https://doi.org/10.3389/fmolb.2022.869192.
39. Huda, M. N., Nafiujjaman, M., Deaguero, I. G., Okonkwo, J., Hill, M. L., Kim, T., & Nurunnabi, M. (2021). Potential use of exosomes as diagnostic biomarkers and in targeted drug delivery: Progress in clinical and preclinical applications. *ACS Biomaterials Science & Engineering, 7*(6), 2106–2149. https://doi.org/10.1021/acsbiomaterials.1c00217 From NLM.
40. Kang, P., Mechref, Y., Klouckova, I., & Novotny, M. V. (2005). Solid-phase permethylation of glycans for mass spectrometric analysis. *Rapid Communications in Mass Spectrometry, 19*(23), 3421–3428. https://doi.org/10.1002/rcm.2210 From NLM.
41. Urzì, O., Olofsson Bagge, R., & Crescitelli, R. (2022). The dark side of foetal bovine serum in extracellular vesicle studies. *Journal of Extracellular Vesicles, 11*(10), Article e12271. https://doi.org/10.1002/jev2.12271 From NLM.

Characterization of Post-translationally Modified Proteins in Extracellular Vesicles

Jyoti Singh, Marco Hadisurya, Zhuojun Luo, and W. Andy Tao

1 Introduction

Extracellular vesicles (EVs) are lipid-bilayer-bound particles released from cells into the extracellular space. They are heterogeneous in size and composition and can be classified majorly into small extracellular vesicles (sEVs) and large extracellular vesicles (lEVs) which are further classified into various subtypes, including exosomes, microvesicles, and apoptotic bodies, based on their biogenesis, size, and content [39]. The biogenesis of EVs involves a series of tightly regulated processes, initiating with the contribution of intracellular membranes, such as those of the endoplasmic reticulum and Golgi apparatus, to the formation of early endosomes. These early endosomes mature into multivesicular bodies (MVBs) through the inward budding of the endosomal membrane, which encapsulates various bioactive molecules. MVBs subsequently fuse with the plasma membrane, leading to the release of small vesicles known as exosomes into the extracellular space. Alternatively, microvesicles are formed directly from the outward budding and fission of the plasma membrane. The precise mechanisms and regulatory factors controlling EV biogenesis are crucial for determining the composition and functional roles of the released vesicles [24, 39]. EVs are pivotal in cellular communication, serving as carriers of a diverse array of bioactive molecules, including proteins, lipids, and nucleic acids. By mediating the transfer of these molecular cargos between cells, EVs facilitate the modulation of cellular processes and signaling pathways, influencing immune responses, tissue homeostasis, and developmental programs [7]. They play critical roles in numerous physiological and pathological processes, including immune responses,

J. Singh · M. Hadisurya · Z. Luo · W. A. Tao (✉)
Purdue University, West Lafayette, IN, USA
e-mail: taow@purdue.edu

J. Singh
Department of Biochemistry, The Maharaja Sayajirao University of Baroda, Gujarat 390002, Vadodara, India

R. K. Marcus and D. Lubman (eds.), *Exosomes and Extracellular Vesicles*,
https://doi.org/10.1007/978-3-031-89270-7_13

tumor progression, and tissue repair. In the context of disease progression, particularly in cancer, EVs can alter the tumor microenvironment, promote metastasis, and mediate resistance to therapies by delivering oncogenic factors and modifying recipient cell behaviors [98]. Additionally, EVs hold significant promise in biomarker discovery, as their molecular content reflects the physiological and pathological states of their cells of origin. This makes them valuable for non-invasive diagnostics and monitoring of disease progression, offering insights into disease mechanisms and aiding in the development of personalized therapeutic strategies.

Post-translational modifications (PTMs) involve biochemical alterations that occur on proteins after their synthesis. These modifications arise through the covalent addition or proteolytic cleavage and are essential for modulating protein conformation, subcellular distribution, and molecular interactions [72]. Regulatory factors such as enzymatic machinery, cellular signaling pathways, substrate specificity, cellular context, and protein binding interactions influence the decision of a protein to undergo PTM [64]. Notably, PTMs diversify protein roles, facilitate signaling processes, ensure protein quality, regulate gene expression, and enable adaptation to environmental changes. PTMs significantly enhance the functional diversity and complexity of the proteome, which is essential for maintaining cellular homeostasis and ensuring proper physiological functions [63]. Enzymes catalyze specific modifications based on cellular cues and the structural features of target proteins. Cells harbor a diverse array of enzymes responsible for these modifications, including kinases, phosphatases, acetyltransferases, methyltransferases, ubiquitin ligases, etc. [3]. The activity and specificity of these enzymes play a crucial role in determining which proteins receive modifications, thereby influencing their biological role. For example, protein kinases phosphorylate target proteins at specific serine, threonine, or tyrosine residues, while acetyltransferases add acetyl groups to lysine residues. Signaling pathways regulate enzyme activity in response to stimuli (including growth factors, hormones, environmental stressors, and changes in nutrient availability), and these signaling pathways ensure that enzymes modify the right proteins by recognizing specific sequences or structures in those proteins [4]. Additionally, changes in cellular metabolism, stress responses, or developmental processes can alter PTM patterns in response to changing cellular needs [83]. Proteins with specific amino acid residues or involved in key cellular processes, such as signaling, transcriptional regulation, and cellular cross-talk, are particularly prone to PTMs, which modulate their activity, localization, and interactions [97]. Dysregulation of PTMs is implicated in diseases like cancer, neurodegenerative disorders, and other pathologies like inflammatory, autoimmune, and cardiovascular diseases [17, 30, 82]. Understanding PTMs is vital for targeted therapies against several diseases and can lead to new therapeutic targets and biomarkers, offering potential advancements in disease diagnosis and treatment. These PTMs extend beyond cellular boundaries, as they are present in EVs released by cells, facilitating intercellular communication and the transfer of bioactive molecules (Fig. 1).

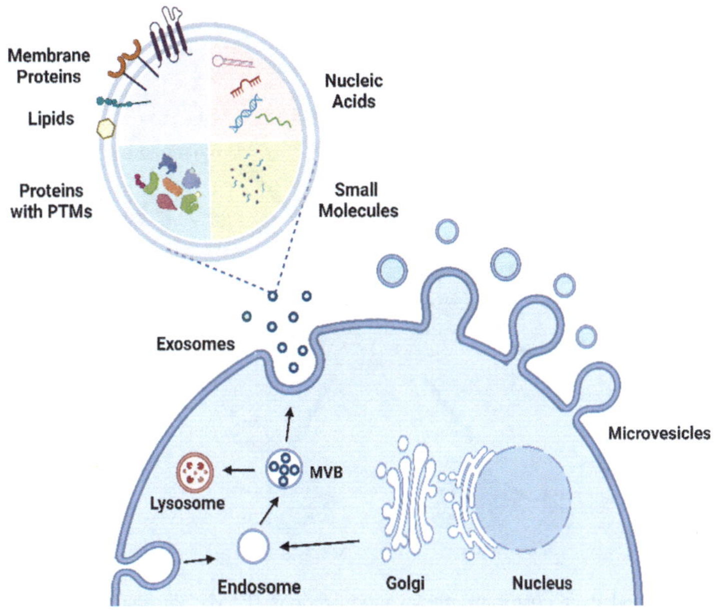

Fig. 1 Packaging of post-translational modified proteins into extracellular vesicles for intercellular communication

2 Post-Translational Modifications in Extracellular Vesicles

Common types of PTM include phosphorylation, glycosylation, ubiquitination, acetylation, methylation, sumoylation, palmitoylation, nitrosylation, and neddylation, each of which involves specific chemical modifications at distinct sites within the protein (Fig. 2). It occurs at various cellular locations, influencing protein function and interactions. Phosphorylation, acetylation, methylation, and ubiquitination regulate cytoplasmic proteins in the cytosol. The nucleus hosts modifications such as histone methylation, acetylation, and phosphorylation, affecting gene expression and chromatin structure. Membrane-associated proteins undergo palmitoylation, myristoylation, and glycosylation, often processed in the ER and Golgi apparatus before reaching the cell membrane. Mitochondria experience PTMs like phosphorylation and acetylation, critical for metabolism and apoptosis. PTMs also occur at the nuclear pore complex, regulating nucleocytoplasmic transport, and in specialized compartments like nuclear bodies, stress granules, and lipid rafts, depending on the specific enzymes and substrates involved.

PTMs in EVs are pivotal in determining their function and biological impact and play a pivotal role in EV biology by influencing EV formation, cargo selection, target and function. Phosphorylation, glycosylation, lipid modifications, and ubiquitination are essential for the regulation of these processes. For instance, phosphorylation of Annexin A2 (AnxA2) at Tyr-23 is essential for its recruitment to plasma membrane

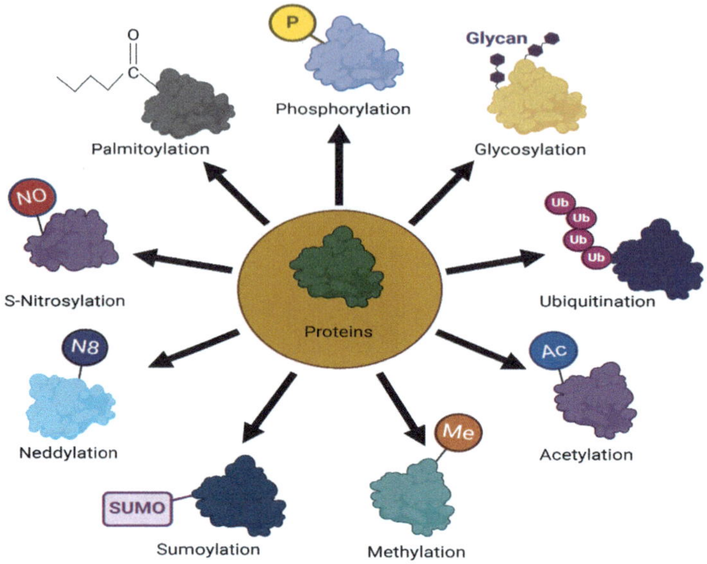

Fig. 2 Proteins and their post-translational modifications (PTM): Phosphorylation, Glycosylation, Ubiquitination, Acetylation, Methylation, Sumoylation, Neddylation, Palmitoylation, and S-Nitrosylation

rafts and subsequent incorporation into intraluminal vesicles (ILVs) within MVBs, which are precursors to EVs [78]. Additionally, proteomic analyses have revealed a high abundance of phosphoproteins in EV cargo, with studies identifying over 10,000 phosphopeptides in plasma-derived EVs [20, 35, 67]. This indicates that phosphorylation affects not only the targeting of proteins to EVs but also the selection of cargo within them. Glycosylation affects protein stability and the ability to interact with cellular receptors, thereby impacting the selective sorting of proteins and lipids into EVs. For instance, the lectin microarray-based profiling of EVs and the parent cell membrane of SkMel-5 has shown an enrichment of specific glycans, such as mannose, α-2,6 sialic acid, polylactosamine, and complex N-linked glycans, in EVs compared to the parent cell membrane [6]. Lipid modifications, such as palmitoylation and myristoylation, are crucial for anchoring proteins to the EV membrane, affecting both the inclusion of these proteins in EVs and the functional properties of the EVs themselves. It has been reported that the incorporation of retroviral Gag proteins into EVs is facilitated by the N-terminal myristoylation of Gag, which anchors the protein to the inner leaflet of the plasma membrane, thus enabling its subsequent incorporation into EVs [9, 26]. Similarly, the role of palmitoylation in EV biology has been highlighted by Verweij et al., who demonstrated that the palmitoylation of LAMP1 at the conserved C78 site is critical for its sorting into EVs [79]. Ubiquitination often tags proteins for degradation or alters their interaction with other cellular components, influencing the types of proteins incorporated into EVs and their subsequent biological activities. It has also been recognized as a crucial PTM for the sorting of proteins into ILVs within MVBs. A proteomic analysis revealed that ubiquitinated proteins constitute approximately 10% of the protein cargo in EVs derived from myeloid-derived suppressor cells, with a significant presence of proinflammatory histones and other proinflammatory mediators. This highlights the integral role of ubiquitination in determining the composition and characteristics of EVs, particularly in the context of immune modulation and inflammatory responses [10]. Collectively, these PTMs ensure that EVs are properly formed, selectively loaded with cargo, and equipped to mediate various physiological and pathological processes, such as cell–cell communication, immune responses, and cancer progression. This dynamic regulation by PTMs is critical for understanding how EVs function and their potential as therapeutic targets and biomarkers.

3 Overall Strategies for Characterizing PTMs in EVs

- **Isolation and Purification Methods of EVs:**

The initial and most critical step in the characterization of PTMs in EVs is the efficient isolation of EVs from biofluids as it ensures that the sample is pure and representative of the true EV population. There are different methods that have been developed and are in use for EV isolation such as differential ultracentrifugation (UC), size-exclusion chromatography (SEC), polymer precipitation, sucrose density gradient, and affinity purification. Each method has its advantages and drawbacks regarding efficiency, purity, isolation time, cost, and instrument requirements [74, 76].

Differential ultracentrifugation is a widely used method for isolating EVs by separating them from contaminants based on size and density by applying a series of progressively higher centrifugal forces. This technique involves multiple rounds of centrifugation, beginning with an initial spin at 300 rcf for 10 min, followed by centrifugation at 2,000 rcf for 10 min to remove cells, cellular debris, and large vesicles. The supernatant is then centrifuged at 10,000 rcf for 70 min at 4 °C to further eliminate large vesicles. Finally, ultracentrifugation at 100,000 rcf for 70 min at 4 °C to pellet exosomes [48]. Although effective in purifying EVs, this method is time-consuming and may lead to EV aggregation or damage. Furthermore, due to the similar sedimentation properties of various extracellular vesicle subtypes, differential ultracentrifugation often yields low quantities of exosomes with insufficient purity [73]. While separating exosomes from cells, cell debris, and large vesicles is relatively straightforward due to their size differences, the major challenge lies in purifying exosomes from small shedding vesicles, which are of similar size. As a result, differential centrifugation is often inefficient and ineffective for separating these closely sized vesicles.

Size exclusion chromatography (SEC) is a gentle and effective method for isolating intact extracellular vesicles based on size as they pass through a column filled with porous beads. This technique utilizes a stationary phase, typically composed of gel polymers, through which the mobile phase flows and is subsequently eluted. This molecular sieve effect separates molecules based on their Stokes radii, resulting in the earlier elution of larger particles, while smaller molecules diffuse through the pores of the resin and are eluted later [75]. By loading samples onto columns with porous beads, EVs elute earlier than other contaminants like protein aggregates and lipoproteins, yielding high-purity fractions suitable for PTM analysis [56]. However, SEC may not completely remove all contaminants, and recovery can vary with column properties and sample composition.

Immunoprecipitation (IP) can selectively target specific proteins on the surface of extracellular vesicles (EVs), enabling the isolation of distinct EV subpopulations. IP with organ-specific antibodies, such as L1CAM for neuronal EVs, offers precise enrichment of EVs from specific tissues. This method involves initial centrifugation to remove debris, followed by ultracentrifugation to pellet EVs. The resuspended EVs are then incubated with antibodies against markers like L1CAM, and captured using beads conjugated with secondary antibodies. After washing, the EVs are eluted for downstream analyses like proteomics or RNA sequencing, enabling targeted study of EVs' roles in health and disease [15]. Similarly, GPC-1 is used for isolating pancreatic cell-derived EVs, crucial for pancreatic disease research (Melo et al., 2015). IP has a risk of non-specific binding, which can affect the purity and yield of the isolated EVs. On the other hand, high selectivity in this process may also result in lower yields, which can lead to incomplete coverage of EV cargo and a loss of valuable biological information.

Attempting to address the shortcomings of UC and SEC, chemical affinity-based solid phase isolation of EVs has been explored. EVtrap (extracellular vesicle total recovery and purification) beads use chemical affinity-based methods to provide an efficient, easy and rapid approach for EV isolation from various biological samples, including cell culture media, plasma, and urine (Charles Jacob et al., 2022) [35]. These magnetic beads are functionalized with aromatic lipophilic and hydrophilic groups, having a high affinity for EV lipid bilayers. The process involves incubation of EVtrap beads with samples for 30 min, in contrast to hours in UC. After bead separation, EVs are eluted with a volatile reagent (e.g., triethylamine) to disrupt the interaction between the beads and EVs, and dried in a vacuum centrifuge for downstream analysis [87, 29]. EVtrap beads have been shown to significantly outperform ultracentrifugation and other standard exosome isolation methods by capturing higher levels of EV markers from plasma, as confirmed by western blotting, silver staining, and LC–MS analyses [35]. This technology based on non-antibody affinity beads routinely identifies over 1,000 unique proteins from EVs from only 0.01 mL of plasma or 0.2 mL of urine, a significant improvement over any prior data [87].

It is a widely recognized fact that EV cargo analysis in particular proteome analysis is highly susceptible to contamination from high-abundance plasma or serum proteins, lipids, and reagents used in polymer-assisted precipitation techniques [85]. Consequently, some EV isolation methods may be suitable for nucleic acid sequencing or immunoassays but not for MS-based analysis.

Automation significantly enhances the isolation of EVs by improving efficiency, reproducibility, and scalability. Automated systems, such as Izon's qEV platform, Millipore Sigma's ExoEasy kits, and Thermo Fisher Scientific's KingFisher Flex, standardize protocols, reducing variability and human error, and enabling high-throughput processing essential for large-scale studies. They decrease hands-on time, minimizing contamination risks and freeing researchers from other tasks. Automation optimizes isolation conditions, improving EV purity and yield, and can integrate seamlessly with downstream analytical techniques. Our lab has established an automated EV isolation workflow from 1 mL urine sample [44, 95]. This scalability is crucial for clinical applications, ensuring consistent and compliant production of diagnostic and therapeutic EVs. Advances in microfluidics and lab-on-a-chip technologies further enhance these automated platforms, making EV isolation more precise and efficient.

- **Protein Extraction from EVs:**

It is well-established that EV membranes are derived from the cell membrane, allowing the application of standard cell lysis and protein extraction techniques. However, it is essential to use a systematic approach that preserves protein integrity and ensures comprehensive recovery. The initial step involves the lysis of the vesicle membranes, which can be achieved using specialized lysis buffers containing a balanced composition of detergents, salts, and protease inhibitors. This combination not only disrupts the lipid bilayer of EVs but also stabilizes the proteins, preventing them from degradation. The workflow for extracting EV proteins utilizes the phase-transfer surfactant (PTS) protocol for efficient lysis and protein extraction. The PTS buffer includes sodium deoxycholate, sodium lauroyl sarcosinate, Tris–Cl, tris-(2-carboxyethyl) phosphine for reduction, chloroacetamide for alkylation, and a protease inhibitor that is essential to prevent proteolytic degradation during the extraction process, ensuring that the protein content reflects the in vivo state as closely as possible. Each sample received PTS buffer, boiled at 95 °C in the dark, and then diluted with TEAB for digestion. A BCA assay is determined for the estimation of protein amount. Using sodium deoxycholate and N-lauroyl sarcosine sodium salt as surfactants improves peptide identification with PTMs and enhances enzymatic activity, reducing missed cleavages.

Alternative methods such as ultrasonication can also be employed, where high-frequency ultrasonic waves facilitate the disruption of vesicle membranes, thereby releasing the enclosed proteins. This technique is particularly useful for samples requiring gentle handling to maintain the native structure of proteins. Following lysis, the extracted proteins can be concentrated and further purified, if necessary, using techniques such as ultrafiltration or precipitation.

This meticulous extraction process is pivotal for subsequent downstream applications, including mass spectrometry analysis, immunoblotting, or other proteomic investigations, ensuring that the EV-derived proteome is accurately represented and amenable to detailed characterization.

- **Characterization of PTMs in EVs:**

Mass spectrometry-driven proteomic approaches have been instrumental in characterizing the molecular composition of small extracellular vesicles from various tissues and cells, leading to the identification of thousands of proteins. Beyond the protein content, MS has also enabled the detection of several post-translational modifications and is generally considered one of the best strategies for characterization due to its unparalleled sensitivity, specificity, and ability to provide detailed molecular information [19]. MS enables the precise identification of PTMs by detecting the mass shifts that occur when modifications are added to proteins or peptides [25]. Through tandem mass spectrometry, researchers can pinpoint the exact sites of modifications within the protein sequence, offering insights into the structural and functional consequences of these modifications [53]. The high resolution and accuracy of mass spectrometry enable the detection of low-abundance PTMs of EVs and the quantification of modified versus unmodified peptides, facilitating a comprehensive analysis of the PTM landscape. When the stoichiometry of modified peptides is low, fractionation and enrichment techniques are often employed before MS to enhance detection sensitivity and yields [25]. Furthermore, MS can handle complex biological samples, making it a versatile tool for studying dynamic changes in PTMs across different conditions and stages of disease. Its ability to analyze proteins and their modifications in a high-throughput manner makes MS an indispensable approach in proteomics for elucidating the roles of PTMs in extracellular vesicles and identifying potential biomarkers or therapeutic targets.

- **Phosphopeptide Enrichment:**

A comprehensive analysis of phosphorylated peptides derived from EVs involves a combination of state-of-the-art techniques, including the essential enrichment of these peptides due to their often low stoichiometry, thereby enabling a deeper understanding of phosphorylation dynamics. To date, various phosphopeptide enrichment strategies have been developed, but the most notable ones for phosphopeptide enrichment in EVs are based on metal ion chelation, including immobilized metal ion affinity chromatography (IMAC), metal oxide affinity chromatography (MOAC), and ion exchange chromatography (IEX) and PolyMAC (Polymer-based Metal ion Affinity Capture). These techniques play crucial roles in phosphoproteomic studies by selectively isolating phosphorylated peptides from complex biological samples [34, 38, 37, 77, 93]. IMAC is based on the strong affinity of phosphate groups on amino acid residues for metal ions, such as Fe^{3+} or Ga^{3+}, to capture phosphorylated peptides. IMAC involves preparing peptides in a weak acid solution to equilibrate their protonation states, thereby facilitating effective binding to the chelating matrix. These interactions are highly sensitive to pH fluctuations and can be disrupted by metal chelators or specific reagents. Following the binding phase, non-interacting peptides are washed away, and an elution buffer is employed to release the bound phosphopeptides, allowing for their subsequent analysis [14]. This approach of phosphopeptide enrichment in EVs facilitated the discovery of potential biomarkers for breast cancer, highlighting the utility of IMAC in enhancing the detection and quantification of phosphoproteins in complex biological samples [20]. Meanwhile, MOAC utilizes metal oxides like titanium dioxide (TiO_2) or zirconium dioxide (ZrO_2) to selectively bind phosphorylated peptides, particularly phosphotyrosine-containing peptides. These methods are essential for isolating low-abundance phosphorylated peptides from complex biological samples, facilitating their subsequent analysis [54,

60, 61, 99]. PolyMAC utilizes polymer-based materials functionalized with titanium ions to bind phosphorylated peptides specifically. This approach offers advantages like high specificity and efficiency in capturing phosphopeptides, minimizing non-specific interactions, and improving detection sensitivity [2, 57]. It has been further developed as the Spin-Tip version to handle a small amount of starting materials including EV samples [87].

Very recently, our group has introduced one integrated strategy termed Extracellular Vesicles to Phosphoproteins (EVTOP). It is a streamlined method tailored specially for EV phosphoproteomics, enabling the one-pot EV isolation, protein extraction, digestion of EV proteins, and phosphopeptide enrichment from trace amounts of biofluids. It utilizes magnetic beads functionalized with Ti^{IV} ions and a membrane-penetrating peptide, octa-arginine R_8^+ [71]. This combination not only facilitates the efficient capture of EVs but also ensures the retention of EV proteins during lysis due to the hydrophilic surface provided by R_8^+. On-bead digestion then transitions the EVTOP surface to Ti(IV) ion-only, facilitating efficient phosphopeptide enrichment for phosphoproteomic analysis.

Phosphotyrosine, phosphoserine, and phosphothreonine-specific antibodies are commonly used to pull down phosphorylated proteins or peptides for mass spectrometry analysis. While these antibodies have been used for immunoprecipitation, the specificity of antibodies against phosphoserine and phosphothreonine is relatively low, making only highly selective phosphotyrosine-specific antibodies suitable for the enrichment of tyrosine-phosphorylated proteins in EVs [21].

Each enrichment method offers certain advantages, and over the years improved enrichment methods, better mass spectrometry speed and resolution, and enhanced data analysis have enabled comprehensive phosphoproteomic analyses, which has extended to the EV research field, providing insights into intercellular signaling pathways and cellular processes regulated by protein phosphorylation.

- **Glycopeptide Enrichment:**

Protein glycosylation is critical in a wide range of biological processes and disease-related pathological progression. Mass spectrometry (MS)-based glycoproteomics offers a robust and high-throughput approach for comprehensive profiling of the glycoproteome on a system-wide scale [18]. Most methods have been developed primarily for the analysis of N-glycopeptides, with serine and threonine O-glycopeptides receiving subsequent attention. In contrast, studies on C-glycosylation, tyrosine O-glycosylation, and S-glycosylation are notably rare [59]. Among these, most of the mass spectrometry-based investigations into the EV glycome have predominantly focused on N-glycosylation.

Various methods have been developed for the enrichment of glycopeptides and glycans, but only a subset has been effectively applied to the glycosylation analysis of EVs. Among these, lectin-based enrichment, which targets specific glycan types through lectin-glycan interactions, offers good selectivity but is hindered by limited universality, low throughput, and high costs. Yamamoto et al. identified a high-mannose-type glycan-specific lectin that can be used for the selective enrichment of glycopeptides from tumor-derived EVs. This method utilizes the specific binding affinity between lectins and glycans to isolate glycopeptides, enabling detailed glycoproteomic analysis of EVs [90]. Alternatively, hydrophilic interaction chromatography (HILIC) leverages the hydrophilicity of sugar chains to achieve high universality and rapidity, though it suffers from low selectivity. This technique utilizes a hydrophilic stationary phase to retain glycopeptides, which are then eluted with a gradient of increasing organic solvent. For instance, Sun et al. utilized a HILIC cartridge to enrich glycopeptides from EVs, identifying fibrinogen beta chain (FGB) and beta-2-glycoprotein 1 (β2-GP1) as potential biomarkers for colorectal cancer by comparing glycoproteins in EVs from cancerous and paracancerous tissues [70]. Another method used for glycopeptide enrichment is size exclusion capture, which relies on hydrophobic and polar interactions with adjustable pore sizes. This technique provides good selectivity and sensitivity but necessitates the complex synthesis of mesoporous nanomaterials. Wu et al. developed a novel material named FeMPC and integrated it with MALDI-MS to analyze glycans in urinary exosomes, successfully identifying 30 and 45 N-glycans from 10 μg of digested urinary exosomes derived from healthy individuals and gastric cancer patients [89]. In the sequence of glycopeptide enrichment techniques, boronic affinity chromatography interacts with diol groups present in glycan structures through boronic acids to capture glycopeptides. The bound glycopeptides can be eluted by altering the pH or using a competing agent. This method offers high universality and mild reaction conditions. However, this method is constrained by its low selectivity and enrichment efficiency. Utilizing this approach, Luo et al. identified 495 glycopeptides from 249 glycoproteins in exosomes extracted from human plasma and introduced a novel concept involving lectin affinity materials for enhanced glycoproteomic analysis of extracellular vesicles [49].

Lastly, hydrazide chemistry, which involves the reaction between hydrazide and aldehyde groups, offers excellent selectivity but is irreversible, potentially leading to the loss of glycan information. Andaluz Aguila et al. refined this approach by developing a workflow that sequentially isolates phosphopeptides and N-glycopeptides, using hydrazide chemistry to enrich N-glycopeptides from EVs isolated from 1 mL of plasma. This technique involves oxidizing glycans on glycoproteins to introduce aldehyde groups, which are then covalently captured onto hydrazide-functionalized beads [2]. This method allows for the specific isolation of glycopeptides which can be released from the beads by cleaving the glycosidic bonds using an enzyme like PNGase F which liberates N-linked glycans [94]. Each glycopeptide enrichment method, including hydrazide chemistry, has unique advantages and limitations, with trade-offs between selectivity, universality, throughput, and procedural complexity, highlighting the necessity for tailored approaches in glycoproteomic studies of extracellular vesicles.

- **Other PTM Enrichment:**

Immunoaffinity-based techniques are pivotal for characterizing many PTMs in EVs due to their specificity and sensitivity. These techniques rely on the use of antibodies that are designed to bind specifically to the modified residues of target proteins, allowing for the selective enrichment and analysis of PTMs within EVs

[25]. By employing antibodies that recognize PTMs, researchers can effectively isolate and study these modifications from complex biological samples, including those containing EVs. This approach is particularly advantageous because it enables the targeted capture of modified proteins or peptides, reducing background noise and improving the accuracy of PTM detection. Pan-PTM-specific antibodies are frequently employed for immunoaffinity enrichment prior to techniques such as LC–MS/MS, as well as for Western blotting, protein microarrays, immunohistochemistry, and flow cytometry. Additionally, antibodies targeting specific modification sites that are known to have significant biological roles are available for certain modifications, such as methylation and acetylation [5, 37, 40]. However, generating site-specific antibodies for PTMs, such as ubiquitination, can be challenging due to issues related to the size or transient nature of the modification or insufficient evidence of site occupancy, which hampers the investment in antibody development [41].

Anti-acetyl-lysine antibodies are used against acetylated lysine residues to purify acetylated proteins/peptides. The strategy was applied to EVs to identify and quantify acetylation patterns. Samples of approximately 100 μg of sEVs derived from MCF10A, MCF7, and MDA-MB-231 cells were collected. The proteins were digested, and acetylated peptides were enriched using anti-acetyl-lysine (KAC) immunoaffinity chromatography to improve the identification of low-abundance acetylated proteins [51]. However, it had been identified that the number of acetylated proteins varied among them. Specifically, EVs from MCF10A cells had fewer lysine-acetylated proteins compared to those from MDA-MB-231 and MCF7 cells.

Anti-K-ε-GG Antibody enrichment technique is used against the K-ε-GG remnant (left after tryptic digestion) to enrich and identify ubiquitinated proteins. Although initially developed for whole-cell lysates, this approach has been adapted for EVs to study ubiquitinated proteins. Burke et al. identified 50 ubiquitinated proteins in exosomes purified from myeloid-derived suppressor cells (MDSCs) by immunoprecipitation with anti-ubiquitin antibodies at the protein and peptide levels. The number of identified ubiquitinated proteins was 10% of the number of proteins identified in MDSC-derived exosomes, and 35 of these proteins had not been previously reported to be ubiquitinated [10], indicating that enrichment enabled the identification of low-abundance ubiquitinated proteins in exosomes. Huebner et al. performed a large-scale analysis of ubiquitination sites in human urinary exosome proteins. The ubiquitinated peptides were immunoaffinity-purified using an anti-diglycine remnant (K-ε-GG) antibody. They identified more than 600 ubiquitinated proteins, which accounted for 15% of the proteins identified in human urinary exosomes. This result supports the model that deubiquitination of protein cargo is not essential during the process of protein cargo incorporation into multivesicular bodies [33].

4 Functional Implications of PTMs in EV Proteins

The functional implications of PTMs in EV proteins are profound, as they orchestrate the structural integrity, selective cargo loading of EVs, signaling capacity, and, facilitating tissue remodeling thereby shaping their pivotal roles in intercellular communication and disease pathogenesis.

EV proteins isolated from cell culture media and biofluids, including plasma and urine have been found to exhibit PTMs [96]. These alterations are not merely chemical modifications, they represent critical regulatory mechanisms that can dictate the functional roles of EV proteins in various biological contexts, which are essential for protein configuration. For example, phosphorylation, glycosylation, and acetylation can introduce bulky or charged groups onto protein residues, affecting their spatial arrangement, and leading to changes in protein structure [44]. These alterations in protein conformation can directly impact protein function within EVs, such as enzymatic activity or binding affinity to target molecules. Additionally, PTMs like ubiquitination and sumoylation can affect protein stability by preventing proteolytic degradation [44] [92]. Increased resistance to degradation can prolong the lifespan of proteins within EVs and ensure their functionality during intercellular communication or other biological processes.

Understanding PTMs in EVs is essential, as it provides crucial insights into EV biogenesis, shedding light on the complex mechanisms governing the formation and release of these vesicles from donor cells. PTMs modulate the activity, stability, and localization of proteins involved in EV biogenesis. For instance, lipid modifications like palmitoylation or myristoylation can facilitate membrane association, directing proteins to specific regions of the EV membrane or promoting their inclusion in EV cargo [58]. Verweij et al. highlighted the critical role of palmitoylation at the established C78 site in facilitating the sorting of LAMP1 into EVs [79]. Phosphorylation of the cell surface protein caveolin-1 (cav-1) at tyrosine 14 (pY14) reveals its scaffolding domain (CSD), enabling interaction with the RNA-binding protein hnRNPA2B1. This cav-1/hnRNPA2B1 complex is subsequently incorporated into EVs, along with hnRNPA2B1-bound miRNAs [45]. It has been reported that SW620 cancer cell-derived exosomes exhibit a notably high prevalence of tyrosine (Y)-phosphorylated sites, with 6.4% of proteins in exosomes compared to just 0.6% within the cells themselves. This significant enrichment of tyrosine phosphorylation in exosomal proteins suggests a potential role in regulating exosome formation and function [96]. Moreover, both phosphorylation and O-GlcNAcylation of cellular proteins have been implicated in the selective incorporation of miRNA molecules into EVs [81]. Phosphorylation plays a critical role in cellular communication, facilitating the transfer of proteins like Sphingosine kinase-1 via membrane vesicles in SK-hep1 and 8701-BC breast carcinoma cells [65] and PTEN through exosomes, typically absent in exosomes from normal cells [27]. Beyond its involvement in signal transduction, phosphorylation also targets proteins for incorporation into EVs. For example, phosphorylation of AnxA2 at Tyr-23 is essential for its Ca^{2+}-dependent recruitment to plasma membrane rafts, leading to its inclusion in the ILVs of MVBs and subsequent EV membrane integration [78]. Glycosylation plays a crucial role in the biogenesis, protein sorting, and uptake of EVs. One of the findings provides compelling evidence that N-linked glycosylation specifically directs the sorting of glycoproteins into exosomes [47]. Emerging evidence underscores the critical role of glycosylation in regulating EV protein sorting. Alterations in glycosylation profiles during malignant transformation can disrupt this sorting process. Specific glycan structures, such as increased high mannose, polylactosamine, α-2,6 sialylation, and complex N-linked glycans, along with the loss of terminal blood group A and B antigens, have been linked to glycoprotein and glycolipid incorporation into EVs [6]. Recent studies revealed that altered expression of the glycosyltransferase FUT8, which adds fucose to N-glycans, significantly modifies the EV proteome in prostate cancer [22]. Additionally, aberrant O-glycosylation has been shown to influence the sorting of proteins such as CD44 into EVs [28]. Notably, the glycosyltransferase ST6Gal1, which catalyzes the α-2,6 sialylation of N-glycans, has also been implicated in modulating EV protein cargo sorting [23]. Conversely, the sorting of certain proteins into EVs remains unaffected by their glycosylation status. For example, Lipocalin 2 (LCN2), a highly conserved protein with a single N-glycosylation site potentially implicated in its trafficking into EVs [8], was shown to be sorted into EVs even when glycosylation was inhibited by tunicamycin in various cell lines. This finding confirms that LCN2's incorporation into EVs is independent of its N-glycosylation status. N-glycoproteins, which adorn the cell surface and are released into the extracellular milieu, play a crucial role in cell-to-cell communication. It has been reported that the EVs are notably enriched in glycoconjugates, including glycoproteins, glycosphingolipids, and proteoglycans, and exhibit distinct glycosignatures [50]. Among the various protein sorting mechanisms, PTMs have been identified as key factors in directing proteins into EVs. Despite numerous studies demonstrating the presence of PTM within EVs, the specific role of each modification in the packaging process remains inadequately understood. For instance, ISGylation of the ESCRT protein TSG101 enhances the aggregation and colocalization of MVBs with lysosomes, leading to MVB degradation and reduced EV secretion [80]. Conversely, SUMOylation regulates the sorting of α-Synuclein into EVs, facilitating the incorporation of both cytosolic and transmembrane forms of this protein [42]. Additionally, ubiquitination has been established as a critical PTM for the sorting of proteins into ILVs within MVBs, a process orchestrated by the ESCRT complexes [31].

On the other hand, PTMs are essential for fine-tuning RNA-binding Proteins (RBPs) functions and maintaining cellular homeostasis [68] where RBPs are key regulators of RNA metabolism and can be transferred to recipient cells via EVs, where they influence cell biology and play a crucial role in sorting coding and non-coding RNAs [69]. PTMs, which can modulate the activation or inactivation of proteins in a context-dependent manner, also play a crucial role in directing proteins into extracellular vesicles [1]. PTMs can influence the interactions of proteins with other molecules within EVs, such as lipids, nucleic acids, or other proteins. For example, phosphorylation can regulate protein–protein interactions by modulating the binding affinity or specificity of proteins for their binding partners [55]. This can affect the composition and function of EV cargo, influencing processes such as signal transduction or nucleic acid packaging [13].

EVs derived from immune cells, such as dendritic cells, macrophages, and B cells, play essential roles in antigen presentation and immune regulation [11]. PTMs on EV proteins, can influence the antigen presentation process by modulating the stability, trafficking, and immunogenicity of major histocompatibility complex (MHC) molecules and co-stimulatory proteins on EV surfaces [43, 52]. PTMs in EV proteins regulate the secretion and activity of cytokines and chemokines carried within EV cargo. For example, phosphorylation and acetylation of cytokines such as interleukins and interferons can modulate their stability, receptor binding affinity, and downstream signaling pathways upon EV uptake by recipient cells [32, 86]. This, in turn, affects immune cell activation, polarization, and recruitment in the local microenvironment. EVs containing PTM-modified proteins can promote immune tolerance by suppressing proinflammatory responses and promoting regulatory T cell (Treg) differentiation and function. Glycosylation and ubiquitination on EV surface proteins may facilitate interactions with immune cell receptors, leading to the induction of tolerogenic signaling pathways and immune suppression [50, 66].

EVs enriched with PTM-modified proteins, such as transcription and growth factors, can facilitate cellular reprogramming and tissue regeneration processes. Phosphorylation and acetylation can alter the activity and stability of transcription factors involved in stem cell maintenance, differentiation, and tissue repair, thereby promoting the reprogramming of target cells toward regenerative phenotypes [62]. PTMs on angiogenic factors such as vascular endothelial growth factor (VEGF) and fibroblast growth factor (FGF) can modulate their bioactivity and signaling pathways, leading to enhanced endothelial cell proliferation, migration, and tube formation essential for neovascularization and tissue perfusion [46, 84]. PTMs in EV proteins regulate the remodeling of the extracellular matrix (ECM) during tissue regeneration [19]. PTM-modified matrix metalloproteinases (MMPs) and tissue inhibitors of metalloproteinases (TIMPs) within EV cargo can modulate ECM turnover and degradation, facilitating cell migration, invasion, and tissue remodeling in response to injury or pathological conditions [13].

5 PTM in Extracellular Vesicles as Disease Biomarker

EVs are released by nearly all cell types and can be found in different biological fluids, including blood, urine, saliva, breast milk, seminal fluid, cerebrospinal fluid, amniotic fluid, bronchoalveolar lavage fluid and pleural effusion. Initially, EVs emerge from the immediate cellular environment or niche, where they reflect the molecular state and signaling of the originating cells. These vesicles then enter the bloodstream or other bodily fluids, allowing them to circulate systemically and carry biomolecular information throughout the body. EVs are known to carry a diverse array of PTMs alongside other biomolecules, such as proteins, lipids, and nucleic acids reflecting their cellular origin. Like other cargo within EVs, PTMs are also protected by the lipid bilayer, which shields them from various environmental stresses, such as freeze–thaw cycles, enzymatic degradation, and oxidative damage which ensures their stability. This protective environment allows EVs to effectively mediate intercellular communication and influence various physiological and pathological processes. This widespread distribution and stability of EVs underscores their potential as biomarkers or therapeutic agents in medical research [24].

PTMs on EV surface proteins, such as integrins and tetraspanins, regulate their adhesion to ECM components and immune cell receptors, facilitating tumor cell dissemination and metastatic colonization in distant organs [12]. PTMs in EV proteins are implicated in the pathogenesis of neurodegenerative diseases such as Alzheimer's, Parkinson's, and Huntington's disease [2]. PTM-modified proteins within EV cargo, including amyloid-beta peptides, tau proteins, and alpha-synuclein, can aggregate and propagate pathological protein misfolding and neurotoxicity, leading to neuronal dysfunction, synaptic loss, and cognitive impairment [91]. EVs carrying PTM-modified proteins contribute to the dysregulation of immune responses and the development of autoimmune diseases [36, 66]. PTMs such as citrullination and glycosylation on EV proteins can induce aberrant immune activation and inflammatory signaling pathways, triggering tissue damage and autoantibody production in susceptible individuals [19, 88]. These reports not only deepen our understanding of PTMs in EVs but also hold promise for innovative treatments and diagnostics. Ultimately, PTMs in EV circulation or body fluid potentially can serve as novel biomarkers for a variety of human diseases. Discovering PTM proteins within EVs is crucial, as they can indicate the physiological or pathological state of the parent cells [19]. Changes in PTMs may signal disease states, offering insights into disease mechanisms and potential biomarkers for diagnosis and therapy.

One of the reports identified 1,453 unique glycopeptides from 556 glycoproteins in EVs, where 20 of them show significantly higher levels in breast cancer patients. A novel glyco-specific reverse-phase protein array was used to quantify these candidates, highlighting the potential of an integrated pipeline for developing breast cancer biomarkers [16]. Another report has identified nearly 10,000 unique phosphopeptides in EVs from small plasma volumes. Label-free quantitative phosphoproteomics revealed 144 phosphoproteins significantly elevated in breast cancer patients compared to healthy controls. Targeted quantitation using parallel reaction monitoring validated several biomarkers, demonstrating the feasibility of using phosphoproteins in plasma EVs for cancer screening and monitoring [20].

Preclinical research has led to the registration of over 50 clinical trials aimed at evaluating the potential of EVs as diagnostic biomarkers for various diseases. EVs are leveraged across various stages of disease management as these EV-based disease markers can be identified long before the onset of symptoms or the physiological detection of an ailment, positioning them as promising candidates for early-stage disease detection, enabling non-invasive diagnostic tests. Analysis of EVs facilitates monitoring of disease progression and treatment response by tracking changes in the cargo content. For instance, EV analysis helps in evaluating how well a patient is responding to treatment. Changes in EV biomarkers can indicate whether the treatment is effective or if there is resistance. This capability allows for the evaluation of disease prognosis and helps stratify patients for personalized therapeutic strategies. However, few in the current clinical studies are based on EV PTMs, highlighting the technical challenges moving forward from lab research to clinical settings.

6 Summary and Future Perspectives:

Studying PTM proteins in EVs holds paramount importance due to their multifaceted roles in various biological processes. PTMs confer several advantages to EVs by enhancing their functional diversity and specificity. PTMs such as ubiquitination, phosphorylation, acetylation, and glycosylation play crucial roles in the selective sorting and loading of proteins into EVs, ensuring that these vesicles carry the appropriate cargo necessary for their intended signaling functions. By influencing cargo selection, membrane association, and intercellular communication, PTMs ensure that EVs can effectively participate in a wide range of biological processes. Understanding the role of PTMs in EVs can provide new insights into their physiological functions and pathological roles, offering potential therapeutic targets for diseases such as cancer. The biological processes that EVs are involved in and their unique values in clinical applications have attracted increasing attention to EV research in recent years. Particularly, recent applications of "-omics" techniques in the field of EVs provided greater information than before, and researchers become aware that EVs undoubtedly contain vital clues essential for a better understanding of the function of vesicles. Among many-omics techniques, proteomics of EVs is one of the most important fields because EV proteins represent actual functional molecules in the cell, and most cellular functions are carried out by proteins. In a biological process where proteins participate, various PTMs finely tune the cellular functions of each protein, and EV proteome may provide snapshots of the process.

The study of PTMs in EV proteins is hampered by persisting technical challenges. Low abundance of PTM proteins within extremely heterogeneous EVs presents significant difficulties in accurately quantifying and detecting their PTMs. Additionally, the EV isolation process can potentially alter these modifications; for instance, ultracentrifugation may disrupt transient PTM interactions. Future research must investigate the effects of various EV harvest and isolation methods on PTM preservation. Furthermore, given that EVs originate from the plasma membrane of their parent cells, the substantial lipid and cholesterol content in EVs may interfere with the detection and analysis of protein PTMs.

References

1. Anand, S., Samuel, M., Kumar, S., & Mathivanan, S. (2019). Ticket to a bubble ride: Cargo sorting into exosomes and extracellular vesicles. *Biochimica et Biophysica Acta (BBA)-Proteins and Proteomics, 1867*(12), 140203.

2. Andaluz Aguilar, H., Iliuk, A. B., Chen, I.-H., & Tao, W. A. (2020). Sequential phosphoproteomics and N-glycoproteomics of plasma-derived extracellular vesicles. *Nature Protocols, 15*(1), 161–180.
3. Aslebagh, R., Wormwood, K. L., Channaveerappa, D., Wetie, A. G. N., Woods, A. G., & Darie, C. C. (2019). Identification of posttranslational modifications (PTMs) of proteins by mass spectrometry. *Advancements of Mass Spectrometry in Biomedical Research*, 199–224.
4. Baker, S. A., & Rutter, J. (2023). Metabolites as signaling molecules. *Nature Reviews Molecular Cell Biology, 24*(5), 355–374.
5. Barski, A., Cuddapah, S., Cui, K., Roh, T.-Y., Schones, D. E., Wang, Z., Wei, G., Chepelev, I., & Zhao, K. (2007). High-resolution profiling of histone methylations in the human genome. *Cell, 129*(4), 823–837.
6. Batista, B. S., Eng, W. S., Pilobello, K. T., Hendricks-Muñoz, K. D., & Mahal, L. K. (2011). Identification of a conserved glycan signature for microvesicles. *Journal of Proteome Research, 10*(10), 4624–4633.
7. Berumen Sánchez, G., Bunn, K. E., Pua, H. H., & Rafat, M. (2021). Extracellular vesicles: Mediators of intercellular communication in tissue injury and disease. *Cell Communication and Signaling, 19*, 1–18.
8. Borkham-Kamphorst, E., Van de Leur, E., Meurer, S. K., Buhl, E. M., & Weiskirchen, R. (2018). N-glycosylation of Lipocalin 2 is not required for secretion or exosome targeting. *Frontiers in Pharmacology, 9*, 426.
9. Briggs, J. A., Simon, M. N., Gross, I., Kräusslich, H.-G., Fuller, S. D., Vogt, V. M., & Johnson, M. C. (2004). The stoichiometry of Gag protein in HIV-1. *Nature Structural & Molecular Biology, 11*(7), 672–675.
10. Burke, M. C., Oei, M. S., Edwards, N. J., Ostrand-Rosenberg, S., & Fenselau, C. (2014). Ubiquitinated proteins in exosomes secreted by myeloid-derived suppressor cells. *Journal of Proteome Research, 13*(12), 5965–5972.
11. Buzas, E. I. (2023). The roles of extracellular vesicles in the immune system. *Nature Reviews Immunology, 23*(4), 236–250.
12. Buzás, E. I., Tóth, E. Á., Sódar, B. W., & Szabó-Taylor, K. É. (2018). *Molecular interactions at the surface of extracellular vesicles, 40*(5), 453–464.
13. Carnino, J. M., Ni, K., & Jin, Y. (2020). Post-translational modification regulates formation and cargo-loading of extracellular vesicles. *Frontiers in Immunology, 11*, 948.
14. Carregari, V. C. (2022). Phosphopeptide enrichment techniques: A pivotal step for phosphoproteomic studies. In *Understanding PTMs in Neurodegenerative Diseases* (pp. 17–27). Springer.
15. Chatterjee, M., Özdemir, S., Fritz, C., Möbius, W., Kleineidam, L., Mandelkow, E., Biernat, J., Doğdu, C., Peters, O., Cosma, N.C. & Wang, X. (2024). Plasma extracellular vesicle tau and TDP-43 as diagnostic biomarkers in FTD and ALS. *Nature medicine, 30*(6), 1771–1783.
16. Chen, I.-H., Aguilar, H. A., Paez Paez, J. S., Wu, X., Pan, L., Wendt, M. K., Iliuk, A. B., Zhang, Y., & Tao, W. A. (2018). Analytical pipeline for discovery and verification of glycoproteins from plasma-derived extracellular vesicles as breast cancer biomarkers. *Analytical Chemistry, 90*(10), 6307–6313.
17. Chen, Y.-J., Ching, W.-C., Chen, J.-S., Lee, T.-Y., Lu, C.-T., Chou, H.-C., Lin, P.-Y., Khoo, K.-H., Chen, J.-H., & Chen, Y.-J. (2014). Decoding the s-nitrosoproteomic atlas in individualized human colorectal cancer tissues using a label-free quantitation strategy. *Journal of Proteome Research, 13*(11), 4942–4958.
18. Chen, Z., Huang, J., & Li, L. (2019). Recent advances in mass spectrometry (MS)-based glycoproteomics in complex biological samples. *TrAC Trends in Analytical Chemistry, 118*, 880–892.
19. Chen, X., Song, X., Li, J., Wang, J., Yan, Y., & Yang, F. (2024). Integrated proteomic, phosphoproteomic, and N-glycoproteomic analyses of small extracellular vesicles from C2C12 myoblasts identify specific PTM patterns in ligand-receptor interactions. *Cell Communication and Signaling, 22*(1), 273.

20. Chen, I.-H., Xue, L., Hsu, C.-C., Paez, J. S. P., Pan, L., Andaluz, H., Wendt, M. K., Iliuk, A. B., Zhu, J.-K., & Tao, W. A. (2017). Phosphoproteins in extracellular vesicles as candidate markers for breast cancer. *Proceedings of the National Academy of Sciences, 114*(12), 3175–3180.
21. Chen, X., Zhu, N., Cai, T., & Yang, F. (2024). Technologies Used for Analysis of Extracellular Vesicle-Omics. In *Extracellular Vesicles: From Bench to Bedside* (pp. 83–109). Springer.
22. Clark, D. J., Schnaubelt, M., Hoti, N., Hu, Y., Zhou, Y., Gooya, M., & Zhang, H. (2020). Impact of increased FUT8 expression on the extracellular vesicle proteome in prostate cancer cells. *Journal of Proteome Research, 19*(6), 2195–2205.
23. Dall'Olio, F. (2000). The sialyl-α2, 6-lactosaminyl-structure: Biosynthesis and functional role. *Glycoconjugate Journal, 17*, 669–676.
24. Doyle, L. M., & Wang, M. Z. (2019). Overview of extracellular vesicles, their origin, composition, purpose, and methods for exosome isolation and analysis. *Cells, 8*(7), 727.
25. Dunphy, K., Dowling, P., Bazou, D., & O'Gorman, P. (2021). Current methods of post-translational modification analysis and their applications in blood cancers. *Cancers, 13*(8), 1930.
26. Fang, Y., Wu, N., Gan, X., Yan, W., Morrell, J. C., & Gould, S. J. (2007). Higher-order oligomerization targets plasma membrane proteins and HIV gag to exosomes. *PLoS Biology, 5*(6), Article e158.
27. Gabriel, K., Ingram, A., Austin, R., Kapoor, A., Tang, D., Majeed, F., Qureshi, T., & Al-Nedawi, K. (2013). Regulation of the tumor suppressor PTEN through exosomes: A diagnostic potential for prostate cancer. *PLoS ONE, 8*(7), Article e70047.
28. Gao, T., Wen, T., Ge, Y., Liu, J., Yang, L., Jiang, Y., Dong, X., Liu, H., Yao, J., & An, G. (2020). Disruption of Core 1-mediated O-glycosylation oppositely regulates CD44 expression in human colon cancer cells and tumor-derived exosomes. *Biochemical and Biophysical Research Communications, 521*(2), 514–520.
29. Hadisurya, M., Lee, Z.C., Luo, Z., Zhang, G., Ding, Y., Zhang, H., Iliuk, A.B., Pili, R., Boris, R.S. & Tao, W.A. (2023). Data-independent acquisition phosphoproteomics of urinary extracellular vesicles enables renal cell carcinoma grade differentiation. *Molecular & Cellular Proteomics, 22*(5).
30. Haro, I., Sanmartí, R., & Gómara, M. J. (2022). Implications of post-translational modifications in autoimmunity with emphasis on citrullination, homocitrullination and acetylation for the pathogenesis, diagnosis and prognosis of rheumatoid arthritis. *International Journal of Molecular Sciences, 23*(24), 15803.
31. Henne, W. M., Buchkovich, N. J., & Emr, S. D. (2011). The ESCRT pathway. *Developmental Cell, 21*(1), 77–91.
32. Hu, X., Li, J., Fu, M., Zhao, X., & Wang, W. (2021). The JAK/STAT signaling pathway: From bench to clinic. *Signal Transduction and Targeted Therapy, 6*(1), 402.
33. Huebner, A. R., Cheng, L., Somparn, P., Knepper, M. A., Fenton, R. A., & Pisitkun, T. (2016). Deubiquitylation of protein cargo is not an essential step in exosome formation. *Molecular & Cellular Proteomics, 15*(5), 1556–1571.
34. Iliuk, A., Jayasundera, K., Wang, W.-H., Schluttenhofer, R., Geahlen, R. L., & Tao, W. A. (2015). In-depth analyses of B cell signaling through tandem mass spectrometry of phosphopeptides enriched by PolyMAC. *International Journal of Mass Spectrometry, 377*, 744–753.
35. Iliuk, A., Wu, X., Li, L., Sun, J., Hadisurya, M., Boris, R. S., & Tao, W. A. (2020). Plasma-derived extracellular vesicle phosphoproteomics through chemical affinity purification. *Journal of Proteome Research, 19*(7), 2563–2574.
36. Jia, X., Zhai, T., & Zhang, J. (2021). Circulating exosome involves in the pathogenesis of autoimmune thyroid diseases through immunomodulatory proteins. *Frontiers in Immunology, 12*, Article 730089.
37. Jiang, Y.-Y., Jiang, Y., Li, C.-Q., Zhang, Y., Dakle, P., Kaur, H., Deng, J.-W., Lin, R.Y.-T., Han, L., & Xie, J.-J. (2020). TP63, SOX2, and KLF5 establish a core regulatory circuitry that controls epigenetic and transcription patterns in esophageal squamous cell carcinoma cell lines. *Gastroenterology, 159*(4), 1311–1327.

38. Jiang, D., Qi, R., Lv, S., Wu, S., Li, Y., & Liu, J. (2023). Preparation of high-efficiency titanium ion immobilized magnetic graphite nitride nanocomposite for phosphopeptide enrichment. *Analytica Chimica Acta, 1283*, Article 341974.
39. Kang, T., Atukorala, I., & Mathivanan, S. (2021). Biogenesis of extracellular vesicles. *New Frontiers: Extracellular Vesicles*, 19–43.
40. Kim, S. W., Hasanuzzaman, M., Cho, M., Heo, Y. R., Ryu, M.-J., Ha, N.-Y., Park, H. J., Park, H.-Y., & Shin, J.-G. (2015). Casein kinase 2 (CK2)-mediated phosphorylation of Hsp90β as a novel mechanism of rifampin-induced MDR1 expression. *Journal of Biological Chemistry, 290*(27), 17029–17040.
41. Van Kruijsbergen, I., Mulder, M. P., Uckelmann, M., Van Welsem, T., De Widt, J., Spanjaard, A., Jacobs, H., El Oualid, F., Ovaa, H., & van Leeuwen, F. (2020). Strategy for development of site-specific ubiquitin antibodies. *Frontiers in Chemistry, 8*, 111.
42. Kunadt, M., Eckermann, K., Stuendl, A., Gong, J., Russo, B., Strauss, K., Rai, S., Kügler, S., Falomir Lockhart, L., & Schwalbe, M. (2015). Extracellular vesicle sorting of α-Synuclein is regulated by sumoylation. *Acta Neuropathologica, 129*, 695–713.
43. Lee, Z.-C., Hadisurya, M., Luo, Z., Li, L., & Tao, W. A. (2023). Hands-free proteomic profiling of urinary extracellular vesicles with a high-throughput automated workflow. *Journal of the American Society for Mass Spectrometry, 34*(11), 2585–2593.
44. Lee, J. M., Hammarén, H. M., Savitski, M. M., & Baek, S. H. (2023). Control of protein stability by post-translational modifications. *Nature Communications, 14*(1), 201.
45. Lee, H., Li, C., Zhang, Y., Zhang, D., Otterbein, L. E., & Jin, Y. (2019). Caveolin-1 selectively regulates microRNA sorting into microvesicles after noxious stimuli. *The Journal of Experimental Medicine, 216*(9), 2202.
46. Li, W., Li, F., Zhang, X., Lin, H.-K., & Xu, C. (2021). Insights into the post-translational modification and its emerging role in shaping the tumor microenvironment. *Signal Transduction and Targeted Therapy, 6*(1), 422.
47. Liang, Y., Eng, W. S., Colquhoun, D. R., Dinglasan, R. R., Graham, D. R., & Mahal, L. K. (2014). Complex N-linked glycans serve as a determinant for exosome/microvesicle cargo recruitment. *Journal of Biological Chemistry, 289*(47), 32526–32537.
48. Livshits, M.A., Khomyakova, E., Evtushenko, E.G., Lazarev, V.N., Kulemin, N.A., Semina, S.E., Generozov, E.V., & Govorun, V.M. (2015). Isolation of exosomes by differential centrifugation: Theoretical analysis of a commonly used protocol. *Scientific reports, 5*(1), 17319.
49. Luo, B., Li, G., Li, Z., He, J., Zhou, J., Wu, L., Lan, F., & Wu, Y. (2021). Construction of a magnetic covalent organic framework with synergistic affinity strategy for enhanced glycopeptide enrichment. *Journal of Materials Chemistry B, 9*(32), 6377–6386.
50. Macedo-da-Silva, J., Santiago, V. F., Rosa-Fernandes, L., Marinho, C. R., & Palmisano, G. (2021). Protein glycosylation in extracellular vesicles: Structural characterization and biological functions. *Molecular Immunology, 135*, 226–246.
51. Minic, Z., Li, Y., Hüttmann, N., Uppal, G. K., D'Mello, R., & Berezovski, M. V. (2023). Lysine Acetylome of breast cancer-derived small extracellular vesicles reveals specific acetylation patterns for metabolic enzymes. *Biomedicines, 11*(4), 1076.
52. Moreno-Gonzalo, O., Villarroya-Beltri, C., & Sánchez-Madrid, F. (2014). Post-translational modifications of exosomal proteins. *Frontiers in Immunology, 5*, 383.
53. Neagu, A.-N., Jayathirtha, M., Baxter, E., Donnelly, M., Petre, B. A., & Darie, C. C. (2022). Applications of tandem mass spectrometry (MS/MS) in protein analysis for biomedical research. *Molecules, 27*(8), 2411.
54. Neville, D. C., Townsend, R. R., Rozanas, C. R., Verkman, A., Price, E. M., & Gruis, D. B. (1997). Evidence for phosphorylation of serine 753 in CFTR using a novel metal-ion affinity resin and matrix-assisted laser desorption mass spectrometry. *Protein Science, 6*(11), 2436–2445.
55. Nishi, H., Hashimoto, K. & Panchenko, A.R. (2011). Phosphorylation in protein-protein binding: effect on stability and function. *Structure, 19*(12), 1807–1815.

56. Nordin, J.Z., Lee, Y., Vader, P., Mäger, I., Johansson, H.J., Heusermann, W., Wiklander, O.P., Hällbrink, M., Seow, Y., Bultema, J.J. & Gilthorpe, J. (2015). Ultrafiltration with size-exclusion liquid chromatography for high yield isolation of extracellular vesicles preserving intact biophysical and functional properties. *Nanomedicine: Nanotechnology, Biology and Medicine, 11*(4), 879–883.
57. Nunez Lopez, Y. O., Iliuk, A., Petrilli, A. M., Glass, C., Casu, A., & Pratley, R. E. (2022). Proteomics and phosphoproteomics of circulating extracellular vesicles provide new insights into diabetes pathobiology. *International Journal of Molecular Sciences, 23*(10), 5779.
58. Picciotto, S., Romancino, D.P., Buffa, V., Cusimano, A., Bongiovanni, A. & Adamo, G. (2020). Post-translational lipidation in extracellular vesicles: chemical mechanisms, biological functions and applications. In *Advances in Biomembranes and Lipid Self-Assembly* (Vol. 32, pp. 83–111). Academic Press.
59. Piovesana, S., Cavaliere, C., Cerrato, A., Laganà, A., Montone, C. M., & Capriotti, A. L. (2023). Recent trends in glycoproteomics by characterization of intact glycopeptides. *Analytical and Bioanalytical Chemistry, 415*(18), 3727–3738.
60. Porath, J. (1988). High-performance immobilized-metal-ion affinity chromatography of peptides and proteins. *Journal of Chromatography A, 443*, 3–11.
61. Posewitz, M. C., & Tempst, P. (1999). Immobilized gallium (III) affinity chromatography of phosphopeptides. *Analytical Chemistry, 71*(14), 2883–2892.
62. Qausain, S., Srinivasan, H., Jamal, S., Nasiruddin, M., & Khan, M. K. A. (2019). Phosphorylation and acetylation of proteins as posttranslational modification: Implications in human health and associated diseases. In *Protein Modificomics* (pp. 69–86). Elsevier.
63. Qian, M., Yan, F., Yuan, T., Yang, B., He, Q., & Zhu, H. (2020). Targeting post-translational modification of transcription factors as cancer therapy. *Drug Discovery Today, 25*(8), 1502–1512.
64. Ramazi, S., & Zahiri, J. (2021). Post-translational modifications in proteins: Resources, tools and prediction methods. *Database, 2021*, baab012.
65. Rigogliuso, S., Donati, C., Cassara, D., Taverna, S., Salamone, M., Bruni, P., & Vittorelli, M. L. (2010). An active form of sphingosine kinase-1 is released in the extracellular medium as component of membrane vesicles shed by two human tumor cell lines. *Journal of Oncology, 2010*(1), Article 509329.
66. Robbins, P. D., & Morelli, A. E. (2014). Regulation of immune responses by extracellular vesicles. *Nature Reviews Immunology, 14*(3), 195–208.
67. Rontogianni, S., Synadaki, E., Li, B., Liefaard, M. C., Lips, E. H., Wesseling, J., Wu, W., & Altelaar, M. (2019). Proteomic profiling of extracellular vesicles allows for human breast cancer subtyping. *Communications Biology, 2*(1), 325.
68. Sternburg, E.L., da Silva, L.A.G. & Dormann, D. (2022). Post-translational modifications on RNA-binding proteins: accelerators, brakes, or passengers in neurodegeneration? *Trends in biochemical sciences, 47*(1), 6–22.
69. Sun, W., Cui, H., Xu, T., Yue, J., Liang, J., You, W., Sun, W., & Yi, Q. (2023). RNA binding proteins in extracellular vesicles and their potential value for cancer diagnosis and treatment. *International Journal of Oncology, 63*(4), 1–14.
70. Sun, Z., Ji, S., Wu, J., Tian, J., Quan, W., Shang, A., Ji, P., Xiao, W., Liu, D., & Wang, X. (2021). Proteomics-based identification of candidate exosomal glycoprotein biomarkers and their value for diagnosing colorectal cancer. *Frontiers in Oncology, 11*, Article 725211.
71. Sun, J., Li, Q., Ding, Y., Wei, D., Hadisurya, M., Luo, Z., Gu, Z., Chen, B., & Tao, W. A. (2023). Profiling phosphoproteome landscape in circulating extracellular vesicles from microliters of biofluids through functionally tunable paramagnetic separation. *Angewandte Chemie, 135*(29), Article e202305668.
72. Suskiewicz, M. J. (2024). The logic of protein post-translational modifications (PTMs): Chemistry, mechanisms and evolution of protein regulation through covalent attachments. *BioEssays, 46*(3), 2300178.
73. Tauro, B. J., Greening, D. W., Mathias, R. A., Ji, H., Mathivanan, S., Scott, A. M., & Simpson, R. J. (2012). Comparison of ultracentrifugation, density gradient separation, and

immunoaffinity capture methods for isolating human colon cancer cell line LIM1863-derived exosomes. *Methods, 56*(2), 293–304.
74. Taylor, D. D., & Shah, S. (2015). Methods of isolating extracellular vesicles impact downstream analyses of their cargoes. *Methods, 87*, 3–10.
75. Thomas, S., Kaur, J., Kamboj, R., Thangariyal, S., Yadav, R., Kumar, K., & Dhania, N. K. (2024). Investigate the efficacy of size exclusion chromatography for the isolation of extracellular vesicles from C. elegans. *Journal of Chromatography B, 1233*, 123982.
76. Théry, C., Witwer, K. W., Aikawa, E., Alcaraz, M. J., Anderson, J. D., Andriantsitohaina, R., Antoniou, A., Arab, T., Archer, F., & Atkin-Smith, G. K. (2018). Minimal information for studies of extracellular vesicles 2018 (MISEV2018): A position statement of the International Society for Extracellular Vesicles and update of the MISEV2014 guidelines. *Journal of Extracellular Vesicles, 7*(1), 1535750.
77. Tsai, C.-F., Hsu, C.-C., Wang, Y.-T., Kim, H., & Liu, T. (2024). A Tip-Based Workflow for Sensitive IMAC-Based Low Nanogram Level Phosphoproteomics. In *Proteomics for Drug Discovery* (pp. 129–140). Springer.
78. Valapala, M., & Vishwanatha, J. K. (2011). Lipid raft endocytosis and exosomal transport facilitate extracellular trafficking of annexin A2. *Journal of Biological Chemistry, 286*(35), 30911–30925.
79. Verweij, F. J., de Heus, C., Kroeze, S., Cai, H., Kieff, E., Piersma, S. R., Jimenez, C. R., Middeldorp, J. M., & Pegtel, D. M. (2015). Exosomal sorting of the viral oncoprotein LMP1 is restrained by TRAF2 association at signalling endosomes. *Journal of Extracellular Vesicles, 4*(1), 26334.
80. Villarroya-Beltri, C., Baixauli, F., Mittelbrunn, M., Fernández-Delgado, I., Torralba, D., Moreno-Gonzalo, O., Baldanta, S., Enrich, C., Guerra, S., & Sánchez-Madrid, F. (2016). ISGylation controls exosome secretion by promoting lysosomal degradation of MVB proteins. *Nature Communications, 7*(1), 13588.
81. Villarroya-Beltri, C., Gutiérrez-Vázquez, C., Sánchez-Cabo, F., Pérez-Hernández, D., Vázquez, J., Martin-Cofreces, N., Martinez-Herrera, D. J., Pascual-Montano, A., Mittelbrunn, M., & Sánchez-Madrid, F. (2013). Sumoylated hnRNPA2B1 controls the sorting of miRNAs into exosomes through binding to specific motifs. *Nature Communications, 4*(1), 2980.
82. Wang, Z. (2012). Protein S-nitrosylation and cancer. *Cancer Letters, 320*(2), 123–129.
83. Wang, C., Brennan, V. C., Gutierrez, N. M., Wang, X., Wang, L., & Yang, W. (2013). SUMOylation of ATF3 alters its transcriptional activity on regulation of TP53 gene. *Journal of Cellular Biochemistry, 114*(3), 589–598.
84. Wang, H., Yang, L., Liu, M., & Luo, J. (2023). Protein post-translational modifications in the regulation of cancer hallmarks. *Cancer Gene Therapy, 30*(4), 529–547.
85. Webber, J., & Clayton, A. (2013). How pure are your vesicles? *Journal of Extracellular Vesicles, 2*(1), 19861.
86. Wieczorek, M., Ginter, T., Brand, P., Heinzel, T., & Krämer, O. H. (2012). Acetylation modulates the STAT signaling code. *Cytokine & Growth Factor Reviews, 23*(6), 293–305.
87. Wu, X., Li, L., Iliuk, A., & Tao, W. A. (2018). Highly efficient phosphoproteome capture and analysis from urinary extracellular vesicles. *Journal of Proteome Research, 17*(9), 3308–3316.
88. Wu, C.-Y., Yang, H.-Y., & Lai, J.-H. (2020). Anti-citrullinated protein antibodies in patients with rheumatoid arthritis: Biological effects and mechanisms of immunopathogenesis. *International Journal of Molecular Sciences, 21*(11), 4015.
89. Wu, Y., Zhang, N., Wu, H., Sun, N., & Deng, C. (2021). Magnetic porous carbon-dependent platform for the determination of N-glycans from urine exosomes. *Microchimica Acta, 188*, 1–8.
90. Yamamoto, M., Harada, Y., Suzuki, T., Fukushige, T., Yamakuchi, M., Kanekura, T., Dohmae, N., Hori, K., & Maruyama, I. (2019). Application of high-mannose-type glycan-specific lectin from Oscillatoria Agardhii for affinity isolation of tumor-derived extracellular vesicles. *Analytical Biochemistry, 580*, 21–29.
91. Yang, J., Shen, N., Shen, J., Yang, Y., & Li, H.-L. (2023). Complicated role of post-translational modification and protease-cleaved fragments of tau in Alzheimer's Disease and other tauopathies. *Molecular Neurobiology*, 1–20.

92. Zecha, J., Gabriel, W., Spallek, R., Chang, Y.C., Mergner, J., Wilhelm, M., Bassermann, F. & Kuster, B. (2022). Linking post-translational modifications and protein turnover by site-resolved protein turnover profiling. *Nature communications, 13*(1), 165.
93. Zhang, H., Cai, Y.-H., Ding, Y., Zhang, G., Liu, Y., Sun, J., Yang, Y., Zhan, Z., Iliuk, A., & Gu, Z. (2022). Proteomics, phosphoproteomics and mirna analysis of circulating extracellular vesicles through automated and high-throughput isolation. *Cells, 11*(13), 2070.
94. Zhang, H., Li, X., Martin, D. B., & Aebersold, R. (2003). Identification and quantification of N-linked glycoproteins using hydrazide chemistry, stable isotope labeling and mass spectrometry. *Nature Biotechnology, 21*(6), 660–666.
95. Zhang, L., Wang, Y., Zhang, W., Hsu, Y.-I., Asoh, T.-A., Qi, B., & Uyama, H. (2022). Robust dual-biomimetic titanium dioxide-cellulose monolith for enrichment of phosphopeptides. *ACS Biomaterials Science & Engineering, 8*(6), 2676–2683.
96. Zhang, Y., Wu, X., & Tao, W. A. (2018). Characterization and applications of extracellular vesicle proteome with post-translational modifications. *TrAC Trends in Analytical Chemistry, 107*, 21–30.
97. Zhong, Q., Xiao, X., Qiu, Y., Xu, Z., Chen, C., Chong, B., Zhao, X., Hai, S., Li, S., An, Z. & Dai, L. (2023). Protein posttranslational modifications in health and diseases: Functions, regulatory mechanisms, and therapeutic implications. *MedComm, 4*(3), e261
98. Zhou, X., Xie, F., Wang, L., Zhang, L., Zhang, S., Fang, M., & Zhou, F. (2020). The function and clinical application of extracellular vesicles in innate immune regulation. *Cellular & Molecular Immunology, 17*(4), 323–334.
99. Zhou, H., Low, T. Y., Hennrich, M. L., van der Toorn, H., Schwend, T., Zou, H., Mohammed, S., & Heck, A. J. (2011). Enhancing the identification of phosphopeptides from putative basophilic kinase substrates using Ti (IV) based IMAC enrichment. *Molecular & Cellular Proteomics, 10*(10).

Index

R. K. Marcus and D. Lubman (eds.), *Exosomes and Extracellular Vesicles*,
https://doi.org/10.1007/978-3-031-89270-7

GPSR Compliance

The European Union's (EU) General Product Safety Regulation (GPSR) is a set of rules that requires consumer products to be safe and our obligations to ensure this.

If you have any concerns about our products, you can contact us on ProductSafety@springernature.com

In case Publisher is established outside the EU, the EU authorized representative is:

Springer Nature Customer Service Center GmbH
Europaplatz 3
69115 Heidelberg, Germany

Batch number: 10704035

Printed by Printforce, the Netherlands